PROSPECTS FOR
INTERSTELLAR TRAVEL

PUBLICATIONS OF THE AMERICAN ASTRONAUTICAL SOCIETY

Following are the principal publications of the American Astronautical Society:

JOURNAL OF THE ASTRONAUTICAL SCIENCES (1954 -   )
Published quarterly and distributed by AAS Business Office, 6352 Rolling Mill Place, Suite #102, Springfield, Virginia 22152. Back issues available from Univelt, Inc., P.O. Box 28130, San Diego, California 92198.

SPACE TIMES (1986 -   )
Published bi-monthly and distributed by AAS Business Office, 6352 Rolling Mill Place, Suite #102, Springfield, Virginia 22152.

AAS NEWSLETTER (1962 - 1985)
Incorporated in *Space Times*. Back issues available from AAS Business Office, 6352 Rolling Mill Place, Suite #102, Springfield, Virginia 22152.

ASTRONAUTICAL SCIENCES REVIEW (1959 - 1962)
Incorporated in *Space Times*. Back issues still available from Univelt, Inc., P.O. Box 28130, San Diego, California 92198.

ADVANCES IN THE ASTRONAUTICAL SCIENCES (1957 -   )
Proceedings of major AAS technical meetings. Published and distributed for the American Astronautical Society by Univelt, Inc., P.O. Box 28130, San Diego, California 92198.

SCIENCE AND TECHNOLOGY SERIES (1964 -   )
Supplement to *Advances in the Astronautical Sciences*. Proceedings and monographs, most of them based on AAS technical meetings. Published and distributed for the American Astronautical Society by Univelt, Inc., P.O. Box 28130, San Diego, California 92198.

AAS HISTORY SERIES (1977 -   )
Supplement to *Advances in the Astronautical Sciences*. Selected works in the field of aerospace history under the editorship of R. Cargill Hall. Published and distributed for the American Astronautical Society by Univelt, Inc., P.O. Box 28130, San Diego, California 92198.

AAS MICROFICHE SERIES (1968 -   )
Supplement to *Advances in the Astronautical Sciences*. Consists principally of technical papers not included in the hard-copy volume. Published and distributed for the American Astronautical Society by Univelt, Inc., P.O. Box 28130, San Diego, California 92198.

Subscriptions to the *Journal of the Astronautical Sciences* and the *Space Times* should be ordered from the AAS Business Office. Back issues of the *Journal* and all books and microfiche should be ordered from Univelt, Incorporated.

AAS PRESIDENT
    John A. Sand   Ball Aerospace Systems Group

VICE PRESIDENT - PUBLICATIONS
    Georganne Brier Thibault   ANSER

SERIES EDITOR
    Dr. Horace Jacobs   Univelt, Incorporated

AUTHOR
    Dr. John H. Mauldin

ASSOCIATE SERIES EDITOR
    Robert H. Jacobs   Univelt, Incorporated

Front Cover Illustration:

The 40 nearest star systems shown in computer-calculated perspective on a grid 32 LY across in our equatorial plane. A simulated astrogational view from 38 LY away from our Sun (at center), from 20° above plane, and from 20 h 40 m R.A. Size shown codes brightness, and color indicates spectral class and temperature, for a total of 5 dimensions of data. See Figure 6.2 in this book for key to stars.

Frontispiece:

61 Cygni. A typical telephoto of a near star system visible in northern sky. At 11.2 LY this pair of K stars is sufficiently separated (by about 0.001 LY or 10 billion km) and sufficiently warm that either or both could have habitable planets in reasonably stable orbits. These stars are somewhat smaller than the Sun and revolve in 720 years. In the sky they are separated about 20"arc. The disks are an optical problem as the images were magnified about 300 times. Even close up from 0.03 LY as in this photo, the stars would be brilliant points of light. They are receding at 64 km/s from us (Courtesy of Lick Observatory).

AN AMERICAN
ASTRONAUTICAL
SOCIETY PUBLICATION

PROSPECTS FOR INTERSTELLAR TRAVEL

by John H. Mauldin

Volume 80
SCIENCE AND TECHNOLOGY SERIES
A Supplement to Advances in the Astronautical Sciences

Published for the American Astronautical Society by
Univelt, Incorporated, P.O. Box 28130, San Diego, California 92198

First Printing 1992

ISSN 0278-4017

ISBN 0-87703-344-7 (Hard Cover)
ISBN 0-87703-345-5 (Soft Cover)
ISBN 0-87703-346-3 (Unnumbered)

*Published for the American Astronautical Society by
Univelt, Incorporated, P.O. Box 28130, San Diego, California 92198*

Printed and Bound in the U.S.A.

Acknowledgements

Appreciation is expressed to the U.S. Air Force Academy Library and staff, to my local Pueblo Library for obtaining various materials, to Jet Propulsion Laboratory scientist T. R. Gavin for discussion of robotic mission reliability, to the JPL Public Information Office for materials, to George Tsao of the Laboratory of Renewable Resources Engineering at Purdue University for material on CELSS, to physicists Janos Kirz and Malcolm Howells for information on fresnel zone plates, to Lick Observatory for special assistance on a photo, to my wife Susan for helping in the library work and for general life support, and to my publisher Horace Jacobs and his editor son Robert Jacobs at Univelt for long-term encouragement and support.

John H. Mauldin

Preface and Guide for Readers

PREFACE

The possibility of interstellar travel has been of increasing interest since early this century. The popularity of science fiction (sf) stories and the production of major sf movies and TV series involving this concept are indicators. Public interest in the space program, in astronomy, and in other phenomena related to space are also part of a diffuse general background that can support human travel to other stars. This interest is likely to continue, although at low priority, until the chance comes to begin construction of a specific interstellar mission. There may always be interest in interstellar travel even if programs or missions are not fruitful. People have looked up and wondered about the stars for millennia, and in this high-tech space age we seem to be making progress (very slowly) toward human visits to them. Fans of sf hope that zooming about the galaxy to explore strange planets, solve complex problems, or defend against tyranny is more than fantasy. Unfortunately, when a theme as common as travel to the stars is taken for granted in a large but fictional literature, the public tends to assume that the concept will become certainty, even that particular methods will be developed fairly soon.

This book is for those who want a detailed introduction to the problems of interstellar travel. It is also for general readers and dreamers who want more background in the subject. Because of the nature of the overall problem, the reader is taken on a journey of informed imagination, guided by the possibilities of known science and technology. Trying to solve such a problem is like playing a vast game with several kinds and levels of rules. Young people, the next generations who set the course for human progress, are especially enthusiastic about new technology and are justified in hoping for interstellar travel. They may get to make decisions on whether resources will be committed to testing its feasibility or even to implementing early projects, and some will be privileged to do the work. All who are interested should be well-informed about the prospects and problems. Space is the last physical frontier for our species, and the stars are very distant but reachable. If they are reached and our galaxy explored first-hand, then in the remote future humans will turn to the even larger-scale goal of visiting the galaxies beyond (as a few grandiose sf stories have proposed).

It is difficult to name a technical problem which has stimulated more passionate study than interstellar travel. Many scientists and engineers since the 1950s (and a few earlier) have had serious interest in the problems of interstellar travel and its implications for the existence of "extraterrestrial intelligence" (civilizations of other beings). Much work has been done at the level of theoretical research, resulting in a large published literature, mostly as journal articles and specialized publications. Whether or not sf can be said to lead science in a field such as this, there is surely a connection. The serious literature is a fascinating mix of fantasy and hard science, and it is very educational even if the suggested hardware is not built. Serious technical reports abound from the 1960s and 1970s (continuing to the present) which all but promised major breakthroughs to "revolutionize" space technology (and other fields) before 1990. Instead, with hindsight, progress seems slower. Humans have not started on their way to Mars more than 20 years after exploring parts of the Moon; indeed the basic R&D has barely started. Since predicting the future is difficult, this book tries to avoid excessive optimism, especially about proposals demonstrated at best on paper and computer thus far.

The study of interstellar travel provides a broad lesson in the methods of science and technology, especially work connecting many traditional fields. To separate the possible from the fantastic, careful scientific methods are needed and should be kept in mind when reading the results. One should try to know where concepts and facts have come from, aided by good referencing, and should trace assumptions and guesses and distinguish them from established science. In this book a mixture of primary and secondary sources is referred to. Primary sources provide more reliable history but not necessarily the latest science. The writer has examined over 90% of cited references in detail. The writer's scientific and technical judgment (a kind of intuition) has been applied on subjects where references are contradictory or provide a range of still uncertain knowledge. Many reported facts and applications of physical law have been checked at a basic level, if only by fresh estimates, to attempt to insure that no major errors on the prospects for interstellar travel are propagated here. Without burdening the text with a great many more reference notes, the writer has attempted in each paragraph to make key ideas and scientific results traceable to cited references and to distinguish them from ideas and other original work by the writer within the same topic of discussion. An apology is offered to any whose published relevant work was not found or cited here, or was unknowingly repeated afresh from basic science.

Study of interstellar travel, before any implementation, is already a large field in itself. This review of the prospects can only set the stage for further thought and work on both the possible achievement and the implications of achieving or not achieving it. Like designing a starship itself or planning a mission, no one person can study and comprehend in detail all aspects of this subject even at this stage. In planning a program for interstellar travel, there are many complex considerations all of which affect each other to a greater degree than has occurred in many other human projects. There are many ways to go astray at both the common sense level and the scientific level. Many nonscientific sources, aside from sf literature, have presented or created information on this subject that may be misleading. The ideas about "flying saucers" are a significant example. Because of cultural implications as well as scientific and educational ones, discussion of travel to the stars should be as accurate as possible. Much room for healthy speculation still remains, and some of the more educational but scientifically and technically far-fetched ideas are discussed here in context. As many basic principles as possible are presented so that readers may decide in many ways the prospects for and implications of traveling to other stars. The subject should be seen not as a distant abstract one but as one that has had and will have effects on us regardless of whether it ever results in actual voyages. Some of the deepest personal and philosophical issues are relevant to consideration of interstellar travel.

A subject such as interstellar travel, which involves many sciences and many kinds of thought and investigation, is an ideal educational vehicle too. The reader is led through much of modern physical science from elementary particles to astronomy, with excursions into communication theory, ecology, and social studies, to name a few pertinent fields. We are seeing at the end of the 20th Century how the sciences and other human endeavors mix and interact, and such multi-modal education should be further encouraged. Travel to the stars is a reasonable although immense problem to pose for solution, if costs are set aside temporarily. It is hoped that a sufficient variety of principles, modes of analysis, and references are presented here that the serious reader will acquire the means for further study in this field or its related subfields or even in other kinds of multi-field problems.

The end result, a workable starship design, is beyond the scope of any present book (although one publication has come close). The form of a starship is likely to be very complex, and no attempts are made to show a complete starship pictorially. Enticing artwork and detailed film models purporting to show starships are widespread but have little scientific basis, especially before working propulsion is identified. There are also many ways to design a book on such a complex subject, and the author will always wish, surely along with some readers, that some subjects were done differently.

This book originated from the writer's studies of physics, technology, and other aspects of interstellar travel while planning a novel that features a realistic interstellar voyage. While no sf can be fully realistic, else it would become ordinary description of what people are actually doing now, much detail can be provided in accord with established laws of nature. Many other fiction writers have attempted this but the results tend to include more than a minimum number of violations of known science. Perhaps no one can make convincing guesses as to the scale and quality of an interstellar voyage, but a writer's goal can be to leave as little as possible to wild guesses and implausible assumptions about nature. The implausibility might possibly be limited to some subtle deep law, a law that might well be corrected later when scientists understand more. Attempted realism enables readers to experience interstellar travel for themselves as the epic adventure that it must be. But sf readers have had much fun traveling across the galaxy with many great writers who bend the known laws of nature more than a bit.

GUIDE FOR THIS BOOK

This review of the prospects for interstellar travel is intended to be accessible to anyone interested in science, technology, space, or the human future. While some prior reading and study of science and math at the high school level is desirable, sufficient background and explanations are given for the main evidence and arguments to stand on their own. It helps for the reader to have slight acquaintance with physical concepts such as units of measure, common chemical elements, atoms in matter, nuclei in atoms, acceleration, gravity and magnetic forces, temperature, electrical charge, and visible light as part of a spectrum of radiation. These concepts can be found in a dictionary or encyclopedia. No mathematics is used in the main text, and almost no symbols (notably c for lightspeed). Verbal explanation in place of the underlying mathematics is given wherever feasible.

The most difficult technical subjects occur in Chapter 3 and parts of Chapter 4. These can be skimmed or skipped without jeopardizing understanding of later chapters. Those interested in the "softer" social and philosophical subjects can skip technical sections, realizing that political and economic decisions can depend on matters of science and technology. For the technically prepared reader, Appendices B through E provide optional physical and mathematical details of some important results, along with further background. Results from physics at the introductory college level are often presented and used without explanation there. To begin self-study at that level, almost any introductory college physics textbook would be suitable. Some advanced results are derived in the Appendices, including those not known to be published elsewhere.

Appendix A describes the units of measurement used (the SI system, commonly called metric) and their prefixes and abbreviations. To aid the reader in gaining a feel for the magnitudes involved, some common, larger, and mixed units are consistently used instead of rigorous SI units. These are: years (Earthly), lightyears (LY, a long distance), tonnes (metric, about the same mass as in a ton), kilometers per second (km/s, a little slower than miles per second, sometimes called "klicks"), and gee (acceleration). These are often used instead of seconds, meters, and kilograms. The traditional degrees symbol ° is kept for temperatures. A table of conversions to some common US and other units is also provided.

Everywhere possible, simple rounded numbers with one digit or just a multiple of ten are used to illustrate the magnitudes involved. Numerical accuracy is not emphasized in this qualitative study and usually not needed. However, constants and conversions are provided with sufficient accuracy for further work. Sometimes numbers are more conveniently expressed in powers of ten, with a small number shown for example as "(10)-7" instead of the usual superscript power. In this example the exponent or power "-7" shows the number to have 7 decimal places, and a "1" can be in the rightmost place thus: 0.0000001. The large number (10)8 has 1 followed by 8 zeros thus: 100,000,000. This notation is included with the table of prefixes.

References are indicated in brackets [ ] with the author's name, preceded by "a" for article and "b" for book. For an author with many works, these are distinguished by year. Articles and books are listed separately and alphabetically for each chapter in the bibliography at the end of the book. Since each chapter constitutes a subject area, the bibliography is thus grouped by subject. A few articles and books are listed for general interest but not referenced in the text. Some works cover several subject areas and may be listed several times. The most general books are listed in the bibliography for the Introduction if nowhere else. Because the number of published articles and books is now so large, listing all relevant references is no longer possible. Popular sources, as distinguished from scientific journals, are thought of as secondary or tertiary sources and have provided many more articles on the relevant subjects than the few listed here. References called "classic" are the first known published papers on key subjects or the first mention of imaginative ideas. Referencing in the text is provided mainly and most carefully for tracing key points. Sources are cited as close after information from those sources as possible. It is very difficult to distinguish clearly information from other sources from original ideas. All paragraphs contain at least one original idea or interpretation in addition to cited material.

Each chapter opens with general and guiding remarks. Other guidance to the book's structure is given in brackets [ ], such as reference to Appendices (which provide optional supplements to the main text and arguments).

AN INTRODUCTORY GLOSSARY

Some general terms are defined here for the beginning reader rather than within the main text. There are no standard definitions for some of these terms, and this list limits usage to that needed for this book. Other special terms and jargon are introduced in quotes at first appearance in the text. See Appendix A for units of measure. Terms are grouped here approximately by subject area.

Many space terms have originated from nautical or aerial usage. Although picturesque and widely adopted, the linkage should be broken where possible and the images of sleek cylindrical vessels intended for water or air disregarded for deep space travel. More accurate and convenient words are not available to replace "ship" "craft" and "vessel" for space use. "Aboard" tends to slip into discussions of starships. "Astronauts" are here to stay, but literally they are "star sailors". "Spaceflight" is almost a contradiction since no flying occurs in the near vacuum of space.

Science: the study of nature to determine common principles and explanations of the way all things work. The careful process required is embodied in part in this book. The results of science ideally have predictive power enabling good guesses to be made as to what we might be able to achieve technically.

Technology/ Technical: the application of scientific knowledge to practical problems. Much testing and correction of actual hardware is needed to make complex artificial systems work.

Laws of Nature: general statements of how nature acts or "is" under specified conditions; even when not explanatory or written mathematically, consist of precise descriptions of what nature does (e.g. Newton's law for gravity, DNA code). Further scientific work can result in correction or replacement of human-discovered natural laws. Other beings should find similar laws.

Space: the vast regions outside atmospheres away from surfaces of planets and stars; almost devoid of matter; a good vacuum but never completely empty.

Interstellar: literally between/among the stars, referring usually to the vast spaces among the stars outside planetary systems and inside galaxies.

Spacecraft: a small vehicle for local (planetary) travel with human occupants, with some amount of propulsive or guidance capability; used only in space; also may refer to any space vehicle for any purpose. Examples are the Apollo and Voyager spacecraft.

Spaceship: implies a somewhat large spacecraft, definitely occupied, possibly for interstellar travel. None have been built.

Starship: somewhat grandiose term but justifiably used here as the most convenient name for an occupied interstellar spacecraft.

Probe: an automatic, usually small, spacecraft designed to carry out a mission without humans aboard; can be directed occasionally by radio commands.

Robotic: referring not necessarily to a mechanical "human" but to any computer-controlled machine which can do complex, almost intelligent operations.

Mission: for this book, any space voyage for a purpose (leaving pleasure voyages to the future); uses a probe, spacecraft, or starship.

Astrogate/ astrogation: preferred term for "navigation" among the stars, determining locations, directions, distances, and speeds. Astrogation occurs in three dimensions whereas navigation was concerned mostly with two.

Rocket: in space, refers more to the method of propulsion than to the vehicle; an "engine" with or without a fuel supply attached which ejects material in one direction to propel a spacecraft in the opposite direction; on Earth the streamlined vehicle that uses the rocket method to travel through atmosphere from ground to space; needs no air for its function.

Drive: a propulsion mechanism or method, not necessarily a rocket engine.

Payload: the mass of structure, supplies, people, etc. to be delivered after any fuel for the voyage has been consumed and any extra parts jettisoned.

Orbit: the path a planet, spacecraft, or other object takes in space under the influence of one or more larger objects with gravity (stars, planets, moons). An orbit repeats (approximately) so that the position of the orbiting object can be predicted in the near future. A common orbit for space missions is a low circular one around Earth, which must be carefully set up to achieve.

Lightspeed: the ultimate speed, nearly 300,000 kilometers per second; the speed only light and other electromagnetic radiation (radio, radar, x-rays, gamma) can travel at in vacuum. Symbol c.

Gee: a unit of acceleration based on the rate of fall we feel near the ground due to Earth's gravity; a rate of increase of speed equal nearly to 9.8 meters per second per second, conveniently rounded to 10 m/s^2); not limited to situations involving gravity and is used to measure acceleration or deceleration due to any cause.

Mass: the amount of material in any object or particle; the greater the mass, the more force needed to accelerate the object.

Weight: the force on an object due to gravity, whether at the Earth's surface or in space not too far from a planet or star. Objects are nearly weightless only if very far from a source of gravity or if they are falling toward such a source, whether "straight down" or in a stable orbit around the source.

ETI: Extraterrestrial Intelligence, better stated as intelligent beings presumed living in civilizations on hospitable planets at some other stars. None are known to exist, but they might. It is chauvinistic to call them "aliens". "ET" is little better, and "beings" seems the safest term here.

SETI: the Search for ETI; real ongoing programs using radio telescopes.

sf: science fiction, a well-known type of literature, film, even opera.

SYNOPSIS OF THE BOOK

The Introduction provides background, including history and scientific method, in regard to traveling to nearby stars, as preparation for the main part of the book. Included are four model mission plans for the comparison of various methods and approaches given throughout the book.

Chapters 1 through 4 cover the mechanics of the problem of interstellar travel, considering distance, time, speed, acceleration, and the force needed to propel a starship. Rocket theory, gravitational energy, and energy handled by various starship propulsion methods or "drives" are other parts of the mechanics. Chapter 1 covers speeds less than those where relativity becomes important (about 0.2c), and Chapter 3 covers travel closer to lightspeed where Einstein's relativity brings in new rules and fascinating effects. In many ways relativity is science fiction made real. Chapters 2 and 4 cover, respectively, advanced practical methods of propulsion, and very advanced methods involving relativity. Various other phenomena for travel and maneuvers in space are included, such as gas and dust, drag, and electric and magnetic fields.

With propulsion methods in mind, overall engineering design of starships and probes is discussed in Chapter 5. Estimates of size, population, and mass are made, and starship subsystems are outlined. Chapter 6 continues design by considering possible destinations, types of missions, and scientific work. Robotic flyby and self-reproducing probes, and colony missions, are included.

In Chapter 7 astrogation poses many interesting problems in not getting lost and in detecting and avoiding dangers. Communication with Earth is not easy either. Opportunities for further scientific observation, both passive and active, are discussed, including relativistic views.

Details of engineering design of a starship are determined by technological requirements and limits discussed in Chapter 8, as well as by the mission. Some limits are subtle and general. Multiple levels of correction and multiple backups are a major part of engineering for long-term reliability. Recycling, computers, radiation, shielding, and transport to planets are included.

Chapter 9 covers biological requirements and limitations for the transport of people in closed systems for long periods of time. Sources of food, water, air, "gee", and other basic necessities need more careful study. Dangers to living organisms include radiation, weightlessness, depletion of vital elements, and system malfunctions.

Social and psychological considerations in the starship are considered in Chapter 10, and then social, political, and economic aspects of Earth and local space. Scientific and technical aspects of these subjects are necessarily projected from terrestrial experience but without excessive speculation. Costs are estimated very loosely. Less tangible needs of voyagers and mission supporters are not neglected.

If interstellar travel were easy, we should find messages, probes, and starships sent here by "others", as well as find garbage left from their landings on Earth.

Chapter 11 reviews the search for other intelligent beings and speculates within the bounds of science on what other beings might do with interstellar travel. If humans go traveling and visit another civilization, some etiquette may be needed to avoid getting caught in "star wars".

If travel to stars proves too difficult at present, areas for deeper research are identified in Chapter 12. Some wilder solutions to interstellar travel are also considered. This chapter summarizes technical and social conclusions from this journey through the science and technology of interstellar travel and shows some hard decisions to be made by humanity starting now if the stars are to be part of our future.

Contents

Introduction to Interstellar Travel

0.1 TRAVEL TO THE STARS

Thoughts of traveling to the stars spread in the early 18th Century after the stars were suspected to be distant objects rather than sky decoration. As soon as other evidence had convinced early astronomers of the correctness of the 16th Century Copernican model whereby Earth revolves around the Sun, the failure to see motion of the background stars was of concern. Parallax is observed as apparent side-to-side motion of a star with respect to other more distant stars as Earth moves from side to side in its one-year orbit. Methods of measurement were inadequate until 1838 when the distances to Alpha Centauri and other suspected nearby stars were measured in terms of their parallaxes. The result for Alpha by Friedrich Bessel was the largest parallax found, about 1" of arc (1/3600 of 1° or less than one-millionth of a circle) [b-Abell]. (Alpha Centauri is named with the first letter of the Greek alphabet for being the brightest star in the southern constellation Centaurus, a Latin name. Other constellation regions also have their brightest stars named Alpha such-and-such.) [See any world atlas or star atlas for constellation maps, e.g. b-Moore.]

Like most astronomical measurements, the accuracy of the result depends on the accuracy of another result. The size of Earth's orbit has been determined more accurately over the centuries, and we can take it as very close to an average of 150 million kilometers (km). Simple geometry was sufficient for Bessel's result for Alpha to indicate a distance of trillions of kilometers that must have astonished scientists and others. The distance corresponding to 1" of parallax is called 1 parsec and is a common astronomical unit. The preferred unit for discussing stars here is the lightyear (LY), the distance light travels in space in one Earth year. This unit of distance is about 9.461 trillion km. It is conveniently rounded to 10 trillion km or (10)16 meters and is equivalent to 3.26 parsecs. [See Appendix A for a table of units.]

Alpha Centauri with parallax 0.76" has indeed been found to be the nearest star system at 4.4 LY (about 40 trillion km) and is often referred to in discussions of interstellar travel because of this fact. Alpha is a star system consisting of two bright stars A and B revolving around one another, separated by about 4 billion km and a distant companion star Proxima, which is 0.1 LY closer to us at present. Barnard's Star at 6 LY is nearest after Alpha and a more likely candidate for a mission because of the greater likelihood of planets. Counting our Sun and the many double and triple stars, there are 88 stars known within 20 LY of us, with an average spacing of about 5 LY [b-Abell, a-Mattison]. [See Chapter 6 for a table of nearest stars.]

Depending on one's perspective, these distances to nearby stars are reasonably small or enormous. As explored in this book, traveling to these places is a tremendous and fascinating endeavor. For many reasons it is regrettable that stars are so far apart (a million times the size of Earth's orbit) in our region of the galaxy. However, for our Sun to have formed, a volume of typical interstellar gas several lightyears in radius must condense. The density of material is on average only about one atom (usually hydrogen) per cubic centimeter of interstellar space, with large variation from region to region. For us to be here, the wide spacing of stars is unavoidable. We can also view the stars as close together simply by looking outside our galaxy at the other galaxies spread across the universe in huge numbers. Each consists of billions to trillions of stars seemingly packed into a small spot of light but actually as far apart as stars in our galaxy. Study of travel to galaxies, typically a million lightyears apart, is beyond the scope of this book.

Interstellar travel is directed more toward the planets that might be found at stars than the stars themselves, although some stars promise spectacular and dangerous phenomena if seen close up. One of the earliest stories to mention interstellar travel and extrasolar planets is the story "Micromegas" by Voltaire in 1752. In this story the giant being Micromegas journeys by lightbeam, comet, and other means from a planet at Sirius (8 LY) to our Solar System, ultimately visiting Earth as one might expect. More recently, astronomers and other scientists have attempted to guess how many habitable planets might accompany other single stars like our own sun and how many intelligent civilizations might have arisen on planets across our galaxy. Attempts to detect planets at nearby stars have not been sufficiently sensitive to indicate more than the possibility of very large planets. No obvious destination is known today for a first interstellar voyage.

Some fraction of the human species has always wanted to explore or challenge frontiers. We are an adventurous species [a-Finney]. On a planet crowded with humans, some can only turn to space for fulfillment of their visions. Jules Verne was an early true sf writer to propose a new adventure, travel to the Moon, as the U.S. Civil War occupied others. A popular science and philosophy book from 1929 by a chemist [b-Bernal] supposes that exotic rockets and solar sails (only acres in extent) will be used to explore the Solar System and that spherical space colonies will someday easily travel hundreds and thousands of years to the stars and populate the universe. Since the 1920s or earlier, popular sf has helped identify the "outward urge" which would take people to the planets and stars if technology permits. The number of enthusiasts may continue to increase, if public interest in the space program, in movies and books about space, and in space toys is an indication. Even if human exploration of Mars, asteroids, and moons of Jupiter and Saturn occupy the adventurous for several centuries, the stars will continue to beckon to later generations as our "last frontier" (to use a popular phrase).

It is exciting to consider fellow humans traveling far from our Sun. Some of us would love to go if technology, cost, and time permit. But other less difficult methods exist to send our presence beyond our Solar System. Probes operating with self-contained automatic (robotic) machines and programs can make substantial voyages and radio back data. We can send radio messages and set up equipment to scan for messages from other civilizations (SETI). These have been tried on a scale too modest to expect results. This "travel" by radio telescope is not directly in the scope of this book, except for some implications. We might also be able to send out specially-designed living spores to drift in space and land on other planets. Radio methods may be the cheapest, but sending robotic hardware one-way to the stars is much cheaper than sending people one-way to start a colony, and that may be much cheaper than sending people who want to return in a reasonable time. At some point economics must determine our goals in interstellar travel.

Four scientific spacecraft, *Pioneer 10* and *11* and *Voyager 1* and *2*, can be viewed as interstellar missions. Voyager has looked back to photograph Earth from outside our Solar System [a-Sagan]. They will travel through our galaxy in rather different directions, possibly for billions of years. After the Sun's gravity slows them a bit more, the Pioneers will be moving about 11 km/s and the Voyagers about 16 km/s thereafter, barring very close encounters with unknown stars. They would pass by the nearest stars in 100,000 to 400,000 years if the stars were fixed. However, the stars are moving in various directions, almost randomly, and one obscure star now about 17 LY away will pass by us and these spacecraft about 40,000 years from now [a-Cesarone]. Our Solar System and all our spacecraft are also moving, sharing a common speed toward a point about 30° "north" of Barnard's Star (which is therefore coming toward us). Failure of power and other systems will render them nonfunctional in a few decades, but the Voyager spacecraft are expected to radio information from the region just outside our planetary system. No substantial hardware or design studies have been directly funded yet for a new mission beyond Neptune, but some interesting phenomena are out there. Scientific probes to these regions are within our technical capability now, given major commitment. Such a mission should not be counted as interstellar travel but would be important preparation for it.

0.2 SCIENCE, TECHNOLOGY, AND INTERSTELLAR TRAVEL

A tendency is to associate interstellar travel with science fiction (sf). This is not necessarily a disadvantage because science fiction writers and readers tend to be forward-looking people who give high support to practical space exploration while dreaming beyond [a-Bainbridge]. Fiction writers and scientists have separately proposed every interstellar mission and technology that they can conceive of. Beyond imaginative use of rockets and sails have come many still incredible ideas. Direct radio "transmission" of material objects, as if they were elaborate television signals, has been a popular one. Early fictitious ideas such as "hyperspace" were often hopelessly vague until scientific study on the same subject produced quite different results (in this case, Einstein's curved spacetime). Travel via "holes", "warps", or "gates" in space and travel faster than light cannot yet be discussed to the extent that other proposed methods can. While Einstein's general relativity, a basic theory of gravity, permits some of these methods, other basic physics also provides reasons why they cannot be used. If such methods were to be found scientifically possible, severe technical problems would then need to be considered. For example, a blackhole is expected to destroy its users. Blackholes probably exist, but using the nearest suspected one is impractical because of the hundreds of lightyears to travel to reach it.

In a realm closer to fantasy which scientists and story writers have abandoned, mystics and fringe media have accepted or embraced UFOs (Unidentified Flying Objects) as something more than the name indicates [see e.g. a-Markowitz]. Some of the public continues to identify them hastily as "flying saucers", allegedly the means by which visitors from the stars arrive. These apparitions violate many well-understood laws of nature, and no verifiable sighting or artifact has ever been found. That aliens might use "other" laws of physics seems highly unlikely. Human engineers do not foresee the development of practical "aircars" to replace our polluting automobiles, much less starships that move from Earth to space as easily as claimed. Allegations about flying saucers are not covered in this book, but the relevant science given here may be applied as an exercise by the reader to their possibility.

An aspect of interstellar travel which seems fantastic but is much more realistic and of deep philosophical importance is its connection with the existence and whereabouts of other intelligent beings and civilizations. Several conferences have been held on communication with and the nature of other civilizations. Many serious books and conference reports [e.g. b-Finney; b-Hart; b-Papagiannis] and articles have been published on such a speculative subject, often extrapolating from our human experience. Tabloid stories aside, there is no evidence that other beings have visited us. We have seen no traces of powerful starship passage through deep space, we have received no clear messages from other beings either at home or enroute, and we certainly have had no invitation to join a galactic club. Yet there are good reasons to believe some of these events should have occurred already. The paradoxes about other beings, and the logical links among travel, the means of travel, the beings who do it, and their goals, are a fascinating field of study. While the lack of "visitors" is a bad sign for the feasibility of travel, many other explanations have been put forward for this observation. It is partly within the scope of this book to explore these arguments, concentrating more on the relevant laws of nature than on trying to guess "their" social motives.

Most scientists have been conservative about interstellar travel. As the space age began with satellites near Earth in late 1950s, a few studies of the feasibility of much more ambitious star travel were published, showing the concept highly impractical [a-Purcell; a-Hoerner; a-Pierce; a-Drake]. Most writers and readers do not seem to have heard of these basic studies, with the unfortunate effect that the public continues to be overly optimistic and the fortunate effect that all interest was not abandoned. People seem astonishingly eager to believe that interstellar travel is possible, mainly for its consequences. To oversimplify into two camps, the "escapists" believe that they might be able to escape Earth's--that is, our species'-- problems and go elsewhere, or to send our excess population away, or to bring in new resources. Other "salvationists" do not want to go elsewhere but believe that superior beings have, or will, come here to save us from our problems with better ideas and technology. Overall, it seems that more modest hopes about interstellar travel would have better prepared us humans to be more respectful of our one limited planet. For these social reasons careful study of the prospects for interstellar travel is also important.

Since early but serious studies, the prospects for interstellar travel have waxed and waned as scientific and technical knowledge has expanded and new problems have been found. Strong theoretical interest continues unabated and a large if not very visible literature has accumulated, most of it rather technical. Scientists and engineers at NASA, at JPL (Jet Propulsion Laboratory, principle designer of planetary missions), at other contracting spacelabs (corporate and academic), and contributors to the British Interplanetary Society (BIS) have studied interstellar-related subjects, mainly as an addition to their more Earthly work, but also in work funded by NASA, USAF, aerospace companies, and others. They are working in at least 10 different nations and have published in assorted space, astronautical, astronomical, and general scientific journals (as indicated in the Bibliography). Publications of the American Institute of Aeronautics and Astronautics, of the American Astronautical Society, of the BIS, and of several international societies for astronomy and space, and the journals *Nature, Icarus, Science,* and *Astronautica Acta* are prominent but not the only sources that have carried serious, refereed reports on what are most broadly called "interstellar studies".

Journals which publish scientifically-sound interstellar studies (including the special subjects of interstellar travel, SETI, extrasolar planets, and exobiology) promote progress in this subject area by carrying on the normal process of science whereby theories and results are exchanged, critiqued, and tested in open literature. Studies have been published not only by physical scientists and engineers but also by biologists, sociologists, anthropologists, and many others. The extent of popular and professional interest is indicated also by a bibliography with some 2700 items in some 70 subject categories, published by BIS in 1980 [a-Mallove]. The output may have accelerated since then. The growing complexity of the problem is revealed by these studies and reports and by other considerations shown in this book.

At the present time no one knows really where to start a major research and development program for interstellar travel. A landmark theoretical study was published by BIS in 1978 as Project Daedalus, where scientific and engineering work was done in impressive detail on hundreds of aspects of design for a robotic mission to Barnard's Star. This project has stimulated much more interest and serves as a baseline against which many small and large studies and reports are to be measured. Reference to Daedalus is made at many points in this book. While the public and scientific interest is strong, the presently supposed approaches to interstellar travel appear at or beyond present technical and economic feasibility. For now, a proposal to build a starship in hardware would be properly rejected as impractical and vague. Further exploration of places much closer, the Moon and Mars, has been long delayed because of much simpler problems with transportation, survival, funding, and political will. Travel to a nearby star would require travel over a distance typically a million times farther with a time period up to thousands of times greater than missions to Mars. Costs could be thousands to millions of times greater than all that has been spent on space thus far. The effort seems so large that no one can make a realistic estimate of when a world society might be ready to plan an interstellar program. The technology deserves to be called giga- or tera-technology, to use the prefixes for numbers in the billions and trillions. Yet the prospects are too fascinating to ignore.

The discussion in this book is limited to the modes of travel within foreseeable technical possibility. While it might seem that no one can guess the future of technology, and virtually anything is possible given enough time and resources, subtle and basic limits are indicated by many parts of our current knowledge. The physical laws that pertain are particularly stringent when applied to this problem, and (as is often the case) our best scientists do not see where crucial new understandings of nature might come from. Interstellar travel is not limited by propulsion methods alone but by many other basic, practical, and human considerations. Travel to and from the stars is surely not easy or we would have had "visitors" by now and seen other clear evidence of it. Even so, travel to nearby stars appears almost routinely feasible compared to the ultra-high technology that some have envisioned: time travel, energy from black holes, planet reconstruction, inhabited rings and Dyson spheres around stars, disassembly of stars, mass migrations among stars, and preventing the collapse of the Universe!

0.3 FIRST LOOK AT POSSIBLE INTERSTELLAR MISSIONS

To gain more appreciation of the size of the technical feat of sending people to a nearby star, a first look at some possible interstellar missions is given here as preparation for the main discussion of later chapters. Full explanations are given later. Since a dozen stars, some that could have Earth-like planets, are within 10 lightyears of us, 10 LY or $(10)17$ meters is a good round first guess for a possible mission distance. This distance is almost a million times farther than the nearest star our Sun. Light, the fastest traveler in the universe, needs 10 of our years to come this distance, as compared to 8.3 minutes to reach us from our Sun. Because of the difficulty of interstellar travel, there is no need to consider farther distances than 10 LY at present. Human explorers will be doing very well indeed to find an inhabitable planet this "close", and visits to more exotic objects such as neutron stars and nebula must wait.

Of course, the nearest set of stars at Alpha Centauri is "only" about 4.4 LY away, but what would be done there? The double star Alpha A and B (not to be confused with Beta Centauri 293 LY away) and the companion Proxima are not known yet to have planets. Planets, habitable or not, are not expected to be found in double or triple star systems. The mission is more difficult to justify if all that can be done is to orbit the star, make observations, look for unexpected planets, and possibly return home. Planning for a 10 LY mission is more realistic. Epsilon Eridani near that distance is the closest single star similar to our Sun but much younger, and Tau Ceti at nearly 12 LY is more similar and old enough to have inhabited planets. If a human mission went to Alpha, it might continue to another nearby star system, preferably without slowing at Alpha. A robotic probe to Alpha would make more sense, because it is sent one-way, need not slow down, and may go on to several other stars.

The journey over 10 LY can take almost any time depending on the speed. Choices in three different ranges serve to set the scale [see Table 0.1]. If the journey proceeds at a speed 100 km/s, only ten times the present capability, then it will require 30,000 years. This duration is far beyond the span of human recorded history and seems ridiculous for so simple a journey. Every effort must be made to raise the speed. Humans might be able to use biotechnology to increase their lifespan, and might even learn to have long-term stable civilizations which can carry out projects requiring millennia. But interstellar missions which 1000-year-old people cannot see the end of are not attractive prospects if faster ones are possible.

If the starship can travel very near the speed of light (300,000 km/s), then observers on Earth need wait only 21 or so years for the first results to be received here. (Less time passes aboard ship, as discussed later, and time is needed to build up speed and slow again.) The roundtrip time for explorers would be about 25 years, including time to explore, and they would return somewhat younger than their friends. A more feasible intermediate speed is 1% of lightspeed (0.01c, using the symbol for lightspeed). But the duration is 1000 years, quite a long journey to plan on both for the voyagers and for scientists back home. Such a long time indicates that a human mission should be a one-way colony voyage. The amount of supplies and equipment needed for surviving a millennium and then starting a colony indicates a very large starship, called a "space ark" by some and a "generation ship" by others because many generations of humans would live during the voyage. Further study of propulsion methods (including Daedalus) shows that 0.1c might be feasible, cutting travel time to 100 years. Still no one might live to see the end of a one-way mission, and a two-way mission becomes doubtful.

Whether the mission is one-way or two-way and whether it stops are vital determinants of propulsive needs [see Figure 0.1]. A flyby of a star does not require a means to slow the starship or probe. A one-way exploration where the explorers become a small colony on a suitable planet, or a probe that stops or lands, requires the same effort to bring the starship to a halt as to accelerate it. A two-way mission requires in addition the means to reaccelerate the starship and, when it reaches our Solar System, a means to slow it again. A full two-way mission is not just four times more difficult than a flyby. The following simple example assumes a 10:1 total mass to payload ratio with fuel as 9/10 of the total. A one-way flyby probe requires 9 tonnes of fuel to send a 1 tonne probe (metric mass, equivalent to 1000 kg or a Volkswagen) at high speed through a star system. Another 9 tonnes of fuel would be needed to halt it there. But to send the extra 9 tonnes (10 total) all the way to the star requires 90 tonnes of fuel from this end (total mass 100 tonnes), and the ratio applies again. If 9 tonnes of fuel are needed to accelerate the 1 tonne probe (say, with samples it has collected) toward Earth after the mission has visited the star, then 100 tonnes must be sent there to be decelerated, requiring 900 tonnes of fuel to start the mission (1000 tonnes total initial mass), another factor of 10 more. Finally, if there is no other means of halting the returning probe here, then it must carry 9 tonnes of fuel for that job, and the original mission must start with another 10 times more fuel here, or 9000 tonnes (total initial mass 10,000 tonnes). For each acceleration and deceleration leg, the total mass to payload ratio is multiplied again.

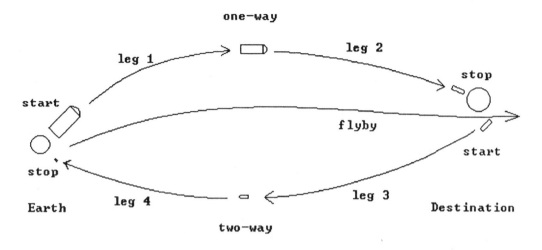

Figure 0.1   Types of interstellar missions.

Leaving Earth and Sun is not trivial, although in comparison to the difficulty of interstellar travel, it is easy. The gravitational pull of the Sun dominates nearby space, keeping all the planets tightly bound to it, as well as any slow starships. There is a minimum speed needed to depart from the Sun permanently from any given position. Escaping from the vicinity of Earth, not its surface, requires 42 km/s, and the starship would end up with zero speed out there. A voyage in interstellar space at 100 km/s would need to start here with about 109 km/s. (The seemingly small extra speed is explained in Chapter 1.) *Voyager 2* leaving the vicinity of Neptune needed about 8 km/s to escape the Sun and had about 19 km/s.

Departing the Earth's surface is a different kind of problem. A starship is likely to be too large and fragile to lift off directly, so it would be assembled in space, perhaps in Earth orbit. Much material would be needed to build it. While a mere 11 km/s is needed to escape from Earth (not from the Sun) and less to reach an orbit around Earth, huge rocket engines have been needed thus far to raise a few tonnes. Perhaps most of the thousands or millions of tonnes of material can be mined from the Moon and wandering asteroids. Either way, major space activity would be needed to assemble a starship. Shuttle or spaceplane flights from Earth would be very numerous and would need to be inexpensive and very safe with the most vital materials and people aboard. If a starship is built near Earth, leaving the vicinity of Earth on the great voyage would be triflingly easy compared to later feats. Just a few kilometers per second are needed to break free of Earth's gravity from Earth orbit.

The human element of missions to the stars provides other and formidable constraints. Of basic importance is the number of people traveling, because this determines the size and mass of the starship, and living space, supplies, and equipment. When the starship mass has been determined, then the required propulsion can be planned. The mass of the engines and fuel may be the largest part of the starship. A preliminary guess in round numbers for mission size is 100 people minimum. Ten people seems too few for a journey of decades or centuries unless hibernation becomes possible, and 1000 constitutes a very large mission. An estimated 100 tonnes mass is needed to support each person. The total payload is then 10,000 tonnes. For a fast trip with less people and less mass per, a 1000 tonne mission might be feasible. If the mission sizes and speeds indicated here turn out to be very difficult, there is still much hope of sending hardware to the stars. The mission can be made much smaller, without people. A probe can be one tonne or less. Since it need not return, or even halt, there is a good chance of finding the means to boost a mere kilogram to near lightspeed at reasonable cost.

There are many possible variations of size and speed for a mission to a star. Here just four model missions are chosen. These are characterized as big slow, small moderate, small fast, and fast probe [Table 0.1]. These are referred to often in further discussion and serve as a basis for comparing various methods of travel and various kinds of problems wherever uniform comparison is possible. Only round numbers are used here. The fast mission at 0.9 lightspeed is about as close to lightspeed as one might reasonably propose, gaining some benefit from the slower passage of time but not too much penalty in the inherent mass increase. Difficulty is estimated numerically in arbitrary units here but is related to the approximate energy involved. The difficulty is much more severe for the fast explorer mission, because it involves four legs near lightspeed. Even so, a very skimpy mass and very lightweight propulsion have been supposed for it, not enough for comfort let alone colonization. The difficulty of the fast mission is not reduced greatly by making it one-way or by refueling at the destination. These models carry humans, but a small moderately-fast one-way flyby probe is also considered. The starship mass is in some proportion to the population size, the duration, and the speed, all of which seem to require extra mass. An optimistic fuel to payload ratio of 10:1 is assumed here and applied to the number of legs of the mission. A colony ship is understood to travel one-way (2 legs), an explorer is to return to Earth (4 legs), and a flyby probe needs 1 leg. The mission masses and difficulties change substantially for propulsion methods not involving carrying the fuel and engines.

Table 0.1
MODEL MISSIONS TO STAR 10 LY AWAY

Type of Mission	Duration(Y^*)	Speed	Pop.Size	Mass (tonnes[‡])	Difficulty
Big Slow Colony	30,000	0.0003c[**]	1000	1,000,000	10
Small Moderate Colony	1000	0.01c	100	10,000	100
Small Fast Explorer	44[†]	0.9c	100	1000	10 million
Fast Probe	100	0.1c	0	1	0.1

Notes:
* Earth years
† roundtrip; time aboard ship one-way about 10 years; add time for exploration
‡ metric tonnes (1 tonne = 1000 kg); payload mass as arrives at star; fuel extra
** 100 km/s
Difficulty estimated from total kinetic energy, assuming mass ratio 10:1 per leg.

Chapter 1

Basics of Travel in Space

Present capabilities and methods of space travel are presented here for comparison with the needs of interstellar travel. The US experience with the Apollo missions, Voyager, and the space shuttle provide practical baselines [b-Blaine]. While a starship is unlikely to be launched from Earth, problems of getting off and back onto a planet are also a crucial aspect of any space travel. Many basic concepts of space travel are introduced as part of this discussion, most pertaining to the mechanics of motion and gravity. All are needed to understand the problems of interstellar travel fully. Acceleration, speed, mass, and force are of first interest, followed by energy and its relation to gravity. [Appendix A lists the pertinent scientific (SI) and other units of measure, and Appendix B provides mathematical background for many topics.] [gen. refs. b-Berman; b-Glasstone]

1.1 MECHANICS OF SPACE TRAVEL

The speed of a rocket or any moving object is accumulated bit by bit in a process called "acceleration", and acceleration is caused by application of a force according to the 17th Century law of motion established by Newton. The mass of the rocket acts as "inertia", and the greater the mass the more it hinders acceleration. The units of acceleration are meters per second per second (m/s^2) and indicate that during each second of acceleration, additional speed in meters per second (m/s) is added to the previous speed. In the simplest case for Newton's law, the acceleration is proportional to the applied force and the mass of the object remains constant. Then it has the traditional form F = m a. But this does not apply to rockets and most spacecraft whose mass changes as fuel is depleted. [Appendix B provides basic formulas from physics for calculation of motion for any reader who wishes to check and compare results in this discussion or to explore other ideas.]

Acceleration for many practical space purposes can be measured in "gees", where 1 gee is the acceleration of an object falling toward the ground near Earth (about 9.8 m/s^2). The value 10 m/s^2 is commonly taken as a convenient alternative definition for 1 gee. During one second at 1 gee an object acquires 10 m/s of speed regardless of its mass. The mass of an object (measured in kilograms) is determined by the amount of material in it (how many atoms of each kind, and so forth) and is always the same whether the object is in space or on a planet. About the only time people experience acceleration as high as 1 gee is in falling to hit the ground or colliding in a car. When motion is brought to a halt, the process is sometimes called deceleration and is the negative of acceleration. If there is no bounce and little room in which to decelerate, the magnitude of gee is so high as to be deadly.

The force on an object due to Earth's gravity near Earth is equal to the mass of the body multiplied by 9.8 m/s^2 (1 gee) and constitutes its weight. Force and weight are measured, not unexpectedly, in newtons (N), a unit compatible with meters, seconds, and kilograms. An object is affected by this gravity force whether or not it moves in response. During "freefall", best done in space with no air friction, the acceleration is about 1 gee near Earth and less when farther away. (The behavior of gravity over large distance is covered later.) A falling person in space feels no force or weight because nothing prevents his/her acceleration at 1 gee, but Earth's gravity is pulling on the person and on every object the same as usual. In a sense (thanks to Einstein) acceleration and gravity cannot be distinguished, and this equivalence can be useful in space travel. Acceleration should not be confused with force although this happens readily in discussion; strictly speaking we cannot feel acceleration, only force. In the case of a jetliner accelerating (usually less than 1/2 gee), the passengers do not feel forward acceleration but rather force on their backs from their seats pushing them to keep up with the plane. Some writers define this force as a "gee" force of the same magnitude as the acceleration.

An important force in this book on space travel is "thrust", the forward force on a rocket or spacecraft when burned fuel or other mass is ejected behind. How this works is discussed later. When a mission needs high acceleration, present rocket technology can burn and eject fuel at prodigious rates, producing accelerations of hundreds of gees. This is well beyond what healthy humans can endure: 3 gees for less than an hour, about 10 gees briefly. It has not proven practical to launch slower with less than a few gees because of the nature of chemical rockets. When a heavy payload such as the 90 tonne loaded shuttle is to be accelerated up to speed sufficient to orbit Earth, the acceleration reaches 3 gees after about 1800 tonnes of fuel are burned in about 8 minutes and the boosters and external tank are dropped away. The people aboard need endure the 3 gees for only a minute [b-McAleer].

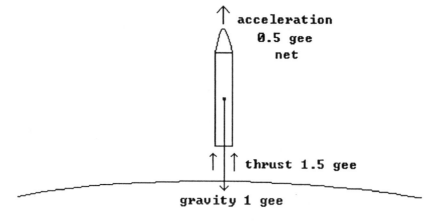

Figure 1.1   Rocket launch under gravity.
(Note: a starship would not be launched from a planet.)

When a shuttle with boosters (or a Saturn rocket) is launched vertically with crew, the actual upward acceleration of 0.5 gee (or 0.25 gee) is the result of the difference between the upward acceleration due to rocket thrust of 1.5 gee (or 1.25 gee) and the 1 gee downward due to gravity [see Figure 1.1]. The initial thrust must exceed the weight of the rocket or it does not move at all. In the first minutes so much fuel is burned and ejected that the rocket becomes much lighter and accelerates faster, soon reaching 3 gees or more. A rocket lifting off does more than accelerate. Its rate of acceleration increases (sometimes called "surge"). The crew feels heavier and heavier because the wall under them pushes upward harder to keep them accelerating and to increase acceleration to several gees while also overcoming the 1 gee downward due to Earth gravity. Rockets do not require much time to pass through 100 km of atmosphere, about 6 minutes at an acceleration that increases from 0.5 to 3 gee. [A later section and Appendix B show how to calculate motion when acceleration increases as mass decreases.]

To achieve low orbit a shuttle must acquire about 8 km/s of speed, a large value to us traveling about on Earth, but rather slow for space travel. Presently only chemical rockets are used for propulsion to leave Earth or to travel in space. These cannot readily achieve speeds as high as 20 km/s because the best chemical fuel, hydrogen, will not burn to produce hot gas moving faster than about 4 km/s. In practice it has been difficult to send spacecraft away from Earth at much more than 11 km/s, the escape velocity for Earth. The Apollo missions to the Moon required huge rockets to acquire about 12 km/s, leaving a mere 1 km/s or so for travel as they approached the Moon. The speed of Earth in its orbit (about 30 km/s) has been important in helping NASA spacecraft reach the large outer planets against the Sun's gravity. There several gravitational boosts by the outer planets have given the Pioneers and Voyagers about 20 km/s for their departures. The Sun's pull will leave the Voyagers with about 16 km/s after a century or so, far slower than our slowest model mission. Thus chemical rockets seem to be out of the question for interstellar travel and other propulsion methods must be investigated. Faster methods would also benefit travel to nearby planets but have not been studied past the laboratory stage.

In space travel as in all other modes of travel, what is speeded up eventually may need to be slowed down. Space missions launched to other planets have sometimes not been slowed. They have kept going into interstellar space, although at low speed. Missions to orbit the Moon, Mars and Venus have used matching of their paths to planetary orbits to minimize the rocket thrust needed in reverse at the destination. Apollo capsules returning to Earth used air friction carefully to dissipate about 10 km/s without any rocket braking. Thrust or other force is used in the direction of travel to accelerate the vehicle, and deceleration requires a force in the contrary direction. In cases of interstellar travel, the same effort would be needed to decelerate the starship as to accelerate it, with one exception. If an enormous rocket were used to accelerate it, most of the mass is fuel and is ejected. The largest tanks and engines can be abandoned along the way (as the shuttle does), leaving a smaller starship which then needs smaller tanks of fuel and engines to provide the thrust to stop it.

   Present achievement in mass lifted to orbit was set by the Apollo program. Apollo's Saturn 3-stage rocket had about 2800 tonnes mass when fully loaded on the ground. It could lift about 125 tonnes payload to low orbit (and 40 tonnes to the Moon). Even with staging the ratio of initial rocket mass to final payload in orbit is about 22:1. The shuttle can lift less (64 tonne orbiter and 29 tonne cargo), but, perhaps not surprisingly, its ratio is the same, about 22:1 for the comparable job. Lest the Saturn and shuttle be thought rudimentary designs, it must be pointed out that chemical fuel with its large amount of mass to obtain high thrust from chemical energy is nearly optimum for launchings, and that rockets were the result of decades of experience and design after the rejection of many advanced methods to be discussed later. (Nevertheless, still other launch methods such as spaceplanes are possible.) A starship may make the 111-meter tall Saturn rocket look like a candle, although it will not be launched from the ground. The minimum payload for a starship (say, the fast explorer at 1000 tonnes) would require at least 50 shuttles to place the important payload mass into orbit before going to work on the really heavy parts, the structure and propulsion and the facilities to build it. These parts could require thousands of shuttle launches, and producing them on Earth is to be avoided.

1.2 ROCKET PROPULSION

   As illustrated by the Apollo program to carry humans to the Moon, travel in local space uses rockets. Rockets work by ejecting mass (material) at high speed from exhaust nozzles [Figure 1.2]. This is called "reaction mass", not necessarily because it came from chemical reactions but because the action-reaction between it and the rocket produces thrust. This law which says that whatever force one uses to push on an object, one feels the object push back with the same force, is also from Newton. The thrust is the force that pushes the rocket to accelerate it as an equal and opposite force pushes burned fuel out. In more detail, thrust pushes on the walls of the combustion chamber, that chamber pushes on the structure of the rocket, and the rocket accelerates. Its parts are well fastened together so that the forces passed along from chamber to structure do not break the rocket apart instead. The reaction on the rocket is the same whether a small mass is ejected at high speed or a large mass at low speed. To conserve fuel mass it is desirable to eject as little mass as possible for a given thrust, and high exhaust speeds are usually better if available.

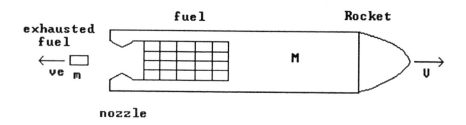

Figure 1.2   Momentum drive moves a rocket.
Thrust is produced from action-reaction.

The physical quantity "momentum" is a special combination of mass and speed (or velocity), calculated by multiplying mass and velocity together. The term "velocity" is more precise than "speed", especially here, because it implies (to a physicist) the direction that goes with speed. A certain amount of momentum must be carried away by the burned fuel mass in the backward direction in order that the rocket gain an equal amount of momentum in the forward direction. The effect is the same whether a large mass at low speed or a small mass at high speed carries the momentum if the product is the same. A rocket engine might be called a "momentum drive" and makes use of a law of physics that states that the net momentum of an isolated system is zero. A rocket on the pad or "motionless" in free space (not easy to achieve) starts with zero speed and therefore zero momentum. As it burns and ejects mass, the rocket body momentum remains equal (and opposite in direction) to the ejected burned fuel momentum. The sum of these momentums remains zero.

Whether the exhaust bounces around on the ground or shoots across the solar system behind the rocket is irrelevant; the rocket moves forward the same. A common misconception is that the exhaust needs something "to push against". Rather, this is a self-contained propulsion system that works in empty space and less well in atmosphere. In contrast to terrestrial vehicles whose exhaust contributes negligibly to motion and is thought of as waste and nuisance, rocket exhaust is the most important result of the functioning of the rocket's most important part, its engine. A jet engine achieves propulsion in a similar way to a rocket engine, except that it pulls air in to help burn the fuel instead of carrying all chemicals needed for combustion.

The theory of rocket propulsion has been worked out in basic physics [shown in Appendix B]. The simple and useful result is the Rocket Equation, which tells what final speed will result when a certain amount of fuel is burned and ejected at a certain exhaust velocity. The term "velocity" is used for the exhaust to remind us that it is ejected in one particular direction, backwards from the intended motion of the rocket. The rocket's mass decreases from some fully loaded initial amount to the mass of a rocket empty of the fuel intended for that "burn" but still consisting of payload, all tanks, engines, structure, and other parts. Its mass has decreased only by the mass of fuel used. No gain is made by releasing unneeded parts unless a new burn is done with the old or new set of engines and further fuel. It has proven difficult to design rockets which do not consist mostly of fuel (and oxidizers). Despite the ferocious burning of the best chemical fuels, their exhaust products do not have sufficient velocity for rapid interplanetary travel.

The thrust on a rocket can be calculated from the rate of mass ejection multiplied by the exhaust velocity. If the rocket is sitting on a planet, it will not begin to move until the rate of mass ejection provides thrust that exceeds the gravitational pull on the rocket. For the shuttle liftoff, the initial mass ejection from boosters and main engine (fed by external tank) is about 8 tonnes per second exhausting at an average 3.5 km/s. As a rocket burns fuel, usually at a constant rate for constant thrust, the exhaust of hot gas is like a controlled explosion. Applied to the enormous initial mass of the rocket (say 2000 tonnes), this formidable thrust can give only a low initial acceleration. If mass could be ejected in a literal explosion to get more thrust, the rocket might leap up uncontrollably and soon be out of fuel if not destroyed. Even for controlled burning, a chamber must be constructed beneath the rocket to carry hot exhaust away and avoid disaster before it lifts off.

Looked at another way, the Rocket Equation states that the proportion of the rocket devoted to fuel mass depends exponentially on the final desired speed. The graph of Figure 1.3 shows this relationship to illustrate what "exponential" means. The dependence is a very strong relationship or "function". M0 denotes the total mass, which consists of fuel mass Mf and final mass or payload M'. If one wants a little more speed, much more fuel mass Mf is required so that M0 must be much larger. Most of that extra fuel is needed to get the fuel moving fast to be used later. As shown, the mass ratio M0/M' to obtain a final speed 3 times the exhaust velocity is 20. This is near the practical limit found with the shuttle, which uses 22 times its payload in fuel (plus the mass of booster shells and tank) to reach about 8 km/s. It is difficult to build a vehicle which consists of over 95% fuel, but by dropping boosters away the shuttle is effectively like a one and a half stage rocket, eliminating some dead mass soon after liftoff. The final speed is about 3 times the typical exhaust velocities for the fuels used (about 2.5 km/s for solid boosters, almost 4 km/s for the hydrogen-oxygen engine).

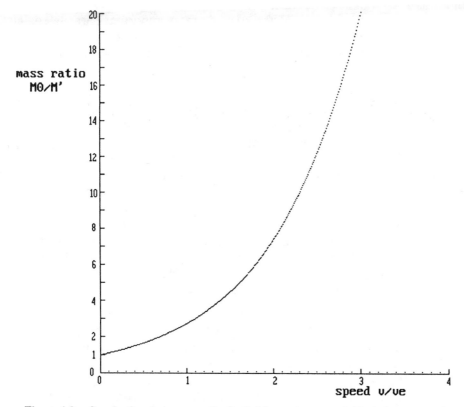

Figure 1.3   **Graph of rocket mass ratio needed for a given speed. Ratio is expressed in terms of initial mass M_0 to mass M' remaining at a given speed. Speed is shown in terms of exhaust velocity v_e. Increase in mass ratio needed is exponential for non-relativistic rocket.**

The Apollo missions needed almost 4 times typical exhaust velocity, or about 12 km/s to go from Earth to Moon. This was achieved with the Saturn rocket whose payload of about 40 tonnes on top was only about 1/68 of the initial launch mass. The ponderous first stage used kerosene fuel which gives less than 3 km/s exhaust, and the upper stages used hydrogen-oxygen. The Rocket Equation calls for a mass ratio of about 68:1, but no single vehicle could be 1/68 payload and structure and 67/68 fuel. The use of several stages permits ratios this large or larger to be effectively achieved. As a stage is used, the speed produced by it is added to the speeds from previous stages, and the mass ratio from before to after the stage is used is multiplied with previous mass ratios. The empty stage (tanks, engines, structure) is left behind so that its mass must no longer be accelerated. Thus practical mass ratios in the range of 4:1 can be used to design stages, each giving about 1.5 times the exhaust velocity, and 3 stages allow reaching about 12 km/s. Three mass ratios of 4:1 multiplied together achieve a large ratio like 68:1. (This example has been simplified by supposing that each stage has the same exhaust velocity. More detailed analysis of a real three-stage mission such as Apollo is complex and not necessary for the purposes of this book.)

Rockets are a very difficult technology because most jobs seem to require more fuel than can be carried. An exponential function such as occurs for the Rocket Equation is unfortunate for scientists and engineers because effects get out of hand very fast. Exponential functions wreak even more technical havoc when travel near lightspeed is considered. Use of stages is only a partial improvement. The Rocket Equation should not be interpreted as absolutely limiting the final speed possible at burnout, but it does show that much more fuel is needed to gain a bit more speed. The Equation works starting from any speed, and it works independently of many details of the motion.

The energy to eject mass in the exhaust comes thus far from chemical reactions of the fuel (although other possibilities are to be shown). The simplest but not the cheapest or safest fuel is liquid hydrogen (H), with liquid oxygen (O) as oxidizer carried in the proper proportion. These substances are difficult to handle on the launch pad but burn to form the most harmless product (steam or water) of any fuel. There are limits to the speed that material can acquire as a result of chemical burning, at best 4 km/s for vigorously burning hydrogen, and therefore limits to its temperature (about 3300°C). Each water molecule formed from reacting hydrogen and oxygen acquires a speed similar to this which represents energy of motion. [See Appendix B for the relation of molecular speed to temperature.] Heat and light radiation is also released. It is important to find fuels that release the most energy for the mass of fuel used. Hydrogen and oxygen are near the top of the list, exceeded slightly by dangerous fluorine. Hydrogen is a good fuel because it is the lightest gas to carry and not coincidentally an energetic burner. Light gases also can acquire more speed than heavy ones. The hydrogen and oxygen reaction produces 13 million joules (13 MJ) of energy per kilogram, a fact to be compared with other fuels. Other rocket fuels that have been used widely and behave well are kerosene with oxygen and a complex mixture in solid fuel rockets. Both produce less than 3/4 of the exhaust velocity that hydrogen and oxygen can.

The energy of motion of water molecules or other reaction products in the rocket combustion chamber constitutes heat energy and spreads around as molecules bounce from chamber walls and hit each other. The chamber is confined partly by a somewhat narrowed exhaust nozzle so that pressure builds up more, and as much of the gas as possible is heated as hot as possible. Indeed the gas gets so hot that chamber walls usually must be cooled to protect them instead of used to help spread the heat. During each collision of a molecule with the chamber wall, it applies a force to the wall. The collisions on the sides do not result in a net thrust, but collisions of molecules moving partly toward the front of the rocket result in thrust as the molecules bounce and then head toward the nozzle and exit from the chamber. Sooner or later each molecule goes out the exhaust. Careful nozzle design is needed to direct as much exhaust as possible backwards and to prevent vibrations and other catastrophes as exhaust gas rushes out into the atmosphere or into space.

1.3 MISSION DESIGN USING ROCKETS

Much must be known about the motion of a rocket for a given method of propulsion in order to predict where it goes and to plan what mission it can do. Newton's law of motion is the basic theory for calculating speed and position in detail as it was for deriving the Rocket Equation that determines the overall mission. The usual "solution" to a given motion problem is a mathematical description of what the rocket does at each point in time. Such a solution is called "analytic", and such solutions cannot often be found for realistic problems. The alternative is a numerical solution where the results are calculated moment by moment from the basic physical laws using a computer program. The solutions are translated into verbal descriptions here, with supporting details in Appendix B. Most interstellar travel is likely to be slow enough to approximate this Newtonian motion. Einstein's relativity theory in Chapter 3 provides replacements for these approaches and new solutions for very high speeds. For the study of interstellar travel, the general problems need not be solved in great detail before estimates of capability can be made and propulsion methods compared. Rocket problems are studied here usually for rockets in free space without the effects of gravity.

The simplest case occurs when acceleration is held constant, but this is not a natural case for most kinds of propulsion. Having thrust which produces 1 gee requires continually reducing thrust in the same way that rocket mass decreases so that the ratio of thrust to mass stays constant (to fulfill Newton's law). Thrust is most easily reduced by reducing the rate of mass burning while the exhaust velocity stays constant. Part of the solution [shown in Appendix B] is that the mass decreases with time exponentially from some initial value until some final value when fuel is gone. The speed increases with time in the usual simple way, and the distance traveled increases with the square of time. Appendix E has tables of speed and distance for selected constant accelerations for this "Newtonian" travel. Even if a miraculous rocket is found that can sustain 1 gee for 1/10 year, this table should stop at about 0.2c where relativity affects the calculations noticeably. When the fuel is gone, the speed is constant. While gravity makes the analysis of this motion more complex, for most cases of interstellar travel gravity can be ignored completely or beyond the outer planets.

When a quantity decreases exponentially, there are few difficulties. The quantity decreases rapidly and smoothly at first, then decreases slower and slower, tapering off very slowly to nothing or in this case to the payload empty of fuel [see Figure 1.4 for a graph of such a function]. The rocket mass decreases ever more slowly because less and less mass is ejected each second, and the decrease is exponential because that is the function that results when the rate of change in mass is proportional to the mass itself. Such a function has a special shape and a characteristic time (called a "time constant") that describes the general rate of decrease. It can be calculated from exhaust velocity divided by the constant acceleration and has units of seconds. If the rate of decrease continued as steadily instead of tapering off, the mass (or other function) would become zero after the elapse of one time constant. For 1 gee this time constant is about 400 seconds for the best fuels such as hydrogen-oxygen and happens to be the "specific impulse" often used to describe fuels in rocketry. After one time constant of 400 seconds has passed, the rocket mass is down to about 37% of its original value; after two time constants it is down to 37% of that or about 14%, and so one for each additional time step.

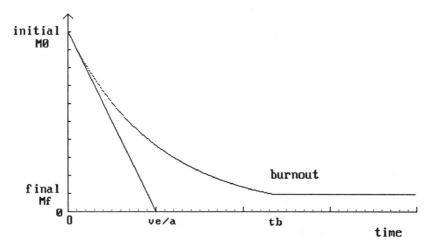

Figure 1.4   Graph of decrease of rocket mass over time as fuel is used.
Decrease is exponential until burnout. Straight line shows initial rate of decrease
and indicates time constant v_e/a for fuel use.

Real experience with chemical rockets has shown that mass ratios of about 50:1 do not provide much speed, only 4 times the exhaust velocity. Suppose that a chemical rocket is wanted for a slow starship leaving the Solar System at 100 km/s, or 25 times the best available exhaust velocity. The mass ratio would need to be about 70 billion. This is not totally impossible if the mission payload is a microprobe of 1 kg and the 70 million tonne rocket is built and fired in space. Huge fragile tanks could be made, no more fragile than some later starship proposals. Except for fragility there is no reason not to use a large acceleration like 1 gee. The time until burnout is the impulse or time constant 400 seconds times 25, or 10,000 seconds. This rocket will travel just 500,000 km before running out of fuel and reaching its coasting phase [see Appendix B]. It will have no problem escaping both Earth and Sun, and neglect of gravity here produces just a small error in these estimates.

Another assumption for rocket motion is that the thrust is constant. This suits better the behavior of chemical rockets. Fuel is exhausted at a steady rate, so many kilograms per second, until gone, and its exhaust velocity with respect to the rocket stays the same, so that the resulting thrust is constant. The rocket mass decreases from the initial value at a steady rate to some small final value. The partial analytic solution for the motion is given in Appendix B, and a short computer program is given in Appendix E which enables step-by-step numerical calculation of the motion as the rocket rises. Figure 1.5 shows a plot of the acceleration a, speed v, and position (height) y for a rocket with modest mass ratio 17:1 lifting off from Earth to low orbit. Gravity is easy to include in numerical calculations, especially near Earth where it is nearly constant. This rocket reaches a high 18 gees before burnout, and, despite being small in mass, represents how a one-stage rocket moves during launch to orbit. More stages are needed to achieve the same results with less final acceleration. Some rockets such as the shuttle main engine can throttle back the fuel burn rate near the end of the burn to avoid the extreme increase in acceleration.

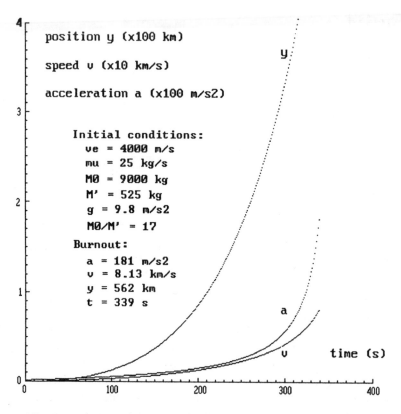

position y (x100 km)

speed v (x10 km/s)

acceleration a (x100 m/s2)

Initial conditions:
  ve = 4000 m/s
  mu = 25 kg/s
  M0 = 9000 kg
  M' = 525 kg
  g = 9.8 m/s2
  M0/M' = 17

Burnout:
  a = 181 m/s2
  v = 8.13 km/s
  y = 562 km
  t = 339 s

Figure 1.5   Graphs of acceleration, speed, and position (height) of one-stage rocket launched in Earth gravity with constant thrust. Note rapid increase in gee near burnout. Speed reaches that needed for low Earth orbit.

For a slow starship reaching 100 km/s at burnout, a payload of 1 kg is cautiously supposed and the initial mass is again found to be 70 million tonnes. No fiddling with the thrust schedule can get around the inexorable conclusion of the Rocket Equation. Any mass ejection rate can be chosen without changing this. If a rate of 1 kg/s is chosen for reasons to be seen, the fuel will last 70 billion seconds or about 2000 years. Again the starship starts in deep space and gravity is ignored. When the analytic or numerical solution is used, the acceleration just before burnout has increased to 400 gees, severe for the hardiest microprobe. The acceleration, speed, and distance are abysmally small for the first centuries and increase steeply in the last years, and it will not work to reduce the mass ejection rate further. A graph of the results is too extreme to give here; the curves would be crammed into the right corner (worse than the human population curve over the millennia). This example shows that a constant-thrust starship can have ridiculous extremes in performance without gaining anything. The absurdity is due mainly to the use of chemical rockets and the resulting huge initial mass. The suggestion that a starship might make the 3000 tonne Saturn rocket look like a candle is supported by these examples. No plan has been made to decelerate it at the destination, much less to bring it back. The mission time is very long, however; a slow starship mission is expected to take 30,000 years.

1.4 GRAVITY AND CIRCULAR MOTION

Gravity is both important and unimportant for interstellar travel. Starships should travel so fast that gravity is irrelevant. But some basic facts about gravity are needed to understand how rockets have been used. And some proposals for starships use gravity to help accelerate them or change their direction.

The principal law of gravity also comes from Newton. Like many laws it comes from much observational evidence. Appendix B provides the mathematical form. Gravity is due to the presence of mass, although scientists are not certain how material really produces gravity. Einstein found an explanation in terms of bending space. Gravity acts like a "field" which permeates all space and is due to all masses present. This is one way to explain why the Sun affects spacecraft moving far beyond the planets. In words the Newtonian law states that the force on an object due to a nearby other object (planet or star, say) depends on the masses of each and on the square of the distance between them. To make the units come out, a universal gravitational constant is also included. This force law describes one of the famous "inverse square" forces which decreases rapidly as the distance between the two objects increases. If the force (weight) on an object is a certain amount at Earth's surface (about 6370 km from the center), then at double this distance (nearly 13,000 km) the force is 1/4 as much. Distances must be compared as measured to the centers of planets and other objects. The Moon is about 60 times farther and feels 1/3600 of the force for each unit of its mass. Because it is so massive, there is still plenty of force to insure that it orbits Earth. Appendix B also shows how to get the value of 1 gee, the acceleration due to gravity at Earth's surface, from the gravity force law.

Newtonian gravity permits special kinds of motion by planets and spacecraft. Simplest are circles but more common are ellipses. Both are properly called orbits, indicating that the motion repeats although not necessarily exactly. Distance remains constant only for circular orbits. For an ellipse the distance between planet and sun varies as the planet revolves around. The imaginary ellipses of the planets move slowly around the Sun because of subtle gravitational effects by the Sun and other planets. If we had two suns, planets would probably never follow circles or ellipses but some other complex ever-changing paths. If the object has too much speed for the local gravitational field, it can escape on a parabola or hyperbola and never return. All these may be called orbits and all are types of the same kind of mathematical curve called a "conic section". Another law (from Kepler) pertaining to orbital motion is that a definite time or "period" holds for each size of orbit, with the period squared being proportional to the size of the orbit cubed [Appendix B]. It is sufficient for this book to know that a larger orbit has a larger period for the particular planet or star being orbited.

Orbital speed can be calculated by combining two Newton's laws, the one for motion telling how much force is needed keep an object accelerating and the one for the gravitational force [see Appendix B]. For circular orbit, the force needed to hold the orbiting object in a circle pulls inward at all times in a simple way. A satellite that orbits Earth is continually falling as usual due to Earth's pull, but it has a high speed sideways (tangential), neither too much nor too little. As it seems to fall it drops around the Earth and never hits. For a circular orbit the required tangential speed is easily calculated to be about 8 km/s for Earth. This is the speed required for a low orbit just above the atmosphere (say 300 km) so that the radius of the orbit (6670 km) is not much larger than the radius of Earth (6370 km). The orbital speed needed depends on the mass of the planet (or star) and on the desired radius of orbit. If the orbital speed is not chosen to match the particular altitude wanted, then the orbit is likely to be an ellipse where the distance from the planet varies during the orbit. Rockets go into circular or elliptical orbits by gradually turning over after passing above most of the atmosphere. If the speed is greater than 11 km/s (for Earth), then no closed orbit is possible, and the rocket has enough speed to depart on a hyperbolic curve that never permits return. The mechanics of orbits is a large subject not fully needed to study starships at a basic level.

Planets in our Solar System follow ellipses near enough to circles that circular orbits are assumed for discussion here. The acceleration needed to follow a circular path depends on the square of the speed, and inversely on the radius of the path [as shown in Appendix B]. This holds for any fraction of a circular path that might approximate part of the path taken by a starship. The circular acceleration is at right angles to the motion, so that a moving object cannot change speed in a circular path. In absence of a source of gravity, a sideways force by other means must be applied to make a starship change course, circular path or not, as shown later.

Actual mission planning in detail requires accounting for the exact shape and orientation of the orbits of the several largest planets involved or that may come near to the spacecraft. Each planet has a near-circular orbital speed in accord with its distance from the Sun. Earth moves around the Sun with a tangential speed of about 30 km/s. Every object launched from Earth starts, in reference to the Sun and Solar System, with this extra and large speed. (Launched spacecraft also start with the speed of Earth's rotation, about 0.5 km/s near the equator.) We on Earth do not notice the Sun's gravity here (about 0.0006 gee) but it is strong enough to control Earth very well. Beyond the Moon, Earth's gravity has diminished to the same weak level as solar gravity, and at that distance one is beyond the gravitational vicinity of Earth. Beyond Earth's vicinity the Sun rules, and to leave the Sun permanently from the vicinity of Earth requires a speed of about 42 km/s. As to be shown, the escape speed is always 1.414... (square root of 2) greater than the circular orbital speed at any given position. Thus a spacecraft leaving a planet will already have about 71% of the speed needed to escape the Sun.

1.5 ROCKETS AND ENERGY

Another basic way to study space travel is to consider the energy of motion, technically called "kinetic energy" (KE). KE and all other energies are measured in joules (J), a small unit. KE is proportional to the square of the speed [see Appendix B] and increases more rapidly as speed increases. This is good and also bad, because as speed increases it becomes harder to increase speed further. The distinction between velocity with direction and speed without it is lost when KE is considered. KE has no direction despite being associated with moving vehicles. High-speed vehicles or rockets are difficult to turn (change direction), but it is the momentum which remains in one direction until force is applied to change it. Drivers have a difficult time learning about KE here on Earth. A car moving 60 mph (about 27 m/s) has four times the energy of motion of one at 30 mph. It is four times more difficult to stop with brakes which convert KE into heat through friction, and it makes four times more violent impact in collision as the KE is dissipated in tearing and heating metal. No matter whether the speed and KE are acquired quickly (high acceleration) or slowly (low acceleration), the KE is the same. KE is also proportional to the mass of the moving object.

The kinetic energy given to a spaceship becomes a liability once the destination is near. The energy must be removed if the ship is to slow down sufficiently to approach the destination. Since friction is usually negligible in space and no brakes are available, the KE must often be removed by turning the spaceship over and pointing the rocket engine in reverse. The same amount of rocket energy is required to cancel the KE as to produce it, a double but unavoidable waste. Thus for a mission that does not just fly by, much fuel must be carried for the end. Just how much energy of motion might be involved for a starship? Our medium-sized mission with 10,000 tonnes payload and carrying the fuel to stop itself might be massing a million tonnes. Traveling at 0.01c (3000 km/s), it carries almost 10(22) joules, a number for which there is no simple scientific name yet. This is about the same as the amount of energy processed in all human technical activity in one present year and about ten million times larger than released in the Hiroshima nuclear bomb. The asteroid that may have caused the Cretaceous-Tertiary extinction would have carried about 1000 times more energy. This starship could not destroy more than a tiny moon in a hit.

There are other kinds of energy, notably heat energy that represents a turmoil of motion, radiation energy which can be disorderly like sunlight or orderly like laser light, and various kinds of potential energy, each one due to a force such as gravitational, electric, and so on. The various kinds of energy can be converted into each other. If all the electro-chemical energy available in a kilogram of H and O (13 MJ) could be converted into KE (mainly as exhaust), the resulting speed would be 5.1 km/s, noticeably greater than the actual exhaust velocity of about 4 km/s. A universal law about energy is that when all kinds are accounted for and added up, the total energy (TE) is conserved. This is to say, energy cannot appear from nothing nor disappear to nowhere (with certain unusual exceptions discussed later). The law of conservation of energy has come from a very wide collection of evidence. If a technical project is planned, and the energy to do it is not available, then there is no way to do the project. We have enough confidence about the physics of energy to make these kinds of claims for different parts of the Universe and even for other supposed civilizations.

Near a planet or star the gravitational pull makes a kind of potential energy seem to exist. As one tries to move away from Earth with a spacecraft, one finds that thrust must be used or that the spacecraft must already have some kinetic energy. To escape Earth, the KE needed corresponds to a speed of about 11 km/s. Each kilogram needs about 63 MJ of KE to depart forever from Earth's surface. (This is not as much energy as it sounds. The energy cost per kilogram, if it could be obtained from the power company, is about 17 kwh, costing less than $2 at present.) An object thrown up is pulled on by gravity, weakly if it manages to get far away. If it starts with less than the KE for escape, gravity wins sooner or later, and it comes crashing back to Earth. If an object far from Earth is motionless (no KE) and let go (ignoring Earth's orbital motion), it will arrive near the ground with the same KE that it would have needed to escape. This discussion assumes that no other energy comes into play, that nothing touches the object, and that it "feels" no other force than gravity. Since by the conservation of energy the object never lost energy, it must have had the same total energy (TE) "up there" as it does down here. Up there it had a different kind of energy, gravitational potential energy (GPE). This potential energy is so-named because the object has the potential to return from there to here and acquire KE along the way. It acquires its KE during falling because Earth pulls on it and accelerates it so that it has increasing speed.

Kinetic energy is always positive. If an object near Earth has 1 joule of KE and fell from above, then up there it had 1 joule of GPE since KE is zero there. And its TE is 1 joule everywhere. But GPE should be numerically larger near Earth because that is where most of the gravity is. Therefore GPE is a negative quantity, reaching its algebraically largest value zero far away and becoming more negative, yet numerically larger, as Earth is approached. Energy-wise, Earth, planets, and stars form "gravity wells", deep imaginary dimples or depressions in space that objects can fall into and from which objects cannot leave unless pushed or thrown out. The Sun's family of planets all orbit at various depths in the Sun's gravity well, as shown in one-dimensional cross-section in Figure 1.6. In this half-slice of the Solar System, the position of a planet below the zero energy horizontal axis is proportional to the square of its escape velocity from that place (but scaled here). Its ordinary distance from the Sun is shown sideways and compressed in scale. The massive Sun forms a very deep well at the center of this system, with 618 km/s being required to escape the "surface" of the Sun permanently if one were foolish enough to go there. Because objects in the vicinity of Earth and Mars must have high orbital speeds, it is very difficult to move inward, more so than outward. No one can just fall into the Sun. The 30 km/s of Earth's motion must be removed from any spacecraft before it could fall to the Sun, just as the shuttle must dissipate its orbital KE as heat in our atmosphere before it can slow enough to land. When orbital KE is counted, the well depth at each planet is not quite so deep, as shown.

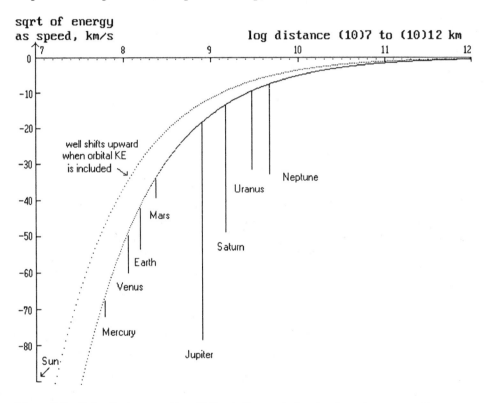

Figure 1.6 **Solar System gravity well due to Sun and planets, shown in cross-section as a graph of energy as a function of distance from Sun. Scales are compressed, with energy expressed in terms of speed to escape (proportional to square root of kinetic energy), and distance on a logarithmic scale. Sun's well is 618 km/s deep, and planetary wells are too narrow to show width here.**

The escape speed can be deduced from comparing the depth of the energy well (GPE) at the surface of the planet (or other position to be escaped from) with the KE needed to end up with zero KE very far away (following a parabolic path). GPE is calculated, using its definition in physics, from gravitational force. The result for gravity is similar to the force except it depends inversely on the first power of the distance and is negative. As Appendix B shows, escape speed from a certain position is the square root of two greater than the circular orbital speed for that position. If KE exceeds in magnitude the GPE (a negative quantity) then the algebraic total or TE is greater than zero. In such a case the object is said to be gravitationally unbound and can escape the planet forever with KE and speed left over (following a hyperbolic path). Near a planet's surface GPE varies so gradually that it can be calculated approximately from the value of 1 gee. GPE is then proportional to the mass of the object and its altitude above ground.

The method of adding energies to find speeds can be introduced here with a numerical example. In the vicinity of Earth a starship needs 42 km/s to escape the Sun. Suppose that we want it to leave forever with 100 km/s and need to know the starting speed. The TE needed is GPE + KE and is proportional to 100 squared. The starship starts in a gravity well with (negative) GPE proportional to 42 squared. The initial KE needed must be greater and is 100 squared plus 42 squared, or about 109 squared. (Note KE = TE - GPE, and GPE is a negative number, so addition occurs.) That extra 9 km/s represents much KE when included on top of 100 km/s, and this is what the mathematics of squaring does for (or against) us.

In review of the energetics of launching a rocket from Earth, the chemical energy in the fuel (a short way to describe the energy stored by electric forces in and among atoms and molecules) is converted partly to KE of the rocket and partly to KE of exhausted fuel, with some energy also lost as sound and radiation. Then, as Earth pulls, the upward moving rocket is decelerated by gravity and by air friction. As altitude increases, KE is converted to GPE, When the burning of fuel is considered, the TE of the rocket is not conserved because much energy leaves as KE in the exhausted fuel. Conserving TE overall would require accounting for all the hot fuel mass spread around Earth and space. But usually TE is considered just for the rocket. If this TE exceeds zero then the rocket can escape Earth. From our viewpoint on Earth, that is all the TE the rocket has. But from the viewpoint of the Sun, the rocket has also KE from Earth's orbital speed and comes out further ahead in its effective TE. Thus, totaling energy is a relative process depending on one's reference frame. To escape the Sun the rocket needs enough extra KE from propulsion to make up the difference between the depth of the Sun's gravity well (now GPE due to Sun) and the orbital KE.

Although not much can be done about it, the poor efficiency of rockets should be mentioned. Not just chemical rockets but any method using rocket propulsion (momentum drive) has the same problem. When a massive rocket lifts off, much energy is lost as hot gases splashing about (and intense sound, heat, and light). The rocket gains very little energy during the initial acceleration, neither KE nor GPE, but the energy stored in its fuel (a kind of TE) is rapidly dissipating. The Saturn rocket with about 2500 tonnes of fuel contains about 75 trillion joules (75 TJ) of chemical energy. After all this fuel is burned, the rocket consists of 125 tonnes payload with orbital speed of 8 km/s, or 4 TJ of KE. The PE of this payload has increased (from ground to low orbit) about 0.4 TJ. Several hundred more tonnes of the rocket are empty shells which reached lower speeds and lower altitudes. While the energy which went to the first and second stages is not negligible, accounting for it will not explain what happened to the original 75 TJ. In some sense it is not fair to discount the early stages because the goal is to have a system which puts as much energy as possible into orbiting the payload. At least 60 TJ is missing after all energy put into KE and GPE has been counted. This rocket has put only about 6% of its original energy into orbiting a payload. Without the use of stages, efficiency would be worse.

The problem with rocket efficiency is traced to the basic physics of what happens to energy and momentum when a chunk of fuel is kicked out of the rocket exhaust nozzle at exhaust velocity. At liftoff a large amount of fuel is exhausted per second, with very little gain in speed of the rocket. The fuel carries away most of the energy in this very unequal interaction. Figure 1.7a shows two extreme cases, when a chunk of ejected fuel is about the same mass as the rocket, and when the chunk is much smaller. In the equal-size interaction, the energy released from fuel is split equally, with the rocket gaining the same amount of energy as the fuel takes away. Both parts have the same amount of momentum. In the unequal-size interaction, the very small chunk of ejected fuel carries almost all the released energy while it carries, again, a momentum equal to that imparted to the rocket. When the rocket speed has risen to much larger than exhaust velocity, the amount of KE that the rocket can gain rises to nearly all of the energy available, but then this is no longer very much of the original fuel energy available since the fuel is nearly gone.

Mathematically the problem is due to the calculation of momentum with the first power of speed and of KE with the square of speed. Conservation of momentum determines the resulting speeds and KE must follow in accord with whatever they are. In the unequal size interaction with tiny chunks of mass ejected from a massive rocket, nearly all of the KE goes to the exhausted fuel because it gains much more speed than the rocket does. Figure 1.7b shows a graph of the fraction f of input energy that goes to rocket KE, as a function of rocket speed v. [Appendix B shows how to calculate this fraction.] When exhaust velocity equals rocket speed, the fraction to the rocket is 2/3. Another way to look at efficiency is the overall proportion of fuel energy that went to rocket KE. A simple well-known analysis [a-von Hoerner; a-Oliver; a-Marx; Appendix B] finds the proportion to be largest (about 65%) when exhaust velocity is used that is about 63% of the desired final speed. The dependence of efficiency on final speed for a given exhaust velocity is graphed in Figure 1.7c. This efficiency falls rapidly as rocket speed is driven well beyond exhaust velocity. When the best ratio of final speed to exhaust is put into the Rocket Equation, the mass ratio to use comes out very near 5. This is said to be the most efficient mass ratio. This ratio turns up later for another optimum situation.

The basic inefficiency of rockets cannot be improved upon much. It is very impractical to design a rocket which kicks out huge chunks at low speed first and smaller chunks at higher speed later when the rocket is less massive. Instead rocket engineers focus on getting the highest exhaust velocity at all times so as to minimize the fuel mass required. If rockets could achieve higher exhaust velocities than possible from chemical reactions, efficiency would be even worse to reach orbital speed. In some ways chemical rockets are the best match to the jobs done thus far.

Chemical rockets handle the most mass per unit of energy, making high thrust good for liftoff (and not much else). Higher exhaust velocity would result in more energy going into the exhaust than into rocket KE. There is no good solution for a rocket, but adjusting the exhaust velocity can provide some improvement in efficiency when the rocket mass does not change much. This is nearly impossible for chemical rockets but not for certain others.

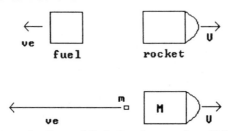

a. Relative momentum of rocket and fuel when large and small fuel masses are exhausted.

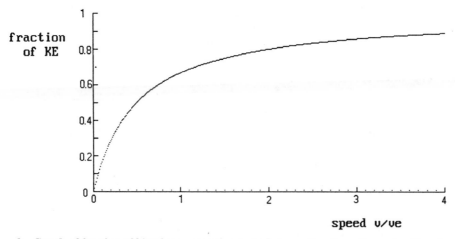

b. Graph of fraction of kinetic energy going to rocket as a function of speed achieved.

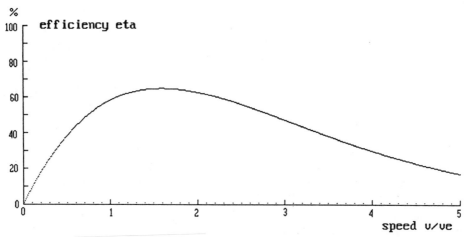

c. Graph of overall efficiency as a function of speed achieved.
Figure 1.7   Rocket efficiency (non-relativistic).

1.6 GRAVITATIONAL ASSISTANCE

The "grand tour" technique of *Voyager 2* used the large outer planets to give the spacecraft additional speed, throw it from one to the next, and then eject it from the Solar System. This acceleration method works because the spacecraft can acquire a substantial fraction of each planet's orbital speed as it is pulled into that planet's gravity well [recall Figure 1.6]. While energy is conserved at each planet, with the spacecraft losing back to GPE what it gained in KE as it accelerated to zoom past the planet, the spacecraft still comes out ahead. As shown in Figure 1.8, it approaches the vicinity of the planet with a modest speed, say 10 km/s, and adds on much of that planet's orbital speed after leaving. Jupiter with orbital speed 13 km/s can provide perhaps another 10 km/s for a total of 20 km/s. This already exceeds escape from the Sun at Jupiter's position. How much speed is added depends on the angle of approach and departure. If the spacecraft is to be directed outward, it should not leave Jupiter tangentially but at an outward angle, taking less extra speed. Energy is conserved in flybys in that the spacecraft has slowed Jupiter very slightly and robbed some of its KE. Jupiter is available for a flyby toward a particular direction once every 13 months. For the Voyager missions, flybys were used to direct the spacecraft from one planet to another, not to build up the largest possible speed.

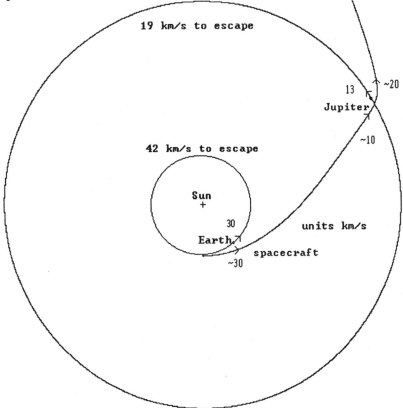

Figure 1.8   Planet flyby path. Spacecraft speeds are approximate.

Planetary gravity assists or flybys are complex to analyze. The results differ depending on the incoming speed and the angles wanted. Lining up more than one large planet for boosting a starship toward a specific distant star requires long waits and careful planning, which we can do. Surprisingly, low incoming speeds are raised more than high ones, so that multiple boosts do not pay off in proportion to their number. And it is not necessary to go in at a planet-grazing altitude. A single good boost at an altitude of 10% of the planet's radius can give close to the largest benefit possible. The most study for starships has been done with Jupiter because of its large mass [a-Jaffe]. For a 10 km/s approach to Jupiter, a final speed leading to 13 to 15 km/s beyond escape from the Sun (a hyperbolic speed) can be achieved in a small range of outward angles. Further improvement is possible with powered flyby, where a rocket engine operates to provide a modest increase in speed during the time the starship is near the planet. Each extra 0.5 km/s added deep in Jupiter's gravity results in an additional 2 km/s outgoing, up to a point. A starship coming in with 10 km/s and adding 2 km/s there will depart the vicinity of Jupiter with 18 km/s, plus the speed of Jupiter, giving up to 31 km/s depending on angle.

To illustrate further the energetics of travel using gravity boosts, consider the effect of the depth of the solar well in the vicinity of Jupiter, where 18 km/s is needed to escape. After a powered flyby at Jupiter a starship might have 31 km/s referenced to the Sun. How much speed will it have after escaping? Conservation of energy determines the outcome. The problem can be solved without considering mass, just KE and GPE proportional to speed squared. When the 18 squared of KE that it lacks is subtracted from 31 squared, about 25 squared remains. Therefore the starship will depart from the Solar System with 25 km/s to spare, still a decent amount of speed considering that much of it was obtained for free. In general, when a spacecraft moves up or down a gravity well and its KE changes accordingly, the resulting speed can be found by adding or subtracting the squares of initial real speed and speed equivalent to well depth.

Saturn is one-third as massive as Jupiter, and moving slower (about 10 km/s), but it is located half as deep in the solar well. After a starship has received gravity assistance from Jupiter it might approach Saturn with say 20 km/s. The faster a flyby is done, the less extra speed can be acquired. The starship might depart the vicinity of Saturn with nearly 30 km/s. To send the starship in a particular direction, Jupiter and Saturn have about the same alignment only every 20 (Earth) years, a disadvantage. (In 20 years Saturn advances about 240°, and Jupiter goes around 1 turn plus 2/3 of another, or 240° also.) The process can be repeated at Uranus with even less gain, but the starship will be farther up the solar gravity well. The required alignment of these three largest planets occurs about every 150 years, too long a wait for most missions. Some of the literature has estimated far larger speeds from two or three gravity assists, but these have been found in error [a-Ehricke; b-Sagan; a-Matloff 1985 p.133-, 1988 correcting self and others]. These references and Jaffe cited above also give many details of calculating the results of flybys. Planetary assists have been popular for studies of very large starships intended for slow multigeneration travel [a-Matloff 1985 p.253-] because large planets are a far cheaper way to boost speed than construction of gigantic rockets. Planets can also be used for braking fast starships. The starship need merely approach a massive planet contrary to its orbital motion and approximately that amount of speed will be subtracted from its incoming speed.

The Sun can be used for a gravity boost, provided that speed is added by other means while the starship is deep in the solar well. Otherwise whatever speed gained as the starship falls toward the Sun is lost coming out, due to conservation of TE. Powered flyby is very effective near the Sun if the starship can take a scorching. One study shows that adding a reasonable 10 km/s on a very close pass to the Sun results in 110 km/s of speed available after escape from the Sun [a-Matloff 1988]. The Sun is a better place to add speed because there is more distance and time available going around the Sun to operate a rocket engine while deep in the well. If one is going to do this, the starship should be dropped from far out and pass as close as possible at "perihelion". The effect is easily illustrated. Suppose the starship starts from beyond Neptune near the top of the solar well and reaches speed 600 km/s as its GPE is converted almost fully to KE. Then 10 km/s is added as it goes around the Sun in a narrow elliptical orbit. Afterwards it starts up the well with 610 km/s. Its original energy was proportional to 600 squared, its new energy to 610 squared. The increase may not sound like much, but it is substantial, and the square root of the difference is 110. The starship has an excess of 110 km/s after it leaves the solar well because a little speed was added on top of a large speed while it was deep in the well.

The extra energy has not come from nothing, nor from the Sun. Before going down the well, the fuel has large GPE too, and had to be transported up there. After being exhausted, this fuel mass remains as far in the solar well as the starship climbs out of the well. As far as the Sun is concerned, Earth is far up the well [recall Figure 1.6]. It would be sufficient to drop the starship from here to gain most of the above benefit, but to pass near the Sun the orbital speed of Earth must first be removed. Inefficient chemical rockets and a flyby of Venus or Jupiter contrary to its motion could be used to do this at reasonable cost. After that a much more efficient propulsion could be used at the Sun to restore much more KE than was lost at Venus or Jupiter. The starship will depart the Solar System at a speed of 100 km/s at the least.

Gravity assistance works best in the orbital plane of the Solar System. As is likely with most planetary systems, ours formed with all planets moving in the same direction and nearly in the same plane (called the "ecliptic"), within a few degrees (excepting Pluto). To reach a star far outside the plane, such as Alpha Centauri 60° "south" or many other nearby stars, Jupiter or Saturn or the Sun can be used to divert the starship away from the orbital plane. The starship is aimed at a point above the ecliptic before its path is bent around the planet toward a destination below. When the bend is extreme, not much speed can be gained at that planet. Other gravitational tricks are possible when two or more massive planets or stars are used along with a starship. In some cases substantial energy can be subtracted from one or both massive objects and given to the starship. The possibilities within our Solar System for significant energy and speed gain have apparently all been found and studied, however.

Here or at other stars some missions and maneuvers might call for using a large planet or star to alter the direction of travel of a fast starship. In terms of physics such an encounter is called scattering because only the conditions of the starship before and after encounter are of interest. To scatter from an attractive force, one need merely loop behind it. What one wants to know is how far from the star (the "impact parameter") the initial velocity must be aimed to achieve a certain new angle of travel after the encounter [see Figure 1.9 and Appendix B; b-Goldstein]. Most of the gravitational interaction occurs in a small region near the star. The result depends only on the mass of the star and the strength of gravity. The path is hyperbolic. For high speeds and small changes of direction the change in angle is inversely proportional to speed squared and to the impact parameter. For a large angle change the starship should be aimed close to the star and/or have low speed. As before, gravity has more effect when the vehicle is in its influence longer. For a numerical example, if speed is 1000 km/s (0.0033c), a starship aimed a close 10 million km from the center of the star would be deflected about 2° from its original course. To veer 10° at Jupiter, a spacecraft moving at 10 km/s must aim about 10 million km from the center of Jupiter.

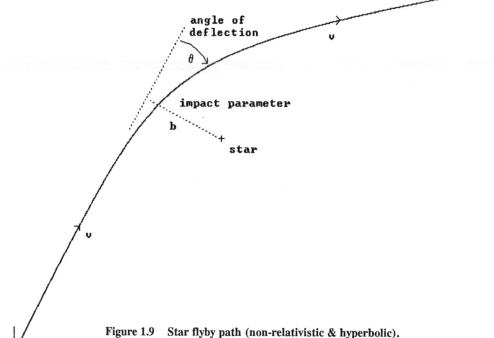

Figure 1.9   Star flyby path (non-relativistic & hyperbolic).

More complex high-speed maneuvers have been proposed at double stars where the pair of stars revolves around each other at high speed [a-Dyson]. The effect is an extension of scattering, where the incoming ship approaches one star nearly head-on in its orbit, makes about one half turn around it, and zooms back out having gained about twice the orbital speed. Ordinary stars revolve slowly and far apart, but compact stars which are small and very dense (e.g. white dwarfs discussed later) might also be close together and moving perhaps 1000 km/s. Then an approaching starship can leave with about 2000 km/s additional. The very high gee is not noticed because gravity acts the same on all atoms of the starship and inhabitants. No such handy binary stars are known near us, but the possibility of such gravity boosts must be mentioned. This system also works in reverse. A starship coming in with moderately high speed can lose its speed at a suitable binary star by catching up with one star from behind and looping around it. The amount of braking or accelerating is adjustable by varying the approach angle.

1.7 FURTHER CONSIDERATIONS FOR INTERSTELLAR TRAVEL

When a very general view of energy is taken, a very perplexing situation appears in regard to interstellar travel. Neglecting the small amount of energy involved in gravitational potentials and the slight drag (friction) in space, the departing starship has approximately the same energy (TE) as it has at arrival. During the trip it builds up enormous KE and therefore TE, but it starts with low KE and TE and ends with a similar low KE and TE. The overall change in energy is negligible no matter how many kinds of energy are included in the accounting. That it moves a little farther from or closer to galactic center is also of no importance. There is no law of nature that prohibits motion of an object from one place to another when no net change of energy is involved. This happens often on the very small scale of atoms, where particles move spontaneously a very small distance from one place to another when energy conservation permits the final result (even if energy is borrowed and returned in between).

Because there is no natural law against it, sf writers cannot be blamed for proposing this effect on an astronomical scale with "instantaneous hyperjumps" of starships from near one star to another. Unlike on Earth where road and air friction convert energy irrevocably into heat, travel in space could be nearly 100% efficient. If a spontaneous jump will not occur, then why not borrow the energy for motion from somewhere, use it to boost the starship to high speed, then return it later? No physical law requires that an amount of energy must be thrown away twice, first to accelerate the starship, and then again to decelerate it. The energy used ends up spread all over nearby space (as fast exhausted fuel molecules) and the starship energy has not changed. If the KE of interstellar travel could be recovered and stored for reuse, such travel would be much more achievable. But thus far this is fantasy within the constraints of basic physical laws. No one has the slightest idea how to manipulate energy in this way on a local or large scale.

Going from the sublime to the ridiculous, a few words should be said about the popular but fictitious notion of "inertia-less" drives in this chapter on the mechanics of travel. Such things seem to be defined as propulsion methods which do not push, but possibly they may pull. The vehicle accelerates but passengers do not feel any gee. Gravity is an inertia-less drive in the sense that we do not feel it pulling us if there is nothing under our feet. People in satellites feel virtually weightless. Physically, inertia seems to be an inherent part of motion. Anything to be accelerated has mass and therefore inertia and requires a force to accelerate it in proportion to mass or acceleration (Newton's law of motion). No new force is known or expected that pulls or pushes on every atom of an object equally so that the object does not feel the force. And inertia cannot be eliminated in any known way. While some elementary "particles" are massless (and can move at lightspeed), we and our atoms are not.

Assuming that better energy resources will be found for interstellar travel, a few other basic aspects of the mechanics of travel should be considered. Given that the mission is not a flyby, it is important to decide whether the starship is to be accelerated for half the trip, then turned around and decelerated with the same propulsion system. This is a good way to provide artificial gee for the passengers because an accelerating or decelerating vehicle feels like it has artificial "gravity" pulling opposite to the thrust. In actuality, the thrust is pushing forward on the feet and seats of the passengers, and the passengers push back in reaction to simulate weight. The effect cannot be distinguished (on a small scale) from real gravity produced by planets. A powerful rocket that develops 0.5 to 1 gee would seem to provide comfortable travel and a fast trip, while preventing loss of strength in people. When speed is nearly as high as it can get (say 0.9c), there is less reason to continue using propulsion (except for a bonus effect, thanks to Einstein).

Using acceleration for artificial gee may be outweighed by the growing energy cost, especially when speed nears c. Another means of artificial gee should be planned since using much lower acceleration to stretch out its duration does not help preserve bones. What acceleration level could be used the whole trip while minimizing the size of propulsion? The answer depends on how long the trip is to take, since any very low acceleration will eventually accomplish the mission. For minimum travel time, the speed should approach lightspeed at midpoint, say 0.8c. To cover half of 10 LY, acceleration should be about 0.06 gee, somewhat helpful and hopefully easy to achieve. The time until turnover is about 13 years, not bad at all. Deceleration is done the same way and the total trip (one way) takes 26 years (Earth time) for 10 LY. If 0.8c is difficult to achieve, perhaps 0.2c is possible at the cost of a four times longer trip, about 104 years and beyond a lifetime. One difficulty in this simple analysis is the assumption of constant acceleration.

In the next chapters propulsion methods are examined which carry more energy per kilogram than chemical fuels can, or which do not carry fuel or reaction mass. Higher speeds become possible and the prospects for travel to the stars improve.

Chapter 2

Advanced Propulsion Methods

Many advanced propulsion systems have reached various stages of paper and hardware design. Most of the basic ideas were suggested in the 1950s or earlier [b-Sänger]. In the search for better interstellar drives, no possibility should go unexamined. For this and the next chapters, a rocket should not be thought of as a chemical burning system with tanks of fuel and engines inherently combined. Any propulsion system should be considered in three component parts: an energy source, the reaction mass, and the drive. Not all of these need be part of the moving spacecraft or starship as was the case for chemical rockets. A general survey of energy sources is given here first, ascending approximately toward the more energetic and exotic. This chapter considers mainly methods which might achieve speeds of less than about 0.2c, avoiding relativity. The next chapters go higher. Methods which make use of the contents and properties of interstellar space are deferred until Chapter 4.

Beyond chemical and gravitational sources of energy are several other source that could be more effective. The characteristics of importance for a source include the amount of energy carried by a unit of mass (its energy density in J/kg, and hence its portability), the amount of energy available, its controllability, and its safety. Taking energy density as KE, the possible exhaust velocity can be estimated. Chemical energy was limited to about 10 million joules per kilogram (10 MJ/kg) for an exhaust of 4.5 km/s, not nearly enough. Energy for gravity is not so easily quantified, but in ordinary planetary systems, the available energy that can be given to a starship was shown to provide speeds not better than tens of kilometers per second.

For each feasible new propulsion method, a first estimate of mission characteristics (acceleration, speed, time, force, mass) is made numerically, keeping as close to one of the model missions as practical. Some general engineering rules should emerge during study of methods and examples. [For some propulsion methods further mathematical physics is relevant at a basic level and introduced in Appendix B. For help with units, see Appendix A. Note the common prefixes k (kilo-) for thousand, M (mega-) for million, G (giga-) for billion, and T (tera-) for trillion.]

2.1 ENERGY SOURCES: SOLAR

The Sun is the first place to look for more energy in space, since about 1400 joules of radiation energy pass through each square meter of surface in space in our

35

vicinity each second. With one joule per second being one watt (w), the intensity (power density) is about 1400 w/m^2. About half of the energy is medium grade heat (infrared), and about half high-grade light energy, with a few percent in the ultraviolet. For heating purposes the difference between heat and light does not matter. For other purposes it might. The high quality of the energy is due to the temperature of its source, about 6000°K for the visible surface. The spread in wavelength from deep in the infrared to well into the ultraviolet (plus some radio and x-ray) is typical of a normally radiating hot object of that temperature at the surface. Given a good collection method, as much high-grade heat and light energy as wanted can be gathered just by making the area of collection sufficiently large. If more intense radiation is wanted, collection can be moved closer to the Sun. Because of the inverse square decrease of radiation intensity with distance, one can find 100 times more intensity by moving ten times closer than Earth's 150 million km orbit. The closest practical place for a solar power station might be at 15 million km from the Sun, well inside Mercury's orbit. There the intensity is about 150 kw/m^2.

It is practical to gather solar energy on Earth only if a very dense storage method is available. Making hydrogen from water is not sufficient. Even the lightest chemical fuels are too heavy to raise out of Earth's gravity well in large quantities. Converting solar energy to microwaves and "beaming" it up from giant transmitters to giant receivers is hardly better. Using it to separate deuterium from water is a little better, since deuterium is a good nuclear "fuel" to be discussed shortly. More practical is to collect solar energy with equipment in orbit around Earth, preferably far away so that Earth rarely shadows or eclipses the collection array. Still better is a collector orbiting the Sun like a planet and much closer to the Sun. Sun-produced energy is not the the only source of "solar" energy. Once a starship has reached another star, similar methods can be used to collect energy there.

Raw sunlight and heat are not directly useful except for solar "sailing" discussed later. It must be converted to a more useful form, either as electricity or captured heat or another form yet to be invented. Electricity is a very controllable form of energy but difficult to store in large amounts. If efficient batteries and other electrochemical and electromagnetic methods still are not available for the urgent needs of electric cars and overnight electric storage, then they are unlikely to be found for starship drives in tremendously larger sizes. Superconductors provide some promise to be explored later. Photovoltaic or "solar" cells and thermal generators are well-developed methods for converting solar energy.

Photovoltaic cells are solid state devices which produce voltage and current when high-quality sunlight in the visible range (not heat) lands on them. The efficiency of conversion has been around 10% for decades with some chance of reaching 20% and slight chances for 30% [a-Hamakawa]. Efficiency is improved by higher intensity, using lenses or mirrors to collect light over a larger area and focusing it on the cell or moving closer to the Sun. The popular and best understood solid state material is silicon, a very abundant element but difficult and expensive to make into pure crystalline solar cells. A thin amorphous version has been developed without regular crystal structure, but it has no more than about 10%

efficiency. The active layer needed to convert light to electricity is only about one micrometer thick and therefore can be low-mass, good news for propulsion if the whole cell can be made thin enough. Solar cells have been proven to operate slow low-mass airplanes. Cells from much rarer materials can be thinner and more efficient. For strength there must be a substrate for amorphous silicon and metallic connections.

The estimated mass per square meter of hypothetical photovoltaic cells (10)-6 meter thick and delicately supported is 0.003 kg. For comparison, the NASA study for a solar cell-powered mission to Halley's comet planned for 70 kw from solar cells spread over 500 m^2 with mass about 500 kg, or 1 kg/m^2, 1000 times more massive than futuristic estimates [b-Friedman]. Energy from the Sun is a good source but spread into space at such low intensity that use of solar cells may lead to very long interstellar travel times.

Estimating solar energy collection requires distinguishing power and energy. Solar cells used for propulsion in the vicinity of the sun will not stay there long. As shown in the next section about 100 GJ/kg can be collected in the time available. However, if photovoltaic cells are somehow illuminated indefinitely, they will collect energy in proportion to the time used and there is no limit on the energy per kilogram. Solar cells should be thought of as collecting power to avoid confusion.

Another way to convert solar energy to other forms is with a thermal generator. The powerplant method should not be confused with thermoelectric generators used with nuclear fission sources and discussed later. In a space thermal generator or powerplant, sunlight is focused by large mirrors onto a boiler to produce high-pressure fluid to turn a turbine and generator much as is done in Earthly power plants with other heat sources and water. Conversion efficiency is about 30% (40% is claimed for a Brayton cycle [a-Aston]), and power density is estimated as high as 10 kw/kg [a-Nordley]. Removing the waste heat in space becomes a serious problem. Space should not be considered as cold or hot, although heat will dissipate by radiation. Despite the 3°K background radiation temperature, warm things cool slowly. Liquid metal and other radiators have been proposed, dissipating 10 kw/kg to 100 kw/kg [a-Nordley; a-Aston]. On a starship scale, very large boilers, condensers, and waste heat radiators would be needed. Conversion of heat to mechanical is a material intensive process. Powerplants are heavy on Earth at 0.1 to 1 kw/kg, and space designs must be much better. Electric generators do operate presently at 95% to 98%. The generator can be lower mass if it uses the new warm superconductors and new light permanent magnets developed for solar car motors [a-Wilson]. While the latter handle about 1 kw/kg, a large version could have lower mass per kilowatt.

Solar collection area as reflectors for thermal generation can be about 1/3 of solar cell area for the same power because of better efficiency for the whole system. Reflecting material similar to that for solar "sailing" can be used, with mass perhaps 10 times less than for solar cells, providing 10,000 m^2/kg. In the vicinity of Earth, collection gives about 10 Mw/kg (only 30% might be used). For a whole system, the mass of collection is low and about 10 kw/kg is indicated from input to thrust. But as to be seen, solar power of the order of 1 Mw/kg is needed to accelerate a starship at 1 gee already moving at 100 km/s, and more for higher speeds. Projected thermal generation methods fall at least 100 times short of having sufficiently low mass to handle the power needed just to accelerate their own mass. Little if anything would be gained by designing very low-mass probes along these lines. A mission at high-intensity sunlight is considered in the next section and gives some improvement.

If heat were needed directly for a starship drive, then the concentrated sunlight could be used without conversion at higher efficiency. For example, solid or gaseous material as reaction mass can be heated in a rocket exhaust chamber by sunlight somehow brought to focus there. There is no advantage unless the material can be heated much hotter than from chemical reactions. Unfortunately, no amount of concentration will generate a temperature much higher than the surface temperature of the Sun, so this method is little better than hydrogen-oxygen reaction and therefore unfruitful. Studies on laser-heated rockets (covered later) also show increasing problems trying to obtain somewhat hotter gases (plasmas).

Another way to convert solar energy to electric is the solar "wind"mill to turn a generator. It has been mentioned as an rotational energy storage device spun up by the Sun [a-Singer; a-Matloff 1985]. It would not help to mount the solar mill on the starship, but it could provide large amounts of local electric power possibly cheaper and simpler than square kilometers of solar cells. The power would be used to propel the starship in other ways to be described. Many solar mills could be used in parallel. The design is similar to solar sailing and is discussed in that later section.

2.2 SOLAR-POWERED MISSION

Using sunlight to generate electricity is a drastic change in propulsion design, since the source of energy is not carried on the spacecraft. A solar source might be best for travel among the inner planets where sunlight is strong. For interstellar travel, a first question is whether adequate speed can be reached before the light is too dim at about 1 billion kilometers out (beyond the orbit of Jupiter). Because of the short distance to build up interstellar speed, the highest "comfortable" acceleration should be used, say 2 or 3 gees. Thrust and acceleration will diminish rapidly as the starship moves away from the Sun because the sunlight intensity decreases rapidly. The two components of thrust, rate of ejection of reaction mass and exhaust velocity, or their equivalents, are affected differently as thrust decreases for different methods of propulsion as shown later.

Setting aside just how the collected sunlight is converted to propulsion, one fact remains (since storage of solar energy is not feasible): the rate of energy usage (the power) for thrust can be large near the Sun but must decrease as the inverse square of distance from the Sun. Even so, not all the power taken in goes to starship acceleration. At low speeds most of the power goes to the exhaust (or reflected light), the worst time to lose the benefit of solar power when it is largest. This is a consequence of action-reaction for a rocket or a "sailing" starship (see later section for "sailing"). Solar gravity also decreases as the inverse square and should be considered. This analysis can be applied to any mission where thrust must decrease with distance in this way. [Appendix B indicates part of the mathematical solution for the motion with gravity assumed negligible and mass unchanging; i.e. negligible reaction mass.]

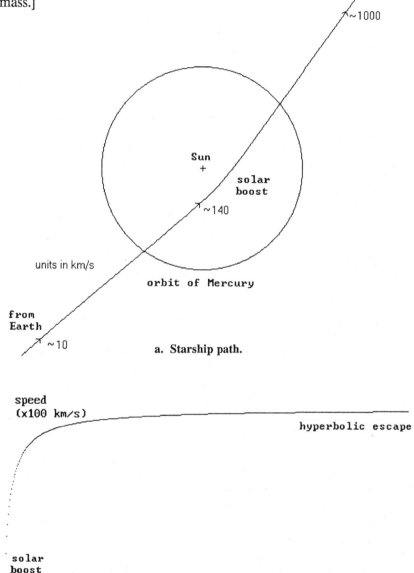

a. Starship path.

b. Graph of speed as a function of radial distance.

Figure 2.1   Solar energy used to boost starship speed (using solar sail or other method). Starship is dropped toward Sun after Earth orbital speed is removed, up to 3 gees of acceleration occurs near Sun, and speed quickly goes well beyond escape velocity (hyperbolic).

Further consideration of energy often can provide exact answers for motion problems where speed and distance over time cannot be solved. As shown in Appendix B and Figure 2.1b, the speed and KE can be found as they depend on distance from the Sun. A numerical example is provided here, expressed per kilogram of starship mass. Solar gravity is 0.06 gee where the starship starts nearest the Sun and remains negligible compared to starship acceleration starting there at 3 gees. Thrust required is 30 N/kg starting at 15 million km from the Sun (1/10 of Earth orbit). As to be seen, a starship starting with about 10 km/s at Earth might have about 140 km/s if dropped to this position [Figure 2.1a]. This initial speed turns out to be negligible compared to the 940 km/s that the starship has after it has escaped the Sun. In fact, any KE added by the Sun going in is lost going out. Most of the acceleration occurs before the orbit of Mercury. This performance is very good and requires only that the solar collection system be low-mass. Removing Earth orbital speed to drop the starship is not easy but whatever solar collector provides the 3 gees can also be used to remove the "small" 30 km/s orbital speed.

It might seem that this mission could leave from Earth orbit, going outward, with about the same outcome but the much larger solar collection area needed to provide 3 gee acceleration turns out to be too massive. Besides removing Earth orbital speed, the solar collector-powered drive could be used to add speed going in if it runs some sort of reaction drive (using mass). Significant gain occurs if 1 to 3 gees can be developed before the starship reaches its closest approach. Suppose that 2 gees are obtained during 10 million km of travel around the Sun (over a time of about 100,000 s). The extra 600 km/s or so would quickly move the starship farther away so that thrust is less. Nevertheless, it appears that a few hundred more km/s can be squeezed out during the solar pass, boosting the final escape speed to about 1000 km/s or about 0.003c. The starship does not travel straight out but on a powered curve, no longer ellipse or hyperbola (until drive ceases). The radius of turn widens as the starship speeds up. The speed gain is so large that the starship departs before making more than a small part of a turn.

If solar cells are used for collection, and solar intensity is about 150 kw/m^2 at 15 million km, then about 0.00005 N/m^2 thrust is obtained at 10% overall efficiency. 600,000 m^2 is needed to provide 3 gees of acceleration for each kilogram of starship mass. A 10,000 tonne starship would need 6 trillion m^2 or an area almost 2500 by 2500 km. This is an unfeasibly large area and indicates that solar energy used this way is not a good possibility. Fastening that area to the starship poses other problems to be seen later. When the gravitational force on 1 m^2 of, say solar cells with 0.003 kg/m^2 is calculated, the result is 0.002 newton, much greater than the thrust obtained from that square meter. Solar cells cannot move themselves unless they become 40 times lower in density.

Some have proposed that acceleration start another ten times closer to the Sun [e.g. a-Matloff 1988], where 100 times less collection area is needed for 3 gees, but one cannot go close enough to gain a major improvement. Moreover, gathering energy so near the Sun is very hazardous. Even with solar collection massing a low 0.00005 kg/m^2 instead of 0.003 kg/m^2 so that it can move against gravity, over 300,000 tonnes are needed for the collector. Both a closer approach and a smaller starship might provide a workable design, say 300 tonnes of collector to drive a 1000 tonne starship, but ultra-thin solar cells handling 15 Mw/m^2 and rejecting most of this as heat do not seem possible. Using lower speed is undesirable, but some systems (such as the sail later) have higher efficiency.

If a solar-powered launch is possible, the same method can be used to halt the starship at a destination. If the star's luminosity is given in terms of the Sun's (a table in Chapter 6 provides this data), then the intensity at a given distance can be found proportionately. The starship need merely approach the star in a tight orbit close enough to develop the 3 or so gees needed to reduce speed to that needed to travel to nearby planets. But even at 1000 km/s these solar powered starts and stops provide for slow interstellar travel. Since mission time is well over 1000 years, the starship must be a large colony, at odds with the low mass required. Whether any solar-powered drive can remain within feasible mass limits remains to be estimated. One other direction for study is the application of solar power to low mass robotic probes which can withstand high gee.

More generally what is the usable energy obtainable per kilogram of solar cells at 10% efficiency, to compare with other methods? An estimated collection time for solar use is about 100,000 seconds, using the above analyses of motion. The solar cells cannot remain in intense sun longer because they accelerate and move out of it (if they have sufficiently low mass). If one kilogram of solar cells has an optimistic area of 1000 m^2, then at least 100 GJ/kg is obtained starting at 15 million km. This is for very intense sun and is 10,000 times better than chemical energy. If solar cells could be sufficiently low mass, speeds of 1000 km/s would be possible, as found above. To obtain this performance, it is necessary to accept that solar cells might be over 90% of the starship mass and payload and structure the remainder, still a better ratio than chemical rockets. Solar cells seem more realistic for local planetary missions. Another unfortunate aspect of solar cells is that they are not used near their full capacity during most of the acceleration period and constitute dead mass the rest of the time. They are likely to be too expensive and valuable to abandon along the way. Indeed they are needed for deceleration. As to be seen in Chapter 4, some have proposed that the solar cells should stay home and the energy collected "beamed" to the starship another way.

2.3 ENERGY SOURCES: NUCLEAR FISSION

Nuclear fission reactors take advantage of the fact that more than a million times more energy per kilogram can be extracted from nuclear "fuels" such as uranium than from chemical fuels. When heavy elements such as uranium with moderately unstable nuclei are exposed to extra neutrons in a reactor, they split into random parts, forming elements of intermediate mass whose nuclei are more tightly bound (have more negative nuclear PE). These nuclei carry away much of the excess energy as KE and are therefore hot. Fission yields about 80 TJ/kg, indicating exhaust velocity over 10,000 km/s if only the fuel mass were to be moved. On Earth the mass of the reactor and associated equipment far exceeds the mass of fuel used per year. The useful output of the reactor is intense heat, limited in temperature by what its materials can withstand. Thus any reaction mass heated by the reactor cannot achieve an exhaust speed much higher than from chemical reactions, or less than 10 km/s. In gas-core designs, only useful for direct propulsion, the uranium fuel is permitted to vaporize and react, and hydrogen gas as reaction mass is heated and exhausted at possibly 50 km/s [a-Cassenti; a-Boyer; a-Vulpetti]. Direct nuclear fission rockets where some of the radioactive products enter the exhaust are a hazardous way to lift off from Earth. Yet work has started on a new system called Timberwind, where uranium carbide pellets heat hydrogen to about three times the exhaust velocity that chemical reaction could in a large rocket engine [a-Aftergood].

The reactor can serve as heat source for any sort of electric generator as high as 40% efficient. With generator included, the combination is called a powerplant, and power outputs discussed henceforth are those obtained as electric power from the generator. Assuming that the drive using the electric power is an impressive 50% efficient, the mass of uranium and power plant needed to provide the KE of a 10,000 tonne starship at 0.01c (3000 km/s) can be estimated. Traditional number names are exhausted, and the KE is expressed as about 10(20) joules. About 2500 tonnes of uranium fuel would be needed if it can be fully fissioned. Unfortunately, experience thus far is that only a few percent of the possible energy is obtained from fission fuel. The fuel mass may need to be ten times greater than the starship mass, leading to design problems.

Nuclear fission reactors and their thermal generators are most efficiently used at maximum safe power level. However, a starship may need low power at early stages of acceleration and high power later. If drive is power-limited, then acceleration can start high and decreases as speed accumulates. If much fuel or reaction mass is used, then starship mass decreases and later acceleration is improved. Exhaust decreases with starship speed if constant power is available. The constant thrust solution [Appendix B] is applicable. As before, at low speed a reaction drive does not put most of the power into the starship but into the exhaust. Any excess is not accepted from the reactor and must be dissipated as heat, a costly process, unless the reactor is damped down. Alternatively, there can be 10 nuclear plants (greater mass but more reliability), and they can be started successively as more thrust is needed. They are dead mass otherwise, never wise for rocket design.

Since the drive is to be used a long time, the initial acceleration cannot be much greater than 1 gee, and the power plants should be sized for producing 1 gee or less when the speed is in a low range. For a 10,000 tonne starship, the power level required to produce 1 gee at a typical starting speed of 10 km/s is one trillion watts (1 Tw). This is the size if not the mass of 1000 typical Earth 1 gigawatt (Gw) powerplants and is unacceptable. The initial acceleration would have to be scaled back to 0.01 gee, requiring 10 Gw of powerplants (say 10 at 1 Gw each). These plants must run 300 years to provide the (10)20 J (at 50% efficiency) required for the mission KE, about 10 times longer than present expected lifetime. Even if terrestrial safety features and containment buildings are eliminated, the basic mass of ten 1 Gw plants is at least 10,000 tonnes.

For propulsion of their Mars mission the USSR announced plans to use four "Topaz" reactors of about 100 kw each to provide electricity at 10% efficiency. The estimated mass is 1 tonne each, giving at best 100 w/kg. NASA's SP-100 100 kw nuclear fission reactor with thermoelectric generator planned for space will mass at least 3 tonnes, for about 30 w/kg [a-Kiernan; a-Aftergood]. Only 4% of the 2.5 Mw of heat from its 200 kg of uranium nitride fuel would be converted to electric power. Scaled up to just 1 Gw, the mass of these complex yet supposedly low-mass no-frills systems would be more than 10,000 tonnes. Since this is about the mass of a starship, this approach seems unfruitful. The power density of a fission powerplant must be more like the 50 kw/kg estimated in another starship propulsion study [a-Aston], or at least 10 kw/kg.

Continuing the above example, an optimistic exhaust velocity from fission-powered drives is 10,000 km/s. Then the exhaust requires 0.1 kg/s mass ejected to provide 1 million newtons thrust for 0.01 gee. If the whole 10,000 tonnes of starship were used as reaction mass, it would last only about 3 years, far short of the 300 plus years needed to reach 0.01c. The power required for this exhaust is about 10 Tw, far greater than the planned powerplant capacity. Again, fission does not seem suitable for moderate interstellar travel.

Not considered yet is the scale of operation for safe refueling of the reactor(s) enroute nor the mass of extra fuel required to decelerate the starship nor the reaction mass for rocket propulsion nor the difficulty and hazards of mining and purifying more uranium on a large scale at the destination, nor the radiation shielding. Since the fuel requirement may exceed the total mass of the starship, fission fuel used in reactors seems unsuitable even to achieve 0.01c. A little improvement in propulsion occurs if the used fuel can be exhausted as reaction mass, but much more than 2500 tonnes is required for that purpose too. Lithium hydroxide makes a low-mass neutron shield as proposed for a fission-powered mission to Pluto and beyond. While fission would make travel to any local planet feasible, other problems should be considered for this or any large-scale use. The preparation of uranium for nuclear reactors and the disposal of radioactive mining and reactor wastes have been of increasing concern on Earth, not to mention the dangers of boosting live reactors to space. Even if assembly is done in space, the heavy uranium may come from Earth.

Because of the difficulties of a fission reactor-driven thermal generator, thermoelectric converters have been used by NASA for planetary missions. The Radioisotope Thermoelectric Generator (RTG) used on Voyager and Pioneer missions starts with 2.4 kw of continuous heat at 1275°K from 4 kg of plutonium spheres (plus many more kilograms of casings and structure) and obtains 450 w of electric power at 18% efficiency. The silicon-germanium solid state thermoelectric devices work over a temperature difference of 700° [a-Heacock] and have a poor power density of about 0.2 kw/kg. The electric power available declines fairly rapidly, to half power in about 40 years.

2.4 ENERGY SOURCE: NUCLEAR FUSION

Controlled nuclear fusion has long been proposed for big power needs such as powerplants and starships. Research has been going several decades on fusion technology without reaching energy break-even (30 Mw for deuterium and helium-3) or engineering design phases [a-Furth; a-Conn; a-Gough]. The goal is to bring the nuclei of one or more kinds of very light elements together when hot enough and dense enough that the nuclei will interact and combine, forming lower-energy nuclei and releasing excess nuclear energy as heat in the form of high-speed nuclei and (unfortunately) some high-energy radiation (gamma rays) and other particles [a-Dushman]. Helium-4 is a very stable and therefore low-energy nucleus and is the endpoint of most useful fusion reactions. Fusion makes use of the strongest known force in nature, traditionally called the nuclear force. It is very strong only when particles are very close together [b-Segre]. Fusion can occur for any light and medium nuclei up to about iron-55, releasing less energy as that most stable realm is approached.

Temperatures needed for fusion are millions to billions of degrees, depending on the elements used. This heat represents particles with KE (speed) sufficient to overcome the electric repulsion between nuclei. The Sun's core is a giant fusion reactor at about 30 millions degrees. The materials or "fuels" there are in a state hotter than gases. Atoms are stripped of all electrons, leaving their nuclei bare. This soup of electrons and positively-charged nuclei is called a "plasma" and can be dense or very tenuous. Some regions of interstellar and local space have this fourth form of matter. Two general methods are being tried on Earth to contain, heat, and compress material for fusion, magnetic squeezing with electric heating of the fusion fuel, and blasting small pellets of fuel with laser or particle beam to heat and collapse them. The Tokamak, a large magnetic torus (donut), is presently the most promising approach. Lasers, electron beams, and heavy ion beams are being tried on frozen pellets of fusion fuel. Sufficient laser pulse power is still lacking, with 25 kJ now achieved at a rate less than one per hour, and 100 kJ planned [a-Cherfas]. In both cases some fusion occurs but controlled energy release greater than the energy needed to start the reaction is yet to be achieved. Various studies show the pellet method more promising for starship drives.

Without knowing the best technology to control fusion, efficiency is difficult to estimate. The fusion energies discussed here cannot be fully used for propulsion, although any losses result in enormous heat loads to dissipate. Fusion fuel consists of light elements such as hydrogen, its heavy forms deuterium and tritium, and lithium. Theoretically, hydrogen used this way instead of chemically would yield over ten million times more energy per unit mass. This is about ten times better nuclear fission. The best reaction in regard to energy would yield 350 TJ/kg, but of course nature has made this one more difficult to start. The more protons in the two components of the prospective fuel, the more they repel and the more heat or KE needed to bring them together close enough to fuse.

Table 2.1 lists some light element nuclear reactions and the total KE released in the products. In nuclear reactions such as these, the resulting energy release is a precise amount and not an average over a range of energy production. When just two particles or nuclei are produced, the KE is shared between them in inverse proportion to their masses. When three or more products occur, each acquires a range of energies. Release of neutrons and gamma rays is to be avoided because they are difficult to control and cause radiation damage. The best reactions produce only charged particles and nuclei (ions) which can be controlled with electric and magnetic fields. Fusion reactions seem to be nicer to work with than fission reactions because of less problems with radioactivity. The fuels are low-mass materials, good for carrying in starships, and some (hydrogen, deuterium, and helium-3) happen to be available at low density all over interstellar space, free for the taking.

Table 2.1
FUSION REACTIONS

Reactants → Products	Mass Units	Energy out: MeV	J/kg	Ignition
p + p → D + e^+ + v	2	0.42	2.0 (10)13	irrel.
p + D → He3 + γ	3	5.49	1.75 (10)14	
D + D → He3 + n	4	3.27	7.8 (10)13	mod.
D + D → H3 + p	4	4.03	9.65 (10)13	mod.
D + H3 → He4 + n	5	17.60	3.37 (10)14	easy
D + He3 → He4 + p	5	18.30	3.50 (10)14	mod.
He3 + He3 → He4 + p + p	6	12.80	2.05 (10)14	hard
p + Li6 → He4 + He3	7	3.90	5.33 (10)13	hard
n + Li6 → He4 + H3	7	4.80	6.57 (10)13	
p + Li7 → He4 + He4	8	17.00	2.0 (10)14	hard
D + Li6 → He4 + He4	8	22.30	2.67 (10)14	hard
p + B11 → He4 + He4 + He4	12	8.80	7.0 (10)13	hard

Notes: p = proton, n = neutron, D = deuterium nucleus, H3 = tritium, He = helium, Li = lithium, B = boron, v = neutrino, γ = high-energy photon. Another convention is to put the nuclear mass number as a preceding superscript. A proton or neutron is approximately the same nuclear mass unit, 1.67 (10)-27 kg. KE released is shown in MeV per single reaction, a common nuclear energy unit, and converted to J/kg for a kilogram of reacting material.

The theoretical exhaust velocity available from using a typical fusion fuel with about (10)14 J/kg (100 TJ/kg) useful output is about 15,000 km/s (0.05c), not much better than fission fuels. The highest exhaust is about 26,000 km/s, nearly 0.1c, from D + He3. These estimates assume that no other significant mass is to be accelerated than the fuel. At this point it is helpful to show the connection between the rest energy of a nucleus or other particle and its speed in terms of c. As to be discussed further in Chapter 3, any object with mass, even when sitting still, possesses a rest energy. For present purposes protons, neutrons, and nuclei made of these have rest energy (RE) of about 1000 MeV (1 GeV, or approx.(10)-10 J) per unit of mass. When the KE of a particle or nucleus is given, such as 4 Mev for He4 from the reaction D + He3, then its speed can be found in terms of c from twice the square root of the ratio of KE to RE; here 4/4000 gives 0.045c.

When pure simple hydrogen (H), the most abundant element, is fused, the nuclei that react are simply protons (p). Unfortunately the reaction of two protons does not involve a true nuclear reaction and is much much less likely to occur than the other nuclear reactions [b-Segre]. The protons repel each other electrically too strongly for the nuclear force to hold them together. Emission of a neutrino, a very elusive particle that carries away significant unusable energy, is a clue that the "weak" reaction has occurred rather than a "strong" nuclear one, as is the small energy yield. Although deuterium (D, also H2), a heavy form ("isotope") of hydrogen with one neutron (n) added to its one proton, is much less abundant than hydrogen here and in space, it yields energy so much more readily than the $p+p$ reaction that this can make up for its scarcity. Tritium (H3, also symbolized t or T elsewhere), is another "isotope" of hydrogen with two neutrons added, but that is one too many because one of the neutrons "decays" or breaks apart leaving helium-3 (He3, with 2p and 1n) in about 12 years. Thus tritium is radioactive and cannot be stored a long time. None is available on Earth or anywhere else unless it is artificially made. Helium-3 is stable but rarer (about (10)-5 of hydrogen abundance). Jupiter and the Sun are much better sources of it than Earth where we would have to make it in nuclear reactors to produce enough. The only reactions in Table 2.1 which use abundant elements and can produce no neutrons by any route are the ones with lithium-7 and boron-11. But they are very difficult to ignite because so many protons are involved.

The most promising fusion reactions involve deuterium, found along with hydrogen in water (1 part in 5000) and in space (about 1 in 10,000). It fuses with itself to yield either He3 and a neutron, or H3 and a proton [a-Powell]. Deuterium is a good compromise fuel because it is not too rare, it produces an adequate amount of energy, and the reaction is not too difficult to ignite. The range of 10 to 100 million degrees (K or C) is sufficient to cause reactions. Some neutron production is unavoidable around hot deuterium, even if He3 is one of the fuels. Unfortunately, neutrons flying out are very unpleasant particles because they ruin other materials and cannot be controlled with electric or magnetic field.

On Earth the fusion of deuterium and tritium has been of most interest because it "ignites" at the lowest temperature, still more than 10 million degrees. One of the highest energies is available, but an obnoxious neutron comes out, carrying most of the energy. Trapping the neutron in lithium shield enables both the capture of that energy for use as heat (only) and the production of more tritium. Tritium is a hazardous radioactive "fuel" which decays in about 12 years, and the supply will not last for a starship journey. The superior reaction D + He3 is not attempted on Earth because there is no significant source of He3 on Earth. It has been suggested that He3 and D be collected from the solar wind, at least for space uses [a-Matloff 1977; a-Parkinson; a-Fennelly]. The concentration is very low, and an electric or magnetic collection system (discussed more later) would need to be 100 km diameter just to collect about 1 kg/yr near Earth. Of interest might be the recent confirmation that Venus has densities of D about 100 times larger than on Earth [a-de Bergh].

Fusion power can be harnessed in four or more ways. The fusion reactor can be a source of heat for operating thermal generators for electric power. Or the energy in the plasma may be extracted more directly as electrical energy with methods using magnetohydrodynamics (MHD). Or the reactor might be designed for direct release of its products as exhaust in one direction, producing thrust like any other rocket. Or fusion can be used more explosively in tiny, small, or large versions of thermonuclear bombs. Since one reactor method involves the nuclear fusion explosion of tiny pellets of deuterium, it can hardly be distinguished from tiny bombs. Various theoretical studies on exploding pellets seem to show this to be the most efficient way to convert the nuclear fuel into propulsion. If the reactor uses magnetic containment, it may need to operate at a constant power level, the plasma density and therefore power output is low, and yet the hardware needed for containment may be very massive. If it uses explosions, then the rate of these can be varied to vary the power output, the fuel is used at a much higher density in small amounts, and the hardware for feeding, ignition, and control appears to be less [a-Bond 1974]. Optimistic power per unit of drive mass is 100 kw/kg.

Another partially-explored way to aid fusion is to use a catalyst. Muons are negative short-lived particles (about 3 microseconds) much heavier (105 MeV) than electrons (0.5 MeV) which will orbit protons and ions very close to the nucleus. If the muon enters the nucleus or comes sufficiently close to the proton, it effectively neutralizes the charge and another such nucleus can be brought close easily [a-Lederman; a-Rafelksi; a-Subotowicz]. The problem with igniting fusion above was mainly one of overcoming the strong electric repulsion of the nuclei. When the charge is neutralized, little extra heat is needed for nuclei to collide often. After the nuclei fuse, the muon is usually released to catalyze another reaction. Nothing gets rid of a muon at the energies involved except its own natural decay (back to an electron and neutrinos). Muons are produced from the decay of other particles (pions) produced in proton collisions at energies much higher than involved in fusion reactors. A special accelerator/generator is needed to supply fresh muons and does so very inefficiently. Muons must catalyze about 10 reactions just to provide the energy of another muon, and possibly another 100 reactions to operate the process of making muons. There is not time to catalyze very many more before it decays. Over 100 deuterium-tritium fusions have been demonstrated per muon but the whole process has not been found to work sufficiently close to break-even to begin powerplant research. Work has been done only on this reaction that produces neutrons. In some stars carbon catalyzes fusion, and other more practical ways to catalyze fusion may yet be found (discussed in Chapter 4).

2.5 FUSION PROPULSION

Like fission or any other drive where fuel energy is carried as mass, fusion propulsion obeys the Rocket Equation. The final speed is typically some 2 to 3 times the exhaust velocity for mass ratio ranging from 7:1 to 20:1. Fusion may enable moderately fast trips (up to 0.1c) to nearby stars, as short as 100 years one way for 10 LY. Higher-speed designs have been proposed. Aside from its use as a power source for other drives, fusion may be used directly in rocket mode [note a-Roth]. Except for the neutrons, the particles resulting from the reaction could be directed out as exhaust in one direction, guided by magnetic and electric fields. A "magnetic bottle" design rather than a Tokamak would be necessary. The exhaust would typically consist of 4 MeV He4 ions with speed about 13,000 km/s. Achievable plasmas are very thin (about $(10)21$ particles/m^3), and thrust would be very low, especially in terms of the mass of reactor needed. The reactor can operate continuously or in pulses or explosions. Continuous and pulsed magnetically-contained reactors are being studied at the hardware stage but the results do not permit yet a description of whether or how they would be used for propulsion, much less an estimate of efficiency or mass. While the problem of perfect confinement does not apply in space where some plasma must be squeezed out of the magnetic bottle to provide thrust, the thrust appears to be low by this method. If 1 million newtons thrust were wanted, mass must be ejected at the rate about 0.1 kg/s. Since the plasma has only about $(10)-6$ kg/m^2, and only about 1/10 of this should be ejected at a time, 1 million m^3 are needed, a reactor about 100 meters across and a very big reaction chamber. The amount of magnetic field needed around it is indicated in the next discussion.

The most efficient way to use nuclear fusion energy may be as explosions, either small ones by exploding pellets of fuel or large ones as from nuclear bombs already developed in efficient forms but rather excessive numbers for military purposes. The thrust is necessarily impulsive, although some absorption of shock can smooth out the abrupt pulses of acceleration. Many designs for small pellet systems have been proposed. Typically pellets are about 1 cm in diameter, mass about 1 gram, and yield about 100 GJ if fully "burned". Since billions of them might be needed, manufacture should be simple and perhaps done aboard ship from bulk stock. Since deuterium and helium are gases, the pellets must be frozen cryogenically (very low temperature about 3°K) to make them stable and to increase their density to that of a solid. As details are studied and tried, however, the number of layers of special materials seems to be growing [a-Ruepke].

Ignition of fusion is the challenge. In contrast to contained low-density plasmas which need a certain temperature, ignition of pellets is better discussed in terms of the energy that must be put into them before fusion starts. Solid or liquid fuel is to be compressed a factor of 100 or more while at the same time heated. This can be done by depositing a pulse of energy in the outer surface of the pellet, which then implodes while heating itself [a-Dawson; a-Ruepke; a-Hyde]. The process is called "inertial" fusion because the inertia of an outer shell of the pellet is used to collapse it after driven to high speed. An estimated 10 MJ or more is needed to produce effective fusion and get back substantially more energy than put in. Lasers, electron beams, proton beams, and heavy ion beams have been studied for the task. Work has been done with the first two already, up to the 25 kJ level [a-Cherfas]. Various present high-power lasers each have several disadvantages, so that the best laser is yet to be developed. Besides lack of energy, improvements are needed in firing rate, wavelength, efficiency, durability, mass, and short pulses. The often mentioned free-electron laser is a novel approach but has yet to reach the millijoule level [a-Kim]. Laser efficiency over 10% has been elusive in any suitable large laser. Any component this inefficient presents enormous problems of dissipating waste heat as well as making inadequate use of precious fuel. Electron and ion beams are likely to be produced in advanced, not conventional, accelerators. Ion beams are expected to be the best way to deliver energy but development has been slow for scientific needs on Earth. Collision of larger pieces of fusion fuel is also being tried [a-Pool].

Because of uncertainty over lasers, Project Daedalus chose an electron beam ignition system [a-Martin 1978; a-Bond 1978, 1986]. This design of a flyby of Barnard's Star in 50 years at 0.12c was intended to show that known human technology could accomplish such a mission. The first and more powerful of two stages is described here. The 4 cm pellets (about 10 grams) of frozen D and He3 (the latter gas obtained from Jupiter in a large sub-operation) are given a superconducting coating. Then they are magnetically shot into a reaction chamber 100 m in diameter at the rate of 250 pellets per second [as in Figure 2.2]. Multiple electron beams deliver about 3 GJ in less than 1 microsecond to each pellet when it passes through the correct place. The electron sources are simple electrodes with very high voltage used to pull electrons out. The pellet is compressed a thousand times in volume. At best 15% of a pellet is expected to fuse, and about 100 GJ is released. Energy for the next pulse is obtained by putting an additional coil around the chamber for the plasma to induce electric current in. The fusion "engine" with all supporting components (but not payload or fuel) has a very high output, about 15 kw/kg. About 1 kg/s is ejected, for a thrust of about 10 million newtons. Total fuel required for the mission is 30,000 tonnes of He3 and 20,000 tonnes of D (for 1:1 usage). Tanks, engines, structure, and miscellaneous are about 2800 tonnes. Starship mass of about 53,000 tonnes is 118 times the payload of 450 tonnes. Theoretical work on the Daedalus-style mission continues [a-Ruepke].

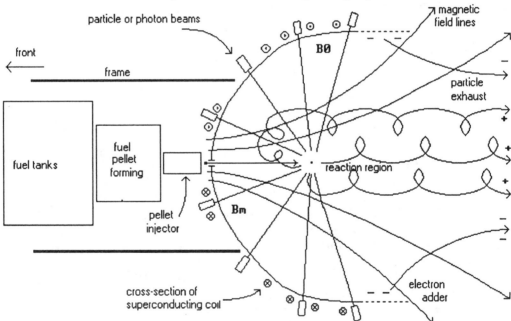

Figure 2.2   Simple hypothetical view in cross-section of a fusion rocket drive. Beam-induced fusion of frozen deuterium-He3 pellets would produce mostly charged particle exhaust.

The next step to get propulsion from small pellet fusion explosion is to contain and direct the reaction products. Daedalus and other designs use magnetic "mirrors" [see Figure 2.2] to reflect fast charged particles one way. Superconducting coils are wound around the reaction chamber to provide for high-strength dense magnetic field Bm at the front end and a background field B0 in the central chamber. B0 ensures that any particles traveling transverse (perpendicular) to the exit be forced into circular orbits and is sized to contain the fastest expected particles, $He4^{++}$ ions moving about 10,000 km/s and having charge to mass ratio 48 MC/kg. [See Appendix B for details on magnetic fields]. A field of about 0.05 tesla is needed to hold them in 4 meter orbits. A current of about 5 million amperes is needed to produce this field in a 100 meter chamber. The particles moving parallel to this field heading toward the front end at about 10,000 km/s begin spiraling around the field lines. They must be stopped and "reflected" within 1 microsecond in the space of 10 m by having the field become much stronger there [b-Jackson; b-Reitz]. The increase in field between the central region and the front must increase about 0.2 tesla in the 10 meters and thus Bm must be about 0.2 tesla. The current to produce this field is about 5 million amperes, because the radius of the chamber is smaller there. These fields are well within present practice, and Bm could be larger to ensure that no ions can touch the structure around the reaction chamber (which may need no material walls, but see a-Bond 1978). Bm causes even the most energetic positive particles which happen to be going forward to reverse course and go toward the rear. This magnetic control of charged particles is seen in other propulsion systems to be discussed. While current for one loop of wire has been estimated, the field is better spread out by using more turns of wire. The current will be proportionally less.

The starship acquires thrust because of the action-reaction between the exiting plasma and the strong magnetic field anchored to the structure by means of the current-carrying superconducting wire. A burst of exhausted plasma carries momentum with it, and leaves the starship with additional forward momentum. The acceleration consists of impulses many times per second, but the magnetic field and mechanical shock absorbers can smooth it somewhat [a-Boyer]. So that the departing positive ions do not leave the starship with increasing negative charge, electrons must be fed to the plasma after it exits to neutralize it. Other "details" are more worrisome. In some designs the reflecting magnetic field for controlling fusion plasma is larger than achieved by our best cold superconductors thus far (about 30 tesla), and the required current density is higher than known superconductors can handle (1 to 10 billion A/m^2). Using a bank of smaller reaction chambers instead of one large one is a way to reduce field and current requirements for each, but other problems such as less protection of hardware from plasma occurs. The superconductors must be protected from heat and radiation and other damage. Some dangerous radiation is emitted and must be controlled or shielded. If a neutron-producing reaction is used, the neutrons must be trapped (in lithium perhaps) and the heat used or radiated away. Inefficiencies in any part of a fusion starship drive would produce huge waste heat loads difficult to dissipate.

Because of the difficulty of igniting pellets of some other fusion fuels, such as boron-11 which is more plentiful and would produce no neutrons but may need 1000 times more energy to start, stages of pellet ignition have been proposed [a-Winterberg 1977, 1979]. The smallest possible pellet of deuterium and tritium would be imploded to release energy sufficient to ignite a larger pellet of the main fuel, with various schemes of magnetic or material reflectors to focus the energy onto the main pellet. Even a third state of still larger pellet is envisioned. If a boron or lithium reaction is used, it might be possible to ignite it with the high-speed protons already coming at the starship in space [a-Bond 1974]. Protons at about 1 MeV are needed (0.02c), so this method does not work at low speeds. Chapter 4 covers the possibilities of gathering some or all of the fuel in space.

The Daedalus mission plan treated the starship as a normal rocket carrying fuel and reaction mass together in small pellets, to be burned by fusion with energy release about ten million times greater per kilogram than chemical fuels. Its travel can be analyzed like any rocket. With all the inefficiencies rolled into a total design, the outcome was 0.12c from a mass ratio of about 120, indicating an effective exhaust velocity about 1/5 of 0.12c or about 7000 km/s. If a mission of this sort were designed to stop by means of fusion power at the destination, what would be the capabilities? Suppose that exhaust velocity is effectively 10,000 km/s and the mass ratio for each leg is 20 (400 overall). The highest speed would be 30,000 km/s or 0.1c. Higher ratios are not worth the small gain in speed. For a 10,000 tonne payload with human colonists, the departure mass is 4 million tonnes, not unrealistic for a space vehicle (several hundred times the mass of an oil tanker and over 1000 times larger than the Saturn rockets). Over 99% of this mass would need to consist of fusion fuel and most of the rest would be engines. Since the trip would require over 100 years and several generations, there is no need to hurry. The acceleration could be moderate and last about 100 years, followed by 100 years of deceleration.

Fuel probably should be exhausted at a constant rate for constant thrust to keep the design and operation as simple as possible, so that acceleration would increase over time as mass is consumed, becoming about 20 times higher near turnover than at departure. Fuel would be used at the rate of about 1 kg/s, which sounds reasonable. Possibly pellets of 10 g each could be fired 100 times per second. Initial acceleration is about 0.00025 gee, rising to about 0.005 gee before turnover, when the whole process is repeated with the same acceleration schedule and 20 times less mass. For some missions there is the possibility of carrying ordinary reaction mass (hydrogen) and feeding it into the reaction chamber during low-speed departure. The drive is more efficient if the mass ejected is larger and its velocity lower when starship speed is low. However, if mass dedicated to fuel is at a premium (as it usually is) then each kilogram of fusion fuel forgone for ordinary hydrogen is a net loss of energy regardless of speed.

The nuclear bomb method of propulsion must be mentioned. An early and incredible proposal to use nuclear bombs (fission or fission-triggered fusion) became a funded project called Orion [a-Dyson 1965, 1968]. The bombs were to be exploded every 3 seconds behind a shock absorber plate, and 300,000 were to be carried to accelerate the starship to 0.03c. About 1/10 the bomb energy can contribute to forward thrust, so this also is not an efficient use of uranium and deuterium. While some scientists and engineers remained enamored of this concept for a long time, the problems not considered are enormous. The environmental problems on Earth from building so many bombs would be even greater than using nuclear fission plants to produce energy on Earth for transmission to space for a starship project. The departure of the mission would leave a trail of hazardous debris in space, although it would be built and depart far from Earth. The starship would have jerky acceleration and difficult steering. Far out in interstellar space might be a good way to eliminate all present nuclear bombs, but this project would have more made. The scale of propulsion with violent nuclear explosions does illustrate the kind of energy needed for accelerating a modest starship. There would be advantage in using much smaller bombs, but fission bombs can only be made in a minimum size, which is a huge explosion about one-tenth the size that leveled Hiroshima. Large pure fusion bombs have also been proposed [a-Bond 1984].

2.6 ELECTRIC ION PROPULSION

Electric ion propulsion has undergone more study and hardware development than any other advanced method of propulsion. Solar-powered ion propulsion was prepared for NASA's "grand tour" mission but cost and reliability questions returned the project to a conventional rocket launch of the Voyagers. Reaction mass must be carried along with the ion drive or "engine", and electric force is used to eject the mass to cause thrust. The energy source can be carried in the spacecraft or supplied externally. Energy can come from the heat of burning chemical fuel, or from nuclear reactions, or from sunlight, and converted to electricity by methods already discussed. Chemical fuel such as hydrogen and oxygen can be converted directly to electricity at good efficiency on a large scale in fuel cells, but fuel cells are inherently massive, about 1 kw/kg plus fuel. Again chemical fuel is out of the question. NASA's ion propulsion program continues as SEPS (Solar Electric Propulsion System) [b-Nicolson], but planned thrust is very low for small spacecraft to travel for months to the planets. Estimated system mass is 1 kw/kg for a short stocky mercury ion thruster getting about 100 km/s using about 15 kilovolts in one stage rather than many stages. Another mission study estimates only 0.06 w/kg for a nuclear-powered ion drive [a-Jaffe].

In ion propulsion, one or more electrons is stripped from each atom of the reaction mass to produce positive ions. Electric fields accelerate the ions between two electrodes or down a tube of electrodes [see Figure 2.3 for one arrangement]. High negative voltages are required to give ions substantial KE and speed. Many electrodes can be used if the field is switched in polarity as ion bunches pass by so that positive electrodes are always behind them and negative ahead. Action-reaction occurs as the ions leave as exhaust and the electric field structure is moved ahead, pushing the spacecraft with it. To avoid charging the spacecraft, the exhaust must be neutral, and an electron injector is added at the end of the system. Physicists are working at ever-increasing scales of ion acceleration for experimental purposes, but the mass of ions moved is still too small to consider for propulsion. Accelerators with heavy magnets to force particles to travel in circles and with small beam currents are not suitable for propulsion. The Stanford Linear Accelerator (SLAC) uses 3 km to give electrons 30 GeV (almost c); heavier ions would be used much slower. Ions of all masses up to that of uranium have been accelerated to nearly lightspeed. When such energetic ions collide with any matter, extremes of temperature, energy, and nuclear interaction are created but no application for interstellar propulsion has been found thus far.

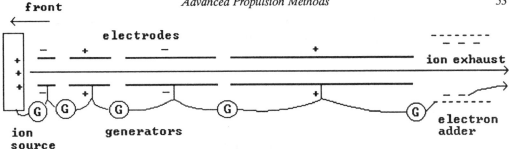

Figure 2.3 Schematic view in cross-section of multi-stage ion rocket drive. Ions exhaust in bunches as generators periodically reverse polarity.

Most elements require considerable effort to ionize, not in terms of energy but in terms of technical method. The energy for ionization is much smaller than the KE to be given to the ion. Intense heat, electric arcs, or other methods are needed to remove electrons, after the material is produced in gaseous form. The working material must become a plasma. Interest has shifted from heavy dangerous metals like mercury and cadmium to a medium mass gas such as argon (40 mass units) or xenon (131 mass units) [a-Matloff, 1983; a-Brown]. Once moving the ions are an electric current like any other, and this current is measured in amperes. With one positive charge the ratio of charge to mass for argon is 2.4 million coulombs per kilogram.

To provide a sample analysis of an ion drive, let one kilogram per second be ejected [see Appendix B for background physics for moving charges]. The current is 4.8 million amperes. If 10 million volts is used to accelerate the ions (mass 20 GeV chosen for simplicity), each ion gains 10 MeV of KE (1/2000 of its RE) and its speed is about 0.03c or 10,000 km/s. The thrust is 10 million newtons. The acceleration of a 10,000 tonne starship would be about 0.1 gee if its mass remains constant (which is unlikely). The KE flow to exhaust is the electric power to the beam, 4.8(10)13 J/s (watts). Additional power is needed to produce the exhaust speed as the starship accelerates, with the extra going to the starship speed via action-reaction. Since most power and energy handled goes (later) to the starship as its speed nears that of the exhaust and efficiency approaches 60%, the power source need be sized only a few times larger than the power sent to the exhaust.

Electric fields are easy to handle up to 100,000 volts per meter, and the upper limit of feasibility with materials, even in vacuum, is a field of about 100 kV per acceleration section. The acceleration tube is divided into 100 electrode sections, each longer than the previous one in proportion to the increased speed. If the first section is 1 m and 100 km/s is gained there, and the last is 100 m, where 100 km/s is added to 9900 km/s, then the total length can be found by adding all the lengths in steps of 1 m from 1 m to 100 m. This simple arithmetic sum gives 2.55 km, not unwieldy for a starship. A high voltage 100 kV generator can be attached between each section but each must be isolated from the others electrically and synchronized. The ion current is switched on and off to provide bunches of ions. One hundred bunches of ions can be handled at one time, one passing through each section. Theoretically an ion drive can eject mass near lightspeed, getting the most thrust out of each kilogram of reaction mass carried, but the technical limit is less and depends on how many kilometers of hardware one wants to build. Moreover, exhaust velocity much greater than starship speed means most energy is wasted. Since both the mass ejected per second and its speed can be varied downward from the design maximum, an ion drive is a very versatile thruster. If used at maximum capability it is a constant thrust drive.

The power used in the above example is about 2(10)10 kw, a very high power. Can this ion drive move itself? An ion drive might be built at 1 kw/kg, so this one masses 20 million tonnes and would accelerate very weakly with about 7 million newtons, less than 0.0001 gee. The mass of the reaction fuel tanks, a possible on-board energy source, and the payload has not been included yet. With travel times of 1000 years or more and 1 kg/s used the whole time, 30 million tonnes of reaction mass is also called for. Clearly this is not a viable interstellar drive if the drive and reaction mass must be so massive. Increasing the mass flow for more thrust is not the answer, because the huge current requires more hardware. If many smaller ion drives can be built that add up to much less mass than one large one, hundreds or thousands could be used in parallel. Mass flow rate could be varied by turning on and off numbers of these. Redundancy of drive units is always good policy. Provided that a low-mass energy source is found, the ion drive might be good for very small probes, say 10 to 100 kg. But missions will still be very slow. A very optimistic ion system has been proposed [a-Aston 1986, 1987] which estimates component masses in the range of 100 kw/kg but still reaches only 0.01c.

An ion drive can be no better for propulsion than the energy source for it. Nuclear fission is a likely power source [a-Bond 1971]. If better sources become available, then they should be used more directly rather than to drive ions. Like most rocket drives, the mismatch of ion speed and starship speed results in low efficiency, and the energy source must supply energy mostly to the exhaust until starship speed increases to near exhaust velocity. With plenty of reaction mass (unlikely) the mass ejection rate could be high, and the voltage and therefore exhaust velocity lowered, during early acceleration for more efficiency. The Sun could be the energy source, in which case the mission is best done by dropping the starship around the Sun while pushing it as hard as is feebly possibly with the ion drive once speed has built up near the Sun. For this power input solar cell voltage, therefore exhaust velocity, remains nearly constant, and current decreases as distance from the Sun increases, reducing the mass flow rate. If the solar source is a thermal generator, the reduction might occur mainly in voltage and result in lower exhaust velocity.

For completeness, another ion method should be mentioned, the plasma/arc thruster or magnetohydrodynamic (MHD) drive [a-Cheng]. The simplest version [Figure 2.4] has two conducting rails with conducting material between them. When high current is sent down one rail and high voltage causes it to arc (in some gas) between the rails so that the current can return, a strong magnetic field occurs around the rails. The arc is a conducting plasma and constitutes a high current flowing across the field. A current in a magnetic field feels a force (Lorentz force) perpendicular to both the field and the current flow. This force is outward along the rails and proportional to the square of the current. The rail spacing is irrelevant.

Magnetic fields are relatively weak, and 1000 amps is needed to produce little better than just one newton of force this way. Thrusters using this principal for the usual momentum drive have been tested in a better design but no better capability. The amount of mass accelerated is small for the hardware because plasmas are thin. The MHD name comes from magnetic force on a conducting plasma causing it to flow and accelerate. There is little hope for this ingenious idea as a large-scale drive. MHD has not taken over Earthly power generation as once predicted.

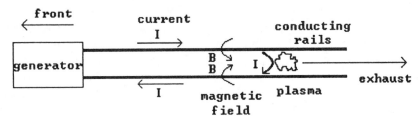

Figure 2.4   Schematic view in cross-section of electro-magnetic plasma (MHD) drive. Supplied gas is ionized to produce exhaust mass.

2.7 ELECTROMAGNETIC MASS DRIVERS AND OTHER MASS EJECTORS

Electromagnetic launching has long been suggested [a-Clarke 1950], and development has been underway on electromagnetic systems for accelerating "buckets" of anything to speeds of 2 km/s or more. The material would be flung from a bucket as reaction mass to propel a vehicle, or to transport the material itself across space. Buckets would be recycled on a continuous path. Rapidly switched magnetic fields are used to repel buckets, which are kept on track also by magnetic fields and supported without friction by magnetic levitation as for experimental trains on Earth. The bucket need not have a power source because the same principle is used of a superconducting loop fastened around the bucket being repelled by a magnetic field. This apparatus is called a "mass driver" [a&b-O'Neill; b-Johnson]. An electric power source, such as solar, would be needed. Success has been shown for kilogram-size loads on a laboratory scale with acceleration over 100 gees (far too high for humans). One gram pellets have been pushed at nearly one million gees over a few meters but reaching only about 3 km/s [a-Rasleigh]. Reaching escape velocity from Earth for one bucket is still a challenge, and this system is unlikely to help boost starships. It has been proposed for auxiliary roles such as moving asteroids and comets around to provide access to their material or to arrange them as anchors or shielding for starship projects. This system is unwise for interplanetary travel and hauling because it would slowly fill the vicinity of Earth with dust and gravel that would be disastrous for all space travel.

Larger scale electromagnetic drivers or launching tracks have been described which operate on the starship itself, or components of it, instead of being part of impulse fusion rocket drive [a-Lemke]. A probe has been proposed to reach 0.33c under 5000 gees over 100 million km. The major advantage is that the power source stays home and is probably the Sun by some means. Big drivers should be far out in the Solar System where gravitational disturbance is less, but their power source is far in, at the Sun. Since approximately constant force is applied to the starship, whose mass is also constant, this propulsion is a constant acceleration type. The simplest analytic solutions for motion would apply. With humans aboard, the acceleration must be limited to about 2 gee. To reach a barely useful 1000 km/s, the driver must be 25 million km long. The required length increases with the square of speed. Reaching 0.1c would require about 900 times longer track, quite unrealistic.

The hardware of a magnetic driver consists of superconducting magnet coils at frequent intervals (perhaps 100 meters) driven by rapidly switched high currents and powered from solar collectors. Figure 2.5 shows a simplified arrangement. The starship has one or more superconducting coils around it carrying high current to make the starship into a strong magnet with field permanently in one direction along the axis of the coil (a magnetic dipole). If currents in starship coil and in fixed driver coil in front are flowing in the same direction, the two attract; the coil behind has current reversed to repel. As the starship advances to the next attractive coil, the current behind is switched to repel it. The effective average force between them depends mainly on the currents in them. Moment by moment the force is impulsive and varies according to the distance between two attracting or repelling coils. Smaller diameter coils would result in smaller forces or require higher currents [see Appendix B]. To accelerate 1000 tonnes at 2 gees, 20 million newtons average force is needed. For 100 m diameter coils 100 m apart the current needed in each is over 4 million amps, or possibly 400,000 amps in the starship coil and 40 million amps in the fixed coils. This is feasible, and multiple turns of wire could be used. The magnetic field strength at the center line is only about 0.06 tesla. Adequate shielding from this moderate field may be possible. A 1 tonne probe launched at 2000 gees requires the same force and therefore current and fields (and can reach 0.1c). Superconductors need no power to keep a steady current flowing, but some is needed for switching and to replace the energy imparted to the starship. An auxiliary coil on the starship can tap power from the drive coils to replenish the starship coil current.

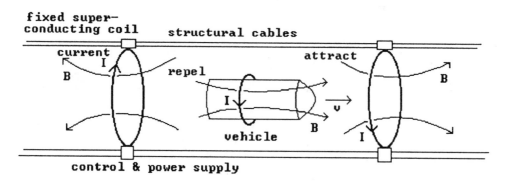

Figure 2.5   Electro-magnetic driver, two of millions of sections.

Considerations for the magnetic driver are applicable to other starship systems using high magnetic fields. The strength of the magnetic field here can be kept within feasible technology [less than estimated by some, e.g. a-Hamilton] but the very high currents needed could be a serious problem for efficient fail-safe electronics. One factor is that any change in field or current results in a voltage being generated to oppose that change (Lenz law). The stored magnetic field energy is very large and dangerous. If superconductivity is suddenly lost ("quenched"), perhaps from a failure at one spot that spreads, backup copper coils are commonly provided to accept the large induced current and dissipate the magnetic field energy as heat. Discussions of magnetic coils for starships often omit this massive extra feature. The strength of the coils and their structure is also a problem. Coils of more than one turn tend to compress themselves longitudinally, and each coil tends to expand in diameter when current flows. The field exerts a high expansive pressure or "hoop stress" on the coil. Coils and structure are severely flexed when current is reversed or otherwise changed.

Each of 1/4 billion or so magnetic coil stations can accumulate its own energy via solar cells and is activated once per launch. Precision timing of the drive system is probably easy, but keeping the magnets in line and stationary during passage of the starship could be difficult. If connected by long cables, alignment may vary, and a magnetic centering system is needed. For economy, the driver coils must be as low-mass as possible and therefore easily moved during action. Propulsion occurs from action-reaction between the starship magnet and each stationary coil. If the magnets can be kept still, then the starship gains maximum momentum.

The magnetic driver is essentially a constant force, length, and diameter device and could be designed to launch a variety of vehicles, from slow massive human missions to fast small probes or fuel for a starship (see later section) by changing its timing. Small fast launches can be done more often but will need about as much energy per launch as slow massive ones, therefore more power input. Instead of a constant coil spacing with ever-faster timing up the line, it may be more practical to increase coil spacing according to the increasing speed, resulting in a much longer track. The driver could be used in reverse to decelerate any vehicle coming through it with a compatible magnetic field, and the KE recovered and stored as electrical energy. The driver probably cannot be in solar orbit but kept "at rest" pointing radially away from the Sun and not near large planets, perhaps not even in the plane of the planets (making access more costly). Effort must be expended to prevent it being pulled toward the Sun.

A proposed long rod to attach the stations together is limited by the gravity tending to pull the closer end apart from the farther end. According to one study, the strength of materials seems to limit a "practical" structural rod to less than 100,000 km long [a-Lemke]. The driver stations may need to be linked in groups that are not connected but kept in place with propulsion. However, four or more cables would be a better choice since gravity keeps them stretched, and there is no need for an inherently rigid connecting structure. The cable size is determined by the impulse it must withstand during launch. Aggregate cable cross section must be about 0.01 m^2 to hold 20 million newtons if tensile strength is a conservative 2(10)9 N/m^2 [see Appendix B for 2.8 and note later caution on strengths of materials]. This is low for "whiskers" but high for bulk material. Total cable mass for a 1 million km set is about 50 million tonnes, for the density of plastics. In a simple model with the cable set near Earth's location and with the force of solar gravity on one half compared to that on the other half, the stretching force in the middle is about 1 million newtons, much less than launch reaction force. A 10 million km set would have about 100 million newtons tension in the middle and still be within the range of the highest measured strengths for materials, about $(10)10$ N/m^2. Since stretching due to gravity decreases as the cube of distance from the Sun, the entire driver could be linked by one cable set if located farther from the Sun. It would be sufficiently massive to hold still during the most strenuous launch. Except for the enormous amount of material and work needed, this magnetic driver could be called feasible.

A magnetically-driven vehicle must be carefully constructed to avoid several kinds of damage. The moderately high field stresses but does not damage magnetic components (does not tear them and their structures apart). But the rapidly changing magnetic field induces current flow in all conductors aboard. Electronic systems must be shielded, and metal structure assembled with insulating sections. Any metal parts will get hot. People and other biological systems may need shielding too. It has been suggested [a-Winterberg] that the induced current be used to heat propellant and augment the electromagnetic driver with rocket exhaust, but this method seems limited to relatively slow and simple probes and not applicable to complex starships.

Another mass ejection system that can use outside power is the laser or microwave heated rocket [b-Dyson]. Serious studies have long been done on ways to accomplish heating reaction mass without carrying the fuel [a-Weiss; a-Kantrowitz 1972, 1987; a-McAleer; a-Nyberg; a-Brown; a-Slocum]. The goal is to heat, for example, hydrogen much hotter than burning it can. Exhaust velocities may be 40 km/s. Hardware tests succeeded over 10 years ago but no vehicles are planned. Since mass ejection systems are limited in the heat that structures can stand, unless plasma containment systems are designed, these systems are unlikely to be used in interstellar travel. For the use of external laser and microwave beams, better ideas have been developed as shown in Chapter 4.

2.8 SOLAR SAILS

The best idea for interstellar travel is not to carry the energy source on the starship because it is inevitably heavy. Since the stars already provide power in the form of light, it might be feasible to harness this source. Light exerts pressure on a surface when it is absorbed by it, and it exerts twice the pressure when it is completely reflected by a shiny surface. Again, the thrust that moves the starship comes from action-reaction involving momentum. The momentum is in the form of light [see Appendix B for mathematical physics pertaining to light]. Physicists have developed an alternative or dual model of light with particles called "photons" as well as waves. A certain color of light is equivalent to a certain amount of momentum and energy carried by each photon, with energy greater for photons/waves of shorter wavelength. Strictly, a photon has no mass, but since it has momentum it can have the effect of mass when it strikes anything. Although photons move at lightspeed, their use for relatively low-speed travel is put here rather than in Chapter 4.

The Sun provides a very large number of photons per second, all traveling radially outward at lightspeed. At visible wavelengths the number passing through a square meter near Earth is about (10)21, but each carries a very small punch. Given sufficient time, they could drive a solar sail or "lightsail" [b-Friedman; a-Drexler 1979, 1982] pulling a starship to near lightspeed. Photons constitute a "reaction mass" as needed for propulsion. Unfortunately, voyages may require long times even if the starship is dropped near the Sun first. Lightsails cannot be indefinitely large even if they are able to accelerate their own mass. They become too fragile to assemble, survive, or hold onto. Sunlight is too weak beyond the orbit of Jupiter to provide enough photons, and any use of sunlight must occur much closer. Light from any star, no matter how bright, decreases rapidly with distance and cannot boost a starship to near lightspeed. Using artificial beams of light is another solution to be covered later. Lightsails are sufficiently feasible for low-mass interplanetary travel (space to space only, no planetary launches) that a contest to sail to the Moon has been started.

At the distance of Earth from the Sun, the intensity of light is about 1400 watts per square meter. If this light reflects from a shiny mirror of 1 square meter area, this intensity results in a pressure on the surface and a force. [Appendix B]. One square meter feels a force of (10)-5 newton, much lighter than the weight of a feather. If the light were simply absorbed by a dark material, heating it, the pressure would be half as much. To get a feel for the amount of light pressure, suppose that a spacecraft for sailing is a 2-seater massing only 1 tonne (Volkswagen size), and is to accelerate at 1 gee. The sail area must be 1 billion m². Its form may be a disk or shallow cone [Figure 2.6] and its radius would be almost 18 km. Accepting much lower acceleration is feasible for interplanetary travel, and the sail becomes manageable in size for a small spacecraft. However, the mass of the sail must be considered.

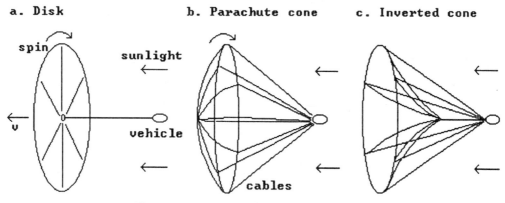

a. Disk　　　　**b. Parachute cone**　　　**c. Inverted cone**

Figure 2.6　Solar "sails" of three general types.

Can the sail be made sufficiently low-mass? Here estimates will usually not be better than a power of ten. Thus all material densities are close enough to that of water to be taken as 1 g/cm³ or 1 kg/m³. One micrometer aluminum reflector, like thin solar cells, has mass about 1 gram per square meter (0.001 kg/m²). The sail above would mass 1000 tonnes, 1000 times too much. A JPL sail design for a trip to

Halley's Comet was 2 micrometers thick, with some coatings [b-Friedman]. The basic material was a very strong plastic, Kapton (DuPont TM). The reflector was aluminum coating less than 100 atoms thick. Plastic film can be much more tear-proof than metal foil. This sail would have been about 10,000 atoms thick. Reduction to 1000 atoms is likely, and to an effective 100 atoms may be possible by perforating the reflector. Reflecting light requires only that the holes in the material be smaller than the wavelength of visible light. The sail could be a microsieve and perform as if solid. The technical literature proposes thinner sails in various ways [a-Matloff 1984] and micrometer thin aluminum foil has been demonstrated [a-Drexler], but for the moment 0.0001 kg/m^2 is assumed, with reinforcing structure included. At this density a sail can accelerate itself (only) at 0.01 gee in Earth vicinity. This is tolerable since solar gravity here would produce 0.0006 gee on it. Ten times closer to the Sun 100 times greater force is possible on the same sail, giving 1 gee. To move payload at 1 gee, the mass of the sail can be about 10 times larger than the rest of the starship, for a "propulsion" to payload mass ratio of 10:1.

The earlier solar-powered mission whereby the starship was dropped around the Sun can be reconsidered with a solar sail instead of solar cells. The power from the Sun is mostly reflected from the sail, not captured, and this propulsion method is very energy inefficient. The photons are traveling about 1000 times faster than the starship will. The starship, via the sail, accepts the small amount of power that action-reaction permits and accelerates. For a 1000 tonne starship, the sail might be 10,000 tonnes and 100,000 square kilometers. Its radius would be about 300 km and initial outward acceleration 1 gee starting at 15 million km from the Sun. The force on the sail is initially 100 million newtons and decreases as inverse square of distance. The solution of Appendix B can be applied, and the speed after escape from the Sun is about 550 km/s (about 0.002c).

This mission merits further study, and some improvements are possible. Lower-density sail can be developed, and initial acceleration can be 3 gees. The most optimistic results hover around a travel time of 1000 years to Alpha at about 0.004c [a-Matloff 1981] if the starship is dropped a scorching 10 times closer to the Sun. This would be a moderately slow colony voyage, and the payload mass needed surely must be larger. The sail should be carried along for purposes of stopping the starship at the other end. The sail is a hindrance going in, tending to retard motion, so it is (somehow) "unfurled" after the starship has reached perihelion and maximum speed (about 130 km/s at 15 million km from the Sun). Then there is a short time to accelerate it.

Many more details of starsail technology should be (and have been) studied than can be covered here. One important aspect is the means of connecting the payload, compact yet probably of lower mass than the sail, to the sail. Only cables in tension are practical structural elements over large areas. The payload can be at the axis or center of a round sail or trailing behind on cables in parachute fashion [Figure 2.6]. The force from acceleration will deform the sail so that the payload is some distance behind the sail anyway, as determined by the tension in the main cables. Cable length must be minimized but tension will be less if the starship trails well behind at a distance about the same as the radius of the sail. Another design looks like an inverted umbrella or curved cone with cables from center and perimeter.

The strength of the cables is a serious problem. For the mission described, the strength needed is said to approach that of pure diamond [a-Matloff 1981]. For a sail of $(10)11$ m^2 so near the Sun as to receive 150 kw/m^2, the force it feels and transmits via cables to the payload is about 100 MN. The choice on cables is few heavy ones or many thin ones. The stress depends only on total cable cross-section area, so the first estimate can be made for one giant cable [see Appendix B]. A slightly conservative value for material strength is $(10)10$ N/m^2, about that measured for iron whiskers and glass fibers, half of that measured for graphite, aluminum oxide, and silicon carbide whiskers, 1/5 of that for diamond, twice that for boron or copper whiskers [b-CRC table, a-Clauser, a-Kelly], about the same as oriented polymers [a-Flam], and 3 times Kevlar (DuPont TM). Whiskers are small needle-like pure crystals with every atom positioned and bonded nearly perfectly without defects and theoretically can have about twice the above strengths. Young's modulus (which measures elasticity and is about 20 times larger still), theoretical and actual tensile (yield) strengths, and allowed working stress are all measured in units of force per unit area but should not be confused. Bulk materials are at least 10 times weaker in tests and are used at about 100 times less stress in actual construction [b-Popov]. Whiskers can only be used in composite materials with some binding material holding the whiskers, passing stress along from one to the next, and halting the growth of cracks. Cable wire cannot be drawn to have the same strength as whiskers.

Using theoretical strength, the cable cross section required is only 0.01 m^2. But the length must be approximately 300 km to match the sail as it is subdivided. With a density of 3000 kg/m^3, the cable mass is about 9,000 tonnes. Clearly a ten times lower mass sail material must be developed, down to $(10)-5$ kg/m^2, giving 1000 tonnes. But then its area can be 10 times less also, since it is 10 times easier to accelerate. The force is reduced to 10 million newtons (10 MN). Sail mass is down to 100 tonnes and radius down to 100 km, and cable length is reduced by about 3. Cross section is down by 10 and cable mass is about 30 times less, or about 300 tonnes. A feasible design is being approached, although cable mass now exceeds sail mass. There is some design room to increase acceleration.

An ideal final speed that can be achieved by this sail can be calculated using conservation of energy as before. Suppose the starship mass is 100 tonnes, just sail, so that 10 gees is obtained to start at 15 million km. The final speed after escape is about 1700 km/s (0.006c). A direct comparison of gravitational and photon force on a piece of this thin $(10)-5$ kg/m^2 sail shows that gravity pulls 1/160 as much as photons push outward. The solar sail can thus acquire almost 13 times more speed than solar gravity can take away from it.

The reasons for putting effort into less sail mass should be clear, since the payoffs are large. On the other hand, if design for a 10 times larger payload were begun, the cable mass would increase about 30. It has been found elsewhere [a-Matloff 1985] that as starship mass and sail size increase, cable mass increases faster and comes to dominate design. In general, cable mass is proportional to sail area, and proportional to sailship mass by a 3/2 exponent. A published mission design attempts a much larger starship, drops it ten times closer to the Sun for a hazardous 100 times increase in solar power, and even specifies a fleet of 16 smaller sailships [a-Matloff 1981, 1986], all in an effort to achieve higher speed and stay within the bounds of feasible cables. Because the trip is slow, a mission must be a colony one, but it must be done with a fleet of about 100 small sailships of 100 tonnes payload each.

Sail structural and orientation stability must be ensured. The sail must keep its shape, particularly for presenting maximum area to the Sun. Photon pressure cannot accumulate like gas pressure and instead will tend to collapse the parachute sail, unlike its earthly cousin. A sail can be rotated at a low rate sufficient that radial cables from its perimeter to its axis are stretched taut. Although the effect is popularly and erroneously called centrifugal "force", the sail material is simply constrained to rotate by the tension in these cables. Payload cables attached to the perimeter must rotate with the sail and connect with a low-friction bearing ring to the starship. Another design, the inverted curved cone avoids spin and uses photon pressure plus inner and outer cable tension to keep the sail expanded [a-Matloff 1981]. A design developed for a mission to Halley's Comet is called "heliogyro" and consists of long strips (vanes) stretched from the axis by rotation. Photon pressure would turn it like a windmill. This design is skimpy on area unless the tens or hundreds of strips are thousands of kilometers long, an extreme size which can cause new problems.

Steering a sail is done by shifting payload position and/or altering the amount of area exposed on one side. Reduction of force can be done by reducing exposed sail area in various ways. The inverted cone sail can be collapsed by shortening the axial cable. All possible maneuvers and braking should be done by sail, for which the energy is free, rather than by emergency onboard rockets or ion thrusters. Manufacturing and assembling the sail would be done in space, to avoid problems of how to get it there. To avoid trying to unfurl a sail rapidly, it has been proposed that an asteroid large enough to shield it from sunlight accompany the starship down to the Sun [a-Matloff 1981]. But the scale of the project is enormously increased to move such an object. Some artificial protection disk would be better, but a sail design that permits rapid "turnon" (rapid presentation of the area to sunlight) and "turnoff" would help both here and at the destination. The sail should be furled during interstellar cruise to avoid damage and drag. There would be less than 1000 seconds to prepare a 100 km sail before the best accelerating opportunity is lost. The precise time and duration of unfurling also has a strong effect on final course and speed. But ways to unfold or to collapse it quickly can surely be worked out.

There are many ways to damage a huge fragile sail. A solar boost at high intensity raises questions of whether the sail material can withstand the heating. No material reflects perfectly, and even 1% absorption could melt a thin sail. Approaching the Sun at a farther perihelion than proposed elsewhere should keep the sail safe without expensive coatings needed. Radiating away 1% absorption or 1.5 kw/m^2 of heat requires about 400°K [Appendix C 4.3]. If the sail is plastic then accumulation of charge must avoided for several reasons. Every section of sail must be electrically connected to the rest.

Sails are so low in density and strength that local drag and damage should be considered. Interplanetary space has about 10 to 100 times more gas density, higher dust density, and many more micrometeors than interstellar space has. Local space dust is in the range of 4 to 40 micrometers size but traveling up to 250 km/s [a-Kerr]. The solar wind consists mainly of hydrogen and electrons streaming out at about 400 km/s. The solar wind pushes dust smaller than the above out of the Solar System and itself comprises most of the gas here. In the vicinity of Earth the density is 1 million to 10 million ions/m^3, a density so low that sunlight provides more pressure and drag [a-Parkinson]. Ion density increases about as fast as light intensity as one approaches the Sun but still provides negligible drag at moderate speeds. Occasional interplanetary dust particles (micrometeors) will puncture the sail and rips could start since tension is unlikely to be uniform throughout its area. Reinforcing strips may be needed all over the sail, increasing its mass. Careful prior study of the dust size and density in the regions of travel is needed. One suggestion has been that the sail could be ruined in five hours [a-Matloff 1981], in which case the starship may never stop again.

There are various ways to boost or enhance a solar sail-powered mission. A small improvement in final speed is gained if the starship speed is increased prior to its passage near the Sun. Earlier references [& a-Matloff 1983] have discussed this pre-perihelion acceleration, but the gain is minimal from a tremendous effort, and the voyage still requires about 1000 years. There are ways to make the Sun shine longer on the sail, such as focusing it with large mirrors. A 100 km mirror made from solar sail material can send significant power a long way beyond the Solar System for propulsion and for operating power. The large area of the solar sail suggests other uses. These and other optical technologies are covered further in Chapter 4 and later chapters.

Solar sailing is a novel, exciting, and different technical approach. Its potential is very good for small but slow missions. It makes more sense than using photovoltaic-powered ion drive even if solar cells can be made thinner than sails. We should not be surprised that some beings somewhere have the ambition, patience, and long lives to sail among the stars. Even if human construction of "solarships" seems far-fetched, demonstrating engineering feasibility on our terms stimulates interesting guesses about the accomplishments of "others" and whether we should be able to observe the results.

A literal spinoff from solar sailing is the solar "wind"mill or solar mill, possibly useful for generating large amounts of power for local use such as operating a launcher. It has "blades" kilometers long, large but not too large, set at a tilt and turned by photon pressure. A seemingly slow rotation speed of about 1.6 cycles per second (angular speed 10 radians/s) is chosen. Materials are not sufficiently strong to design one to turn much faster. The ends of the blades are most effective in obtaining torque, the rotational force [see Appendix B], and are connected to the hub and to an electric generator by long cables. Tension in the cables is much greater than blade force and keeps the blades constrained to rotate. For a site on

Mercury, the Sun provides about 10 kw/m^2. Ten blades each 1 km long and triangular width to make 300,000 m^2 fill the area set by radius 1 km. They have a total force of about 100 N on them. If the effective radius is set to be 1 km, then the torque is 100,000 N m, and the power is 1 Mw. If a blade is made of ultra-thin sail material, density 0.00001 kg/m^2, then it has mass 3 kg. The tension to hold it during rotation is 300,000 newtons. Cable of the strongest material would require total cross section about 0.00001 m^2 and have mass about 30 kg, on the limit of being able to hold during the slow rotation. Other designs should be investigated. The only way to bring solar mill generator size up to a convenient 1 Gw seems to be to place it closer in much higher solar intensity and use somewhat slower rotation. The blades must fill the area of rotation fully and be broad. The skinny designs shown in the literature for gaining or storing power cannot hold together.

2.9 RUNWAYS AND PELLET STREAMS

Using fusion power raises the possibility of sending high-energy density "fuel" ahead of the starship, to be scooped up as needed from a "runway" of fuel [a-Matloff 1979; ramjet discussed Chapter 4]. An acceleration program can be developed which sends pellets or canisters out with any desired combination of speeds, to be intercepted by the starship with minimum mismatch at just the place and time it is needed ("just in time" fuel!). Moving the fuel takes most of the energy, and that is supplied from here, so the starship could be much less massive. While much energy is saved with this system, the final speed is said not to exceed 0.2c. The launching system, probably a magnetic driver, would use solar mirrors and/or large areas of photovoltaic cells to capture solar power. The fuel pellets might be given "cool" superconducting jackets which must be compatible with both launch and fusion. The magnetic field of the driver could repel superconducting pellets forward. Unlike a starship or its launcher, where all equipment must be in one package, the fuel-making and fuel-shipping facilities can be spread all over the Solar System.

A similar idea is the use of a pellet stream for its momentum alone, not necessarily as a fuel source [all a-Singer; followup a-Ruppe]. This sounds dangerous, since avoiding high-speed objects is a major concern in interstellar travel. But there are ways to capture or reflect pellets with pre-determined properties whose paths are known. The propulsion is like solar sailing, but the sail is a much smaller interception device, and the photons are huge. As with fusion fuel pellets, small objects in a steady stream that can withstand huge gees are much more suited to magnetic drivers than human vehicles are. While fuel pellets should be sent at the expected speed of the starship for low-speed capture, momentum pellets should be sent as fast as possible. The proposed million-gee system for 0.25c pellets needs a 280,000 km launcher, or most of the distance, say, from Earth to Moon. Pellet mass is of the order of 0.01 to 0.1 kg, and one may be sent every few seconds. Less ambitious systems are likely and can be shorter in proportion to the square of the speed and longer in proportion to lower acceleration. The problems of aiming and controlling the stream are impressive, but to some extent the starship can move to wherever the stream is coming from. The reaction force that pushes the launcher backwards could be a serious problem. To use the pellet stream, the starship can possess another long magnetic launch system which works in reverse and captures the superconducting pellets, extracting most of their momentum and transferring it to the starship. Or the pellets can hit dust or gas and be converted to plasma which can be intercepted magnetically, partly like a fusion reactor in reverse. The starship will require major radiation shielding from this hazardous process.

An idea not seen elsewhere is to send neutral (or possibly charged) particles across space. This could be a very high-speed system (more appropriately described in Chapter 4) since Earth electromagnetic accelerator technology is well-developed for moving elementary particles to within a millionth less than lightspeed. Particles must be accelerated while charged, but high-speed electrons could be sent along with them at the proper rate to neutralize the beam to prevent spreading. The individual particles might be more useful if not converted to neutral atoms. The particles could be fusion fuel or provide momentum. To control them if neutral, they must be re-ionized at the starship by means to be covered later. As fusion fuel about 1 kg/s might be needed, or about $(10)26$ particles per second, a current (when charged) of about 10 million amps, far greater than the microamps of present accelerators. The KE sent with these particles is much greater than the portion of RE used in fusion. Unless the KE contributes to propulsion, a momentum beam would be 1000 times more efficient. As a source of momentum, the much higher potential of the accelerator can be used (up to 1 TeV thus far) and the current reduced to about 1000 amps. A bank of accelerators in space using solar power can send individual beams in the same direction. Transferring the momentum from this beam safely to the starship is a problem still to be solved.

Stopping a starship with pellets is not necessarily impossible, even if the starship lacks a magnetic launcher. If the starship has engines and burns fusion fuel, then fuel pellets can be sent to it before or during deceleration. If it lacks engines then partial deceleration is possible by placing a slow stream of pellets ahead of the starship and letting it run into them (with careful interception as usual). Such a plan requires time much longer than the mission duration. The reference [a-Singer] provides other uses for pellet streams and notes many advantages of the pellet system over the Daedalus mission and a laser-pushed mission for the same speed and distance. Deciding which system is least exotic will require much more study and some real experiments. These are only theoretical engineering ideas, subject to problems like aiming error and loss of human interest after the starship has departed. Like other missions involving long-term Earth support of a starship, they require an extraordinary amount of societal commitment.

If material is sent across space with a magnetic launcher and received by another magnetic launcher extended so that it can reverse the pellet momentum (instead of catching and demolishing the pellet), then this is the first leg of a theoretically lossless travel system. The starship needs no extra energy source, since magnetically reflecting a pellet sends it back towards the Solar System at a lower speed. (A bit of operating energy can be extracted.) The returning slower stream can run through the launcher in reverse, and energy (and pellets) recovered for another launch. The other leg occurs as the starship sheds the KE that it has "borrowed" from a source back home. A pellet launcher is needed at the destination star to repeat the process. It can recover reflected fast pellets and gain the energy that the home launcher did not recover. Although slower than light, for various reasons a magnetic driver with superconducting pellets or an electromagnetic accelerator with particles seems more feasible. All that seems missing (theoretically) is a way to store titanic amounts of electrical energy, and ways to allow for the motion of stars.

Chapter 3

Relativity and Interstellar Travel

This chapter is partly for the fun of the physics found in travel at speeds so high that Einstein's relativity becomes a dominant law of nature. Many dream and hope about what it would be like to travel so fast although the methods of achieving this are formidably difficult. This chapter and the next cover theory and methods for travel from about 0.2c (60,000 km/s) up to lightspeed c itself (nearly a round 300,000 km/s). Besides being a realm where the effects of relativity are noticeable, speeds over 0.2c would also permit a round trip mission to a near star within a human lifetime. No shorter word has been developed for describing this high-speed regime than the multi-syllabic "relativistic".

Almost the only symbol in the main text is "c" for lightspeed, and most speeds are expressed as fractions of c. Relativity can be a difficult subject, even without its mathematics. Readers can skim or skip this chapter and parts of Chapter 4 and still follow the remainder of the book. Those interested in relativity more fully should be aware that errors are more easily made in relativity and that the best assurance for accuracy is to have several physicists confirm statements and results. Since this book is not primarily about relativity or its applications, very few of the very many available references are cited. [Mathematical details of relativity are in Appendix C.]

3.1 EINSTEIN'S RELATIVITY IN PHYSICS

The physical background for interstellar travel must include relativity. In 1905 Einstein developed his theory of relativity, which includes two new principles, that nothing can ever move faster than light, and that what observers moving at different speeds see cannot differ in any absolute way. Hence "everything is relative", at least in physical theories. This new description of nature is not strictly an invention of Einstein but rather a revelation of deeper aspects of nature which are apparent only at high speeds. Newton's physics still works if one does not make measurements too accurately or at too high speeds. Einstein's physics works at all speeds from zero to c. A sad result from relativity is that interstellar travel is limited to speeds less than c, rather slow for the enormous distances between stars and galaxies.

Many astonishing results follow from Einstein's innocent principles. His theory is not remote, because some results are widely used in technical applications. Some consequences can be seen in daily life if one knows how to look for them. If relativity did not hold, even ordinary matter would act quite differently. Magnetic fields are a manifestation of relativity for the slow speeds of electrons moving as currents in wires. Particle accelerators all over the world follow laws established by relativity to very high accuracy. Noninvasive medical instrumentation uses relativistic particles to make internal pictures of the body. Some astronomical objects in our galaxy revolve about each other sufficiently fast to show the effects of relativity, and the most distant galaxies and other objects are moving near c. Relativistic particles (often called "cosmic rays") from unknown sources in our galaxy strike Earth's atmosphere at a high rate, making showers of other relativistic particles that pass through us. Cosmic particles are of serious concern for any long-term space travel. Relativity helps determine the energy balance of stars and the creation of chemical elements in stars, which are then flung out by supernovas to condense as more stars and planets like Earth.

Many books have been written to explain relativity at all levels [b-Einstein; b-Russell; b-Bondi; b-Born; b-Kaufmann; b-Calder; b-French; b-Taylor; and chapters in many textbooks, e.g. b-Goldstein], but only a few features and results are needed here. The basic approach needed in relativity is to understand that every observer has his/her own reference frame. A starship reference frame moving at any speed is different from an Earth reference frame (which is rotating, revolving around our Sun, and other motions). No reference frame is special or absolute, and all must be considered relative to others. In particular experimenters everywhere should find the same speed for light in vacuum when they measure it, regardless of how they are moving. Observers should find the same results for any other experiment. It should not be possible to tell how fast one is moving because there is no absolute standard. However, the microwave radiation background of space seems to provide a fixed frame for the Universe as a whole. We think we are stationary on Earth, and it rarely matters that Earth is rotating, orbiting more rapidly, its Sun moving still faster around the galaxy, our galaxy moving still faster in another direction, and so on. Out in space the "fixed" visible stars are all individually moving and have a net average motion around the galaxy too. For the study of interstellar travel at relativistic speeds, the motion of Earth, of stars, and of our galaxy can usually be assumed negligible.

In relativity, observers, events, and reference frames are supremely important. Defining and understanding these concepts carefully is at least as important as being able to work out the more difficult mathematics. A "reference frame" is an imaginary but mathematically useful framework that can be thought as attached to an observer [see Figure 3.1a]. Everything that is observed should be measured with respect to the frame. The things observed are all "events" or happenings that occur at a certain place and time, such as a burst of light from a star. The reference frame includes the three dimensions of space that we already know (independent directions such as up-down, north-south, east-west) and clock time as a fourth dimension which we cannot see but can measure. The "background" in which any frame is placed is called "spacetime". Since more than three dimensions are difficult to represent in a diagram, only x for space and t for time are shown when possible [Figure 3.1b]. Most work in relativity, including interstellar travel, can be done with one space dimension x and the time dimension t. The direction of interest is usually that in which the starship is traveling.

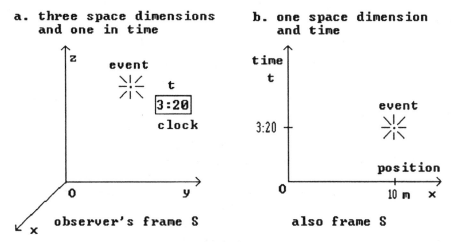

Figure 3.1 Reference frames for relativity.

Since a starship accelerates to reach relativistic speed, its frame of reference requires more thought. Some relativists emphasize that accelerating reference frames are tricky. But all one need do with an accelerating starship is imagine that there is a frame moving alongside it momentarily but not accelerating [see Figure 3.2]. This reference frame is soon left behind, and another faster frame is momentarily moving with the starship. Constant speed frames are called "inertial". All these frames can be made to coincide at the same place and time back near Earth. They have different speeds, and the starship catches up with each one in succession.

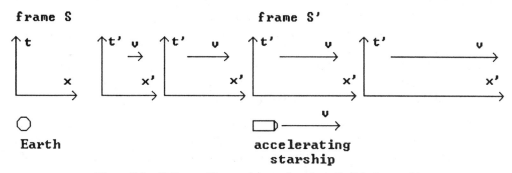

Figure 3.2 Reference frames for accelerating relativistic starship.

For studying interstellar travel, two frames are important, a frame S' momentarily traveling with the starship (but not accelerating with it) and a frame S representing another point of view usually Earth or the Sun [see Figure 3.2]. The S frame often represents the point of view of those who remain at home such as us studying the hypothetical voyage, but it could be assigned to any planet, satellite, or star. The Earth frame is used mainly to describe the journey to Earthlings, although Earth-based observations are limited by the long time which light and other signals require to travel from starship to Earth. Another possible difficulty with the Earth frame is that it is moving, in a substantial and complex way at about 600 km/s when Earth rotation, Earth orbit, sun's motion around our galaxy, and our galaxy's motion are taken into account [a-Muller]. The starship shares the latter motions, and astronomers and space mission controllers have long had to take account of Earth rotation and revolution.

The crucial methods for relativity are that an observer be motionless in his/her frame and that an event be measured as a "point" in spacetime with four numbers or "coordinates", of which x and t are of most interest. A different observer, in S' moving with respect to the preceding observer in S, will measure different coordinates for the same event. To observe something as simple as the length of a meter stick, at least two events must be measured, the two ends of the stick. Only an observer in a frame where the stick is not moving has a chance of having these two events occur at the same time ("simultaneous") and get a proper answer, one meter. Other moving observers do not find these events simultaneous and get a variety of conflicting answers, seemingly paradoxical.

3.2 EFFECTS ON MASS, LENGTH, AND TIME

If lightspeed is really the ultimate speed for everything, there must be some mechanism for limiting all speeds to less than 299,792 km/s. Otherwise a 1 gee rocket engine could keep going for a year and exceed the limit. One barrier is that the effective mass of an object increases as its speed increases. One way to define mass is in terms of the force needed to move it. More and more force is needed to push the increasing mass of a fast-moving object to slightly higher speeds. People aboard the starship do not notice objects becoming more massive, but outside observers knowing the thrust of the rocket would observe that it was not speeding up as fast as expected. This is most unfortunate, because interstellar travel is already plagued with finding sufficiently powerful drives to reach high speeds. Doubling a speed already requires four times more energy. Einstein's results make it even more difficult to reach a speed like 0.9c, and infinitely difficult to accelerate to lightspeed. No object with initial mass can be pushed to the speed of light. Only photons (and possibly neutrinos), massless if they could be stopped, travel at c, where they act as if they had some mass (not infinite).

The mass increase does not become noticeable until about 0.2c or higher. At 0.2c the mass is increased by about 2%, and at 0.9c the increase is a factor of about 2.3. The factor which describes the mass increase is called "gamma" (from the Greek letter used as a symbol for it) and can be calculated from speed as shown in Appendix C. Figure 3.3 gives a graph showing how gamma depends on speed. Note that gamma must be near 1 at ordinary low Newtonian speeds and equal to 1 at zero speed. As speed increases, gamma reaches about 1.1, for a significant 10% effect at 0.4c, and increases greatly thereafter. The regime where gamma is greater than about 10 (0.995c) is called "ultra-relativistic" because relativity dominates there. Relativistic mass is calculated by multiplying the known rest or "proper" mass by gamma. Rest mass is the mass found by measuring it in a frame where the mass is not moving. Masses measured in a relativistic starship would give normal results. Scientists aboard a fast starship cannot tell it has greater mass, but they can notice that the atoms in interstellar space are hitting very hard when the starship runs into them. The hydrogen gas of space, nearly motionless in our S frame, has high speed and even larger momentum and KE due to its mass being increased by gamma, as determined in the starship frame S'.

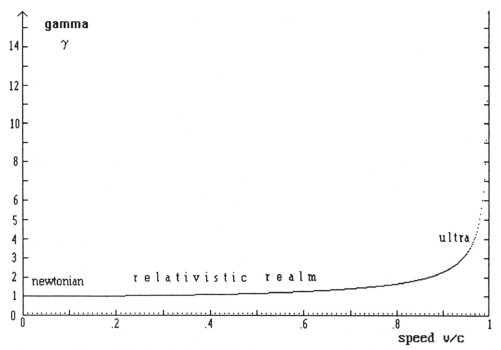

Figure 3.3   Graph of relativistic factor gamma as a function of speed.

Einstein's results should not be interpreted to imply that people aboard a fast starship cannot tell that they are moving. They need merely look out a porthole every year to observe that some stars have changed positions. They know that stars did not move especially for them, therefore the starship moved. They could determine starship speed from careful observations and prior knowledge of star locations in space. But if they had only a few star position measurements and no special knowledge, then there really is no way to determine what is moving and what is not. Scientists have had no opportunity to attempt to measure the mass of a starship or other relativistic object flying by, although the small mass change of a satellite could be measured. Similar observations occur everyday on a large scale at accelerators which work because moving particles have exactly the larger mass that they are supposed to. Can the accelerators be proven motionless and the particles moving? From the particle's point of view, the heavy machine is unaccountably rushing and whirling around it in an astonishing way. No absolute reference can be established despite the crazy viewpoint. (Accelerators have been called "ponderators" because they add mass to particles and hardly increase their speed.)

Besides mass, relativistic speeds affect length, time, and other physical quantities. Rapidly moving objects are shorter in the direction of travel, an effect called "length contraction". And the rate of passage of time aboard them is slower. Lengths are shorter by gamma; the normal, proper, rest length is divided by gamma to obtain a smaller length. Time periods are longer with proper time intervals being multiplied by gamma for an effect called "time dilation". From the viewpoint of a micro-observer traveling on a high-speed particle, our world appears foreshortened. Particle accelerators must be designed with the equipment farther apart than Newton would specify, so that as particles rush by they "see" the components at the correct spacing for their speed. Length contraction and time dilation are connected, being two aspects of the same effect in spacetime. A particle which would decay after a microsecond and is rushing along about 0.9c "sees" length passing by twice as fast and finds it has traveled further. But physicists in lab frame S find the particle lived twice as long and had its lifetime dilated by a factor of gamma. With the great accuracy available from atomic clocks, time dilation has been observed for systems as slow as satellites and passenger jets [classic a-Hafele].

Length contraction is very important for relativistic travel because the spacing between particles and stars conveniently shortens from the viewpoint of the travelers. If their speed is near c, they find their travel time to be shorter by gamma and ascribe it to closer stars. Earthlings, analyzing messages that the receding starship sends back, would find that the radio frequency is lower and thus that starship time is dilated with time units lasting longer. (This effect is masked by another larger doppler effect discussed later.) If Earth astronomers could see the starship, they would note also that the starship appears shorter, but they would see nothing wrong with the star positions.

A favorite story in relativity involves two starships, named Intrepid and Dubious, passing each other at high speed. Paintbrushes or some other definite marking device on Intrepid are activated to mark 10 meters on Dubious as it goes by. (This marking is a standard procedure to establish length and can be done simultaneously in Intrepid's frame.) Intrepid expects that since Dubious is moving fast and therefore contracted in length, the marks it received would measure as farther apart than 10 meters. Dubious has a bonafide metric ruler laid out along its side. Dubious expects that, since Intrepid was contracted, its markers are closer together and mark less than 10 meters. When the marks are examined and reported, perhaps by radio, the marks are found more than 10 meters apart. Under questioning Dubious admits that the marks were made at different times from its point of view and that "possibly" starship Dubious advanced a few meters during the marking process. What at first appears to be hopeless paradox in Einsteinian physics is readily resolved when observers refresh themselves on their college courses in relativity. If they do some calculations on events in the way Einstein prescribed, they can work out their differences. It will continue to bother many possessors of common sense that time can act this way, but at high speed it can. If it did not, light would not have the same speed everywhere.

If different observers obtain different results for their observations of the same phenomena (e.g. the two events needed to measure length), how are they supposed to see the same physics? The answer is to use what is naturally called the "proper" frame of reference, the one in which the events are not moving. The "proper" length of the starship will not usually be of as much interest as the "proper" time will be. Proper time is the time as measured in the starship with any clocks and represents how the voyagers age. The human body itself serves as an indisputable long-term clock. When everyone does special calculations to transform their observations into how things look in the "proper" frame, the results are all in agreement. Moving observers must convert their results according to rules found by Einstein (and named the Lorentz transform). The calculations are not needed explicitly here to study relativistic travel.

3.3 OPTICAL EFFECTS

Time dilation is experienced by travelers, not by seeing their own clocks run slower, but in noticing that others do, including natural clocks such as stars and gases emitting light. Atoms and molecules everywhere emit precise colors of light ("spectral lines", measured as frequencies or wavelengths) when stimulated to do so in hot gases and by energetic starlight. Blue and violet light has higher frequency (and shorter wavelength), and ultraviolet still higher. Red and infrared has lower frequency (and longer wavelength) and radio still lower. The spread in visible light frequencies and wavelengths is about two to one. Deep red has about half the frequency of violet light. Some stars provide other measures of time. Pulses from rotating collapsed stars called "pulsars" are often very regular and provide precise clocks observable across the galaxy. Whatever the frequency, from radio to x-rays and beyond, the effect of relativity on electro-magnetic radiation from a moving source is the same, and often called an optical effect. The pattern of frequencies or "lines" emitted by a moving star or other object is unchanged while the individual frequencies are uniformly shifted downward a small amount. Regardless of frequency or wavelength, visible light or not, the downward shift is said to be a "reddening".

The color of light emitted by stars that a starship passes crosswise [see Figure 3.4a] is observed as redder due to time dilation. This effect occurs regardless of what direction a starship is traveling with respect to the star. But a larger "doppler shift" also occurs when a star is approached or receded from [Figure 3.4b & Appendix C]. Light from stars being approached appears bluer and light from stars left behind is redder. This shift is more easily observed at lower speeds than the time dilation effect. Approaching a typical star at 100 km/s (some stars such as Barnard's are moving toward us that fast), the shift would be 1.0003, or an increase in frequency or shortening of wavelength by 0.03%. Passing to the side of a stationary star at this speed shifts its light frequencies by 5 parts in 100 million. Astronomical instruments at present can determine frequency shifts well enough to detect relative speed toward or away from an object as small as to 1 part in 10 million of c. The largest (red) shifts observed are by a factor of about 4, for the most distant observable objects (quasars) running away as the Universe expands.

Doppler shift in terms of wavelength occurs in the opposite way, with light from an approaching source having shorter wavelengths as the waves of the light arrive more often, and light from a receding source having longer wavelengths as they arrive less often. At high speed the time dilation and regular doppler effects combine noticeably in the relativistic doppler shift, giving an enhanced red shift and a reduced blue shift. In the general case the starlight arrives at an angle to the line of travel of the starship, and the overall doppler shift depends partly on the angle of observation [Figure 3.4c & Appendix C; a-Sänger; b-French; a-Moskowitz]. In the starship frame the star being passed appears to move from near 0° forward to near 180° behind as its colors shift from the blue end to the red end.

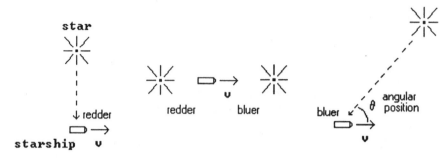

Figure 3.4   Relativistic doppler-related effects. Note: starship and star cannot truly be shown simultaneously in one picture, and angular position also changes.

The angle at which a star is observed is also affected by high-speed motion to produce an effect called "relativistic aberration" [see Figure 3.5a & Appendix C; b-French; a-Moskowitz; a-Sheldon]. Like most effects, this occurs at any speed but is most apparent at high speed. At a speed of 0.9c a star that was located at, say, 45° from ahead appears at about 11°. The original positions of stars seem to be squeezed more and more toward the direction of travel as speed increases, so that most of the stars in the sky, even those "behind" the starship are seen forward. Thus even "forward" and "behind" become relative! As shown in Figure 3.5b, photons moving forward from a star are partially overtaken by the starship, and the starship runs into them as if they were coming from in front. Because most of the light from a star appears to come from ahead when the star is being approached at high speed, its apparent intensity is also greater [Appendix C; a-Weisskopf]. The forward light can be so concentrated that it appears to come as a beam whose width is inversely proportional to gamma. With these effects, the very temperature of a star has been shown to appear different [a-Stimets], so that the population of stars of varying color, position, temperature, and other characteristics all seems different from the starship frame. The possibility of seeing stars spread in a rainbow of colors ("starbow") has been disproven, and these phenomena along with matters of astrogation are covered further in Chapter 7.

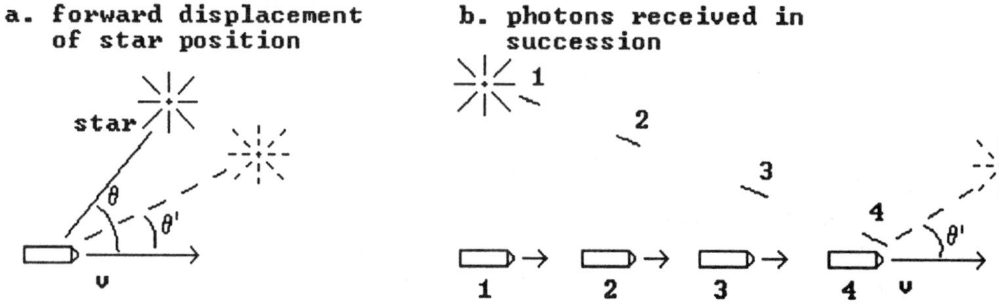

Figure 3.5   Relativistic aberration in apparent position. (See note for Figure 3.4).

For observers at Earth, relativistic effects on light and radio waves can be observed if instruments are sufficiently good. Signals sent from a receding starship to Earth would arrive at a lower rate for two reasons: first they are stretched apart by ordinary recession (doppler), and secondly, the frequency received from the radio transmitter is time-dilated to a lower frequency than the transmitter produces. The only laboratory objects which can be made to move relativistically are elementary particles, and they carry few instruments with them. They emit radiation when accelerated, which shows frequency shift and concentration in direction and intensity that has been measured in the lab frame to agree with predictions from relativity. The aberration of star light due to Earth's motion has been well observed also.

If a very fast starship or other object could be photographed, it would appear redder or bluer in color but not changed in shape. There has been dispute as to whether length contraction and therefore distortion could be observed directly, but the starship should appear rotated as light from different parts of it comes to an observer at the same time [a-Weisskopf; a-Scott 1965; a-Terrell]. The apparent rotation and distortion is likely to remain of academic interest because regular objects may never be made so large that known shapes could be observed when far away and moving at relativistic speed. However, the distortion of star constellations could be observed [a-Scott 1970].

3.4 APPARENT PARADOX OF TWINS

If voyagers return to Earth from a relativistic journey, they would be found not to have aged as much as Earthlings have. Their slower shipboard time would be verified. But the comparison can be turned around to ask why Earthlings do not appear younger to the voyagers. From the starship point of view Earth has been moving at high speed away from them and then came back, with clocks running slower nearly all the while. Earthlings cannot be both older and younger than voyagers, and only one result can be true. Hence the "paradox" here is apparent and not real. (True paradoxes in nature may be non-existent.) It is possible to make physical comparison during and after a relativistic journey and sort out what effects are real. If differing aging is unbelievable, the incredibility is due to traditional intuition not accepting relativity. An Einsteinian explanation is needed here. The problem is often expressed in terms of the Earthbound twin brother or sister of a voyager. Twins are paradigm examples of two systems which evolve similarly and parallel in time and serve as clocks. The so-called "twin paradox" has long been a watershed between understanding basic relativity and not, for physicists as well as ordinary people, with a vast literature [e.g., a-Bondi; a-McMillan; b-French; b-Taylor]. In numerical terms, a journey that occurs at 0.866c (gamma of 2) results in the flow of time at half-rate during the journey. A twin on Earth might have aged 60 years while the voyager aged 30 years.

There is a crucial difference between the two parallel histories of the twins. The people on Earth kept on in their usual routines. But the starship people underwent at least one serious change in direction (an acceleration) during their journey, or else they could not return to see their friends and twin brothers and sisters. In their orbital motions Earth people also accelerated, but this was nothing compared to reversing course between two high-speed voyages. Earth people, from the viewpoint of the starship, seemed to reverse course too, but this time it did not really happen. On Earth no effect was felt, but on the starship effects were felt, no matter how gradual the accelerations and the turning around [see Figure 3.6]. The overall effect for the starship only was two high-speed journeys in nearly opposite directions. One cannot blithely measure time on clocks in relativity, especially clocks not in one's frame. People on the moving starship can deduce that time intervals on Earth are dilated (stretched) but the starting points of the time intervals are not simultaneous with measurements aboard the starship. Back on Earth there is no event midway in time that corresponds to the time the starship turned around. A large time mismatch develops, in the point of view of the starship, as it changes from an outgoing reference frame to a returning one. When this mismatch is added to the dilated ages of Earthlings, advanced ages are obtained.

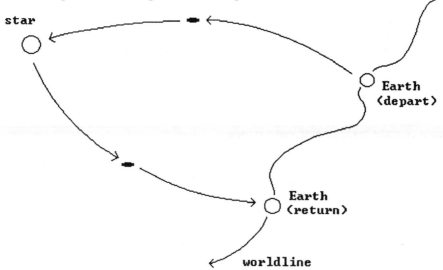

Figure 3.6   Apparent twin paradox for high-speed starship making round trip from slowly moving Earth.

Relativity thus provides one unusual benefit for interstellar travel. The flow of time is variable and can be used to the advantage of the voyagers. The speed required for significant effects is high, about 0.9c to halve the flow of time. The effect is most useful for long high-speed trips. Travel at 0.99c, covering 100 LY, occurs with the passage of about 14 years aboard ship while 100 years pass on Earth. The universe seems to rush at the travelers much faster through the apparent shortening of distances by a gamma of about 7. The benefit is that long trips come within reach of human lifetimes. If (that should be a big IF) one could accelerate at 1 gee as long as desired, in about 10 years aboard ship the voyagers could leave the edge of our galaxy and a few years later pass by the big Andromeda galaxy.

The unfortunate side effect of substantial time dilation is that for the voyagers time passes much faster on Earth. If they were to return to Earth, they would arrive in a future many years later in proportion to the number of light years they traveled. The effect is worse for longer faster trips. Once the regime of large gamma is entered, their Earth is gone forever. There is no way to return to their decade or their century. Another time question arises in sf as a result of supposed technical progress on Earth. Soon after the first starship is sent out at high gamma, a better one might be developed which reaches higher gamma. Its speed can be only slightly greater, but its passengers experience much less time during the journey. Can it "pass" the first starship and have a colony set up before the original voyagers arrive? No, the second ship arrives second as seen from Earth, but its passengers are younger than the first voyagers.

While problems with time produce most apparent paradoxes in relativity, situations can be arranged where lengths seem paradoxical. For example, consider a 1 km starship passing at relativistic speed through a 1 km tube. One kilometer here refers to the proper lengths of each object, each in its own frame. But the two frames are moving and observers on tube and starship will make contradictory claims as to which was longer, tube or starship. Each sees the other object as shorter, and if each is able to measure simultaneously both ends of the other object, each will confirm that the other is shorter. This is a fact of nature. When slow speed is used, dilated clock periods, contracted lengths, and augmented masses return to proper. But the effects of time dilation remain. The people and equipment which went on the starship journey are younger than people and equipment of originally the same age which did not go. No further travel can undo the effect. However, no one will come home shorter and with more mass (aside from over-eating and under-exercising).

3.5 RELATIVISTIC MOTION

Although the preceding effects of relativity are interesting and affect mission planning in many ways, they are secondary to the core problem: can propulsion be found which will send a mission 10 LY in reasonable "time" and despite increasing mass of the starship. The basic analysis of relativity that gives length contraction and time dilation also can give the motion that results from acceleration to near lightspeed (or any speed). Distances, speeds, and travel times are not as easy to calculate as for the Newtonian mechanics of Chapter 1, but analytic solutions are often still possible. A fresh approach is required [b-Sänger; a-Purcell; a-Ackeret; a-Sagan; a-von Hoerner; a-McMillan; b-Belinfante]. Some physical fact is required to begin solution, such as the assumption that the starship gives itself constant acceleration. Also, gravity is assumed negligible for the speeds involved. The mathematical details in Appendix C rarely appear in textbooks. Relativity problems are usually easier to solve in proper time, and solutions mix Earth-based coordinates (frame S) with starship proper time (frame S'). The relation of time between S and S' can be easy to obtain. Acceleration is assumed constant in the starship frame for the first set of solutions, but this is unrealistic for any practical drives. A drive must use increasing amounts of fuel energy to hold acceleration constant.

Relativity is one subject where the hyperbolic functions "sinh", "cosh", and "tanh" seen in tables and on calculators play a basic role. The astonishing numerical results for constant acceleration are summarized here in Table 3.1 for selected calculations in rounded numbers at 1 gee. [See Appendix E for a more complete table for several values of gee.] These results for distance attained, shipboard proper time, time passed on Earth, speed achieved, and gamma reached are for the first leg of a relativistic journey. While full experimental confirmation of these results is not foreseeable, even with elementary particles, there has been negligible scientific doubt about them for at least 40 years. To apply these results to a journey which stops at a star, use one half the distance to find the speed and time at turnover. For example, to find the proper time to stop at a star 10 LY away, find 2.4 years to travel 5 LY, then double this for a total time of 4.8 years.

Table 3.1
TABLE OF DISTANCES, TIMES, AND SPEEDS
FOR CONTINUOUS 1 GEE IN RELATIVISTIC MOTION

Distance (LY)	Time (PY)	Time (EY)	Final Speed	Max Gamma
1	1.3	1.7	0.87c	2.0
2	1.7	2.8	0.95c	3.1
3	2.0	3.8	0.97c	4.1
4	2.2	4.9	0.981c	5.1
5	2.4	5.9	0.987c	6.2
6	2.6	6.9	0.990c	7.2
7	2.7	7.9	0.993c	8.2
8	2.8	8.9	0.994c	9.3
9	2.9	9.9	0.995c	10
10	3.0	11	0.996c	11
20	3.6	21	0.9989c	22
30	4.0	31	0.9995c	32
50	4.5	51	0.9998c	53
70	4.8	71	0.99991c	73
100	5.2	101	0.99995c	104
200	5.8	201	0.999989c	207
300	6.2	301	0.999995c	312
500	6.7	501	0.999998c	512
700	7.1	701	0.9999990c	702
1000	7.4	1001	0.9999995c	1024
10,000	9.6	10,001	0.999999995c	10,323

Notes: PY = proper years; EY = years elapsed on Earth (both measured in Earth years); all values rounded to accuracy shown.

The time spent at slower speeds can be seen by examining the Earth time for the 10 LY journey to stop at a star. The time elapsed on Earth is about twice 5.9 or nearly 12 years. The speed at turnover may be high but the beginning and end of the journey are slow and add about 2 years to the trip. At 1 gee and higher accelerations, the first year or two aboard the starship is not very productive. After 2 proper years, time is passing about 4 times faster on Earth. Speed reaches almost 0.99c just before turnover and the maximum gamma of about 6 indicates that yellowish stars are seen as powerful ultraviolet sources ahead and faint dull red stars behind. For longer journeys it should be no surprise that the time passage on Earth in years is just a bit more (about 1 year) than the distance covered in LY. Observers on Earth would see the starship move almost as fast as light, if they could see it.

After speed has built up, benefits accrue rapidly. After 20 LY of nonstop acceleration, only another 0.4 proper year is needed to go 30 LY. The same holds for covering 300 LY versus 200 LY. Now the starship is really moving, with speed just a bit less than c and gamma about 300. The value of gamma comes to resemble the distance in LY in the cases given here only because of an accident of numbers where c^2 divided by 1 gee acceleration is nearly the same as the number of meters in a lightyear. In general, large gammas are approximately proportional to the distance covered, and to the Earth time elapsed [see Appendix C]. Speeds nearer c can be estimated too. Note in the Table that for each increase in distance by 10, two more "9"s and a "5" are added to the speed in terms of c. As distance increases, the difference between the speed and c decreases inversely proportional to distance squared [see Appendix C].

At an unceasing 1 gee, incredible distances can be covered in proper times of only 10 to 20 years. Again, because of the units used, the proper time (in years) needed to reach a given distance (in LY, without stopping) is approximately the natural (e) logarithm of twice the distance [see Appendix C]. The center of our galaxy (about 50,000 LY) can be passed in about 11 years and the Andromeda galaxy (M31, about 2 million LY) in about 15 years. The log function is so insensitive that it makes only 1 proper year difference whether one goes 1 million or 2 million LY. Thus the time to accelerate halfway and then decelerate to stop is a little greater than twice the time to fly by. The "edge" of the visible Universe (about 10 billion LY) can be reached in about 24 proper years. A popular sf novel by Poul Anderson has described the terrible plight of a starship whose voyage cannot be stopped. Forced to continue at 1 gee, the worried crew passes clusters and superclusters of galaxies and even witnesses the end of the Universe after a few years of their time. With gammas in the billions nothing could be seen, or perhaps the Universe would destroy the starship. Even for gammas in 1000s, no stars could be seen except as an intense beam of hard radiation from dead ahead.

Relativistic motion can be solved for other more realistic requirements such as constant thrust [see Appendix C; a-Anderson 1971, 1974; a-Pomeranz; a-Kooy]. All solutions for speed should result in a speed that approaches c as time goes on. If mass is used for energy and exhaust, it decreases steadily for constant thrust. When thrust or force is constant, speed increases roughly as an exponent of proper time, depending on the exhaust velocity, until fuel is gone. No simple solution for speed can be found as a function of Earth time. Studies of hypothetical starships with engines where exhaust velocity can be varied indicate no significant benefit obtained [a-Powell]. A constant power case, where KE increases linearly in time, has also been studied [a-Powell 1977, unfortunately one of the small number of papers using unorthodox symbolism very difficult to follow or translate]. No analytic solution of relativistic motion is provided and may be impossible.

3.6 RELATIVISTIC ROCKET EQUATION

A Rocket Equation can also be found for the relativistic case [Appendix C; a-von Hoerner; a-Pierce; b-Belinfante; a-Ackeret; a-Sänger; a-Shepherd; a-Kooy; a-Pomeranz; a-Huth; a-Oliver; b-Goldstein]. Again, the results are in terms of the ratio of initial mass M0 to final mass M' of the starship, but with several forms possible. It is difficult to describe the results briefly without assuming mathematical background. One form assumes that all fuel is converted into exhaust with speed c ("total conversion" TC), a goal very difficult to achieve. If conversion uses all the fuel but produces exhaust slower than c, the needed mass ratio rises dramatically (exponentially again). For example, if exhaust velocity is only half of c, then the mass ratio is squared. One form of Rocket Equation gives the ratio as an exponential (e) of the proper time and acceleration. Because of the values of gee, c and Earth years in SI units, the ratio is conveniently, to a good approximation, the exponential of time in proper years (PY). The ratio becomes the Newtonian form if the speeds are slow. The ratio is also approximately given by twice gamma if exhaust velocity is c. Then it can be read directly from Table 3.1 by doubling the gamma reached. As before, the Rocket Equation can also give the final speed as a function of mass consumed during acceleration. For a large mass ratio, the final speed differs from c by twice the square of the inverse ratio (or less according to lower exhaust velocity.) This is a stronger dependence on mass than the logarithmic function for the Newtonian case.

For the 10 LY mission with speed reaching 0.987c and gamma reaching 6.2 at turnover (at 2.4 PY) and an ideal exhaust velocity of c, the mass ratio needed for any kind of rocket propulsion that carries all its own fuel energy and reaction mass is about 12. To halt the starship the same ratio is applied again, so that the initial mass ratio required is about 144. If exhaust velocity is only half of c, then the mass ratio for the first leg is about 150 and for a one-way mission is 150 squared. This is the price to be paid for the time dilation benefits of a gamma of 6. Inadequate exhaust velocity makes performance much worse than an ideal mass ratio of 10 or less per leg, but an exhaust of c would almost achieve it (if fuel conversion efficiency is perfect too). Estimating the fuel needs for an ideal rocket to go 1000 LY at 1 gee in only 7.4 PY is an exercise in fantasy. The high gamma reached (1024, from Table 3.1) indicates a mass ratio of about 2000 for just one leg. Relaxing the exhaust velocity to half c pushes the mass ratio into millions for one leg. Ways should be sought not to carry the fuel, whose mass increases by gamma.

Because of the exponential dependence of mass ratio on acceleration, using higher acceleration is unthinkable, but lower acceleration improves the ratio considerably. Then the journey takes much longer and there is no benefit from a large gamma. As the relativistic table in Appendix E indicates, a starship accelerating at 0.01 gee for 30 PY (30.5 EY) reaches 0.3c (gamma 1.05, insignificant) and covers 4.7 LY. Then the Rocket Equation shows that a mass ratio of about 2 is needed for a total conversion rocket with exhaust at half c. This is very good performance. The two-leg journey of 10 LY requires about 60 years, 25 times longer than the 1-gee case, when acceleration is 0.01 gee. If acceleration of 0.1 gee is used for the same mission, 9 PY are needed to go 4.5 LY, with gamma reaching about 1.5 at turnover, a mildly relativistic value. The mass ratio must be 6.4 for one leg and 40 for two legs, halting after about 20 years. Maximum speed of nearly 0.8c is acceptably close to lightspeed, and pushing to 0.9c would require a major increase in mass ratio. Chapter 4 explores at least one propulsion method which might achieve this performance, with exhaust velocities approaching the half c assumed.

Many studies have been published on relativistic rocket efficiency, taking into account exhaust velocity, fraction of RE converted to KE, fraction of RE lost as non-propulsive radiation, fraction of fuel mass that goes to exhausted mass and to exhaust KE, and more [a-Marx; a-Oliver; a-Huth; & see Appendix C]. In some cases efficiencies over 50% are possible. Popular and simple is the case of photon propulsion with exhaust velocity c. These studies are interesting but academic. Brief use is made in Chapter 4.

3.7 RELATIVISTIC TRANSFORMS OF SPEED AND OTHER QUANTITIES

Some further background in relativity is needed at this point, although study thus far indicates that starships are unlikely to reach extreme relativistic speeds. Thus mathematical rigor is not needed for estimates of interstellar travel. Since many elementary and advanced textbooks teach the transformation of spacetime coordinates between frames, this is covered here only briefly [ref. eg. b-French & Appendix C]. Such calculations would be vital to starship navigation. High accuracy would be needed for navigation so that Einsteinian physics should be used at quite low speeds, say 0.001c, where the difference from Newtonian physics is a small but vital half part in a million.

The Lorentz "transform" is a calculation which can convert the description of an event in starship frame S' to Earth-based or other reference frame S. S' is taken to be moving at a certain speed with respect to S, and the x-coordinate is taken to be in the direction of that motion. The other space coordinates are then unchanged, that is, the same in either frame. The transform can also convert events known in S, such as planet and star positions, to events that observers aboard the starship (S') would see and measure. The transform includes a factor of gamma and mixes space and time coordinates together, unlike the way we experience low-speed phenomena where space and time stay strictly separate. The transform does not permit any events moving at different speeds to be called simultaneous, but it does permit rigorous comparison of events to avoid apparent paradoxes.

The need to know how to compare speeds in different frames arises when a fast (say 0.8c) starship shines a laser forth. According to relativity, the photons from the laser do not acquire speed 1.8c, they move at c in all frames. Therefore speeds cannot simply add. The same holds for any material object sent at high speed from a fast starship. Since direction is important, it is better to use the term velocity for most discussion henceforth to imply a directionality. Given a velocity in one frame, voyagers might want to know what it is in another frame moving at some other velocity? The mathematical rule for "adding" velocities is another transform as shown in Appendix C. It can best be described by saying that at low speeds, speeds add normally but as speed approaches c, the "addition" is shortchanged so that two high speeds add to just a little more than one of the speeds. Unlike the transform of the spacetime coordinates of an event, "adding" velocities is complex in all directions. One sort of effect occurs in the direction that the frame S' is moving, and another effect at right angles to that important direction. At right angles, only time dilation is important, and velocity is reduced by gamma accordingly.

Lest it be thought that errors have been made in considering exhaust velocity from relativistic rockets, what matters most to the starship is what happens in its own frame. If it is moving at 0.8c and exhausts material at 0.8c, the material is left motionless in space according to starship observers but not according to others. What others find for exhaust speed is almost irrelevant, since the starship has properly (to risk a pun) accelerated itself. Transforming speeds can be important in starship operations whenever some object is sent from or observed by the starship. A robotic probe sent from the starship on a mission follows a path determined by how speeds add in relativity. If the starship is to intercept, or avoid intercepting, some foreign object, radar and other sensing must be correctly interpreted. Should missions become so ambitious as to have two starships passing each other at high speed it would be important for each to know what the true speed of the other is in its own frame.

Because of relativity, if for no other reason, interstellar voyagers cannot have an intrinsic speedometer and cannot tell directly their own speed (inertial guidance systems are possible but unfeasible). But they can make measurements on some known star ahead or behind to determine the doppler shift in the light received. Most of the possible frequencies of light that atoms in stars and gases can emit are precisely cataloged, and any observed pattern of frequencies can be compared (preferably by computer) with the catalog and a match made. Then the velocity causing the total doppler shift is easily calculated. More navigational details are in Chapter 7.

If velocity depends on reference frame, what about acceleration, force, and other quantities? What matters to the starship is the acceleration in its own frame, or at least in a frame S' moving at constant velocity which happens to be alongside the starship just before another impulse from ejected mass speeds it up a bit. Therefore how the acceleration would appear to outside observers is an academic question. For the rare cases where it might be important to know, acceleration is reduced by a factor of gamma cubed in the starship direction as measured (somehow) by outside observers in S. Any transverse acceleration is reduced by gamma squared (a time dilation only effect again). The third factor of gamma occurs due to length contraction in the direction of motion. A starship momentarily at a gamma of 2 and accelerating at 1 gee would be observed to accelerate at 1/8 gee by Earth observers. If acceleration is constant aboard, it is not constant as seen from Earth, since gamma varies with speed, and speed varies with time.

Force is unchanged by relativity for the starship. Ejected mass leads to the same familiar thrust at the moment of action-reaction (after the forces have finished acting to produce an impulse). If the engine started out producing 1 gee, it continues to, and the crew feels normal weight as usual, regardless of speed. There is no hope of relativity affecting what gee is felt. Although the starship is more massive by gamma (to outside observers), its exhaust is also more massive, and the starship and its crew do not know it is more massive. They do note that the gas and dust in space is hitting them more often and harder. The mass, momentum, and KE of the gas ("stationary" in frame S) is augmented by gamma as seen from the starship frame, and it is denser by length contraction. If the voyagers did not know about relativity, then the effect to them would seem at first to be as if the starship accelerated steadily toward c and possibly beyond, going faster and faster to hit gas molecules harder and more often. However, if they measured these effects carefully, they would find them to differ from Newtonian predictions, with less and less increase in speed.

Viewed from another frame, force is transformed and changed in a complex way involving power, if the source of the force is moving in S'. Otherwise only transverse force is reduced by gamma. Because force is not transformed the same way as acceleration in both the direction of, and transverse to, the motion, force need not be in the same direction as acceleration, in contrast to the Newtonian case. Strictly, action does not equal reaction at high speeds. Action and reaction are mediated by forces, electro-magnetic, nuclear, or both as in the case of a fusion rocket engine. Nuclear forces expel charged and uncharged particles, the charged ones act and react with a constraining magnetic field, and the end result is mass exhausted one way and equal momentum acquired by the starship. Force felt over a distance, even a short one, is not a viable model because different frames cannot agree on when a force arrived at an object or what distance was covered. The field model for force enables logical explanation compatible with relativity, since the field carries the force from its source in a nuclear or charged particle to another particle, taking time to do that and being subject to relativity all the while. For a short or long while, the field is carrying any momentum being transferred.

Electric and magnetic fields, such as those that might be used in an electro-magnetic accelerator for pellets or starships, are changed by speed. Electric field due to charge is reduced in effect by gamma squared in a transform. This is due to the length contraction of the distance to the charge. There is no relativity effect on the value of charge itself (unlike mass), only on the location, spacing, and density of charges. When magnetic field is transformed, it is found to be changed into an electric field. In relativity magnetic and electric fields are mixed and better called an electro-magnetic field. Indeed, stationary charge in one frame producing electric field is moving charge or current in another producing magnetic field.

3.8 RELATIVISTIC ENERGY AND MOMENTUM

Energy conservation is somewhat different in relativity, although all forms of energy must still be accounted for. Physicists cannot be certain that all forms are known. Whenever energy has not been found conserved in an experiment, a new form has always been discovered to make up the difference. Nuclear energy could not be explained without Einstein's equivalence of mass and energy, derived from his theory. Even if other forms of energy are yet to be found, Einstein established a total energy TE for any object that can never be exceeded. His most famous result defines the rest mass-energy RE in terms of mass and lightspeed squared. It is quoted here in correct form: $RE = m_0c^2$. RE is the energy in the rest frame and should not be confused with TE; subscript 0 denotes the rest or proper mass.

RE is the intrinsic energy an object has when not moving. It can include internal heat and electro-magnetic (chemical), gravitational, and various sorts of nuclear potential energies bound into the object. TE represents the total energy of a moving object and is RE plus KE. TE can be found by multiplying RE by gamma [see Appendix C]. KE can be found from RE by using gamma minus one. KE reduces to the Newtonian form when speed is much less than c, using an approximation for gamma when speed is low.

Einstein's relation gives very large numerical results for energy because of the c-squared. For example, the RE of a Volkswagen is 100 trillion times greater than the kinetic energy KE it has running down the freeway at 30 m/s. The heat and chemical energy stored in all its materials is also small. The available nuclear energy that could be released from the nuclei of all its components is much less than one thousandth of its RE. ("Much less" because much of its mass is iron which has the most tightly bound nuclei and therefore virtually no available nuclear energy.) These results are in accord with the fact that chemical fuel of the best kind stores very little available energy compared to nuclear, and nuclear stores much less than the theoretical maximum. If hydrogen burning releases only 13 million J/kg, then only about one part in 10 billion of the mass has been converted into energy. If all the mass in one kilogram of matter could be converted to energy in a fabled process called "total conversion" (TC), almost $(10)17$ J/kg would be released.

Einstein also found while doing other physics that light (photons) must carry definite quanta of energy proportional to frequency (color). This energy is total energy (TE). Being massless, light cannot be said to have RE or KE, but the fact that light carries energy and momentum indicates that in some ways it acts as if it has mass. Since light moves at speed c, according to relativity it cannot have rest mass. When the speed is c, gamma is infinite, and the mass increase would be infinite if there were any rest mass at all. The idea of stopping light has never found physical favor. Some other particles are suspected of being massless, which would mean that they can travel at lightspeed. When the supernova was observed in the Large Magellanic Cloud in 1987, a burst of elusive and probably massless neutrinos arrived so soon after the light burst that they must have been very very near lightspeed for the past 169,000 years.

Studying starship drives often involves studying energy in the microphysics realm where particles interact and are broken up and recombined. Some RE is released to become KE, and other RE must be acquired from KE to produce new particles. For relativistic particles such as protons and electrons, TE, RE, and KE are often expressed in units of keV, MeV, and GeV. These units of energy come from the thousands (k), millions (M), or billions (G) of volts of electric potential used to accelerate the particle to the stated energy. A proton's rest mass when expressed approximately in energy terms is 1 GeV, which is about $(10)-10$ joule. [See Appendix A for more accurate values].] About 1 GeV is the energy that would be obtained if all the mass in a proton was converted. The best accelerators now give protons a trillion electron volts (TeV), or a thousand times its rest-mass-energy, and 20 TeV is being planned for the Superconducting Super Collider.

When a particle is said to have some energy greater from its RE, the amount stated is understood to be its TE. A 1 TeV proton would have 1000 times its RE and be moving fast with KE of 999 GeV. Gamma is 1000. The speed in terms of c can be found easily from gamma [recall above and Appendix C for 3.2]. The result here is 0.9999995c, with two "9"s for every power of 10 in gamma, then a "5". When a particle has energy less than RE, this is understood to be its KE, as with the 4 MeV helium nuclei obtained after fusion. The speed can be found by considering the ratio of KE to RE. In this case a 4 MeV helium nucleus has RE of 4 GeV and the ratio KE to RE is 1/1000. This is also equal to gamma minus one, the portion of TE that is KE, and gamma is then 1.001. The speed in terms of c can be found exactly from gamma, or can be approximated as the square root of twice this ratio. The 4 MeV helium nucleus is moving at a speed the square root of 0.002, or about 0.045c.

Momentum is found in relativity by using gamma to augment the rest mass, and otherwise is the same as the Newtonian concept. A transform changes energy (always TE) when viewed from a different frame and may bring in momentum [see Appendix C]. Energy is increased by gamma when viewed from another frame but any momentum involved must also be included. Likewise momentum is transformed with a gamma, and any energy involved with it must also be included.

As speed of a starship (or particle) approaches c as seen from an outside frame S, its KE cannot increase much through increasing speed. KE can increase without limit through increase in mass by gamma. Voyagers in S' do not know that they have any KE' or speed except by deduction from the acceleration they feel, from ever more fuel usage, and from observing increasing doppler shifts of starlight. If they use relativity to calculate their TE' from their RE, they find that it is constant (or that starship portion of TE' decreases slightly as fuel is exhausted). Outside observers also have no direct way to measure starship energy but could measure its speed (by doppler) and use relativity to calculate gamma and TE from known RE. They must look at the whole system including exhaust and find TE increasing as gamma increases due to starship acceleration. For constant acceleration as measured on the starship, an earlier approximation of relativity (for long time) gives that gamma increases proportional to observer time in S. TE increases as rapidly. These two views differ by a transform from one frame to the other, reconciling the need for physical law to be the same for S and S'. Also, the time scale for the starship is slowing and its acceleration is diminishing, as seen from S.

There are two relations of momentum to energy, and the one for light has been needed to find the force that photons can exert. For particles with mass the relation is not needed here but mentioned in Appendix C. To tell a bit more of the nature of relativity, momentum and energy add as if they were at right angles. Just as 3-dimensional space and time combine to spacetime with 4 dimensions, other physical quantities such as momentum and energy combine and mix in relativistic physics. Another example is the electro-magnetic field. Understanding such "4-vectors" is vital to those who want to go beyond special relativity as covered here and solve problems in general relativity or gravitational physics involving curved spacetime.

Relativity makes energy a serious problem through the limits imposed to prevent speeds greater than light. Relativity also offers tantalizing solutions: the slowing of time and total conversion of mass to energy. How closely propulsion might approach TC is explored in Chapter 4. One could hope to find a way to travel without the action-reaction rocket method--no exhaust, no acceleration, little travel time, no deadly beams, no titanic low-mass energy source--but these are still mostly dreams from sf. Thus far it is not surprising that "visitors" from other stars have not appeared recently nor left their garbage laying about. They also must contend with what their Einsteins discover about interstellar travel. If visitors were to arrive, one of the first facts we would want to know is "how did they do it?".

Chapter 4

Relativistic Drives and Problems

Several choices of drive remain for attempting relativistic speeds (greater than about 0.2c). Because of the benefits of near lightspeed travel, the methods covered here have received very much general and scientific interest and have been widely studied and written about. They include using light (photons) more actively than ,just catching sunlight in a solar sail, and ways to obtain near total conversion of matter to energy. The complex case of the ramjet is to be considered. Some methods carry the propulsion energy in the starship, and some use fuel or electric and magnetic fields found in space itself for power or maneuvering or protection. If near-light speeds are achieved, space provides new hazards as well as benefits. A summary of propulsion also gives a first look beyond known methods. A summary on propulsion also takes a look beyond known methods. [Mathematical physics for these topics is in Appendix C.]

4.1 SUNBEAMS

Sunlight on giant sails provides a good speed for the short time that the rapidly accelerating starship can remain near the Sun or any star. What if a beam of sunlight could be collected and directed at the starship over a longer time and distance? A large curved parabolic section mirror can collect sunlight to form a tight beam which can travel a long distance before spreading too much [see Figure 4.1; a-Matloff 1986]. The focal length of the mirror is the distance to the Sun's surface and might be millions of kilometers. With a diameter of hundreds of kilometers, the mirror would be very nearly spherical. Rotation might help form its slightly curved shape. Since the mirror does not travel, it need not have low mass, and there are reasons to prefer large mass. The Archimedes method of using many small, perhaps flat, mirrors instead would not work unless they could all control the multitude of light waves from the Sun to a fraction of a wavelength. (Mirror performance limited by diffraction is discussed shortly.) It is not sufficient to add up many parallel beams of light. Their relation must be precisely controlled as a whole. The mirror(s) should be close to the Sun to increase intensity and minimize size. A good first guess is 15 million km from the Sun, 1/10 of Earth's distance, where the mirror would gather 100 times more sunlight than here. The mirror's radius of curvature would be twice the focal length or 30 million km.

Design for this idea begins with estimating the amount of power needed. If the mission could be a fast one, then the starship could be about 10,000 tonnes with little other mass needed for propulsion. The sail that must be carried surely could be kept to about 1000 tonnes. The power sent to the sail should be nearly constant for a lightyear or more, providing constant thrust. If a speed of 0.5c could be reached after 1 LY of acceleration, the needed acceleration is about 1 m/s^2 or 0.1 gee. The force needed is 10 million newtons. A mirror at 15 million km from the Sun can collect an intensity of 150 kw/m^2. If light is beamed from the Solar System at this intensity without much spreading, it can provide a pressure of about 0.001 N/m^2 after some losses. The collection area needed is therefore about 10 billion m^2, or about 100 km diameter. This mirror would differ from flat by less than 1 meter, extreme but perhaps feasible. A bit more acceleration and speed might be obtained by fine tuning this approach, after small losses are accounted for. Acceleration and presumably deceleration would take about 5 years each, and coasting over the intermediate 8 LY about 16 years (all Earth time). The one-way mission requires about 26 years, so reasonable that one should wonder what has been overlooked.

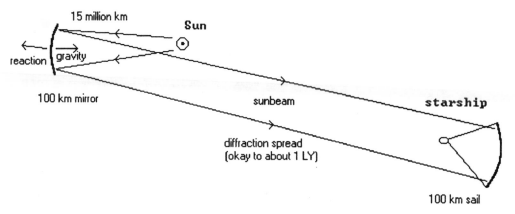

Figure 4.1   Solar mirror providing sunbeam out to 1 LY.
Curvature and size of mirror are much exaggerated.

A critical element is that this intensity be transmitted across about 1 LY, and preferably further, without spreading much. The wave nature of light causes it to spread angularly, depending on wavelength and the size of the focusing mirror or lens, in an effect called "diffraction". Longer wavelengths spread more. The longest wavelength that might be used from the Sun is in the near infrared (10)-6 m. A 100 km near-perfect mirror can form a beam of light tighter than about 200 km until a distance of about 1 LY is reached [see Appendix C]. Most of the light is still within a 100 km beam. Another critical component is the sail. A 100 km sail with generous mass density 0.0001 kg/m^2 has mass 1000 tonnes, plus cable mass.

Some difficult details remain, and a few are mentioned here. The mirror must be held in place, or move in such a way as to maintain beam direction. An orbit about the Sun in a plane perpendicular to the travel direction would prevent eclipsing by the Sun, but brief interruptions of beam are not a problem. More serious is the energy cost of overcoming the reaction force on the mirror. As much propulsion could be needed to hold the mirror as would be applied directly to the starship (without a large fuel mass). Perhaps an arrangement of two or three mirrors can be invented which uses sunlight pressure to stabilize all mirrors involved. Or Sun's gravity can pull back on a mirror aimed near the Sun (over its shoulder) at the starship [Figure 4.1]. A mirror of density about 0.001 kg/m^2 would have gravity and photon reaction forces almost balance at 15 million km. Or at great expense a large asteroid can be moved to the appropriate orbit to serve as a holder for the mirror. Or it can be anchored to planet Mercury. An orienting system is needed to keep the mirror aimed at the invisible starship. As long as down-time for repairs is just a few percent of total time, the starship mission can be kept on schedule. The mirror(s) must be smooth to a fraction of a wavelength to keep focus within the very narrow angular requirement, and active devices may be needed all over its back to control the shape. Both mirror and sail must withstand the heating resulting from reflectivity that may be no better than 99%. At 99% about 2 kw/m^2 is absorbed and must be radiated from the back, but only about 400°K is needed [see Appendix C for 4.3 for photon radiation law].

The largest difficulty is stopping the starship at the other end. The sail and other proposed braking systems are far too weak to slow it from the tremendous speed this approach permits. Sails can decelerate on unconcentrated starlight just as they were accelerated. But at much higher speeds there is not enough time near a star to decelerate even the sail's own mass from near relativistic speed. The speed limit is about 1000 km/s for systems where the sail is as thin as feasible and it and cables comprise most of the mass. Possibly a much larger sail could be used, but very heavy cables are needed to carry the large braking force very near the star. A powerful rocket engine could be carried for braking, but the design mass would increase greatly. Lower speed could be used, and the mirror and sail could be somewhat smaller. This system seems excellent for fast flyby (probe) missions. Ideally this sunbeam system would operate for two-way traffic with a second mirror system at the destination star. A low-mass high-speed mission might be dedicated to carrying a mirror system to the other end. The mirror itself can provide sufficient braking for its arrival. But an automatically aimed and maintained system of this size seems difficult to accomplish.

4.2 LASER AND MICROWAVE BEAMS AND LARGE-SCALE OPTICS

Some have gone further [a-Marx; a-Redding correcting Marx; a-Forward 1984, 1986; a-Norem] and proposed that a giant laser or bank of lasers send a lightbeam from orbit about the Sun. As with the mirror, the reaction force on the laser from its lightbeam is about as large as the propulsion force needed at the starship. As much energy might be expended keeping the laser in position as to energize it. The laser might better be mounted on a massive airless planet such as Mercury and made to swivel. Lasers were first proposed for propulsion by heating reaction mass in conventional rockets [Chapter 2]. A theoretical proposal is to use a big laser to send energy across space to a starship to heat reaction mass [a-Jackson 1978], but aside from low exhaust velocity it seems more feasible to omit mass from the starship and use propulsion provided by light reflection.

The use of a laser to send a momentum beam across space is quite a challenge. Most laser designs produce parallel lightbeams that spread slowly as determined by the laser aperture. The coherence and singular wavelength of laser light are irrelevant for most applications, but such light is more easy to control if a lens or mirror must be used in the system. Lasers at best appear to be about 10% efficient, hardly the 95% to 99% possible with a mirror. Thus a ten times larger solar collector is needed to collect the energy. The laser power for the sample mission above would need to equal and replace the sunbeam, about (10)12 kw continuous, or about a trillion times larger than present continuous lasers (not to be confused with progress in very high-peak power-pulsed lasers). It would be possible to assemble a bank of perhaps a million megawatt lasers gradually as a slower mission progressed, or turn them on one by one, because full power and aperture would not be necessary until near the mission nears full speed.

The lasers could be small if operated at higher intensity and formed into a larger beam with a giant lens or mirror. The lasers alone will surely be more complex and expensive than an active mirror. A laserbeam must come from an aperture larger than 100 km diameter if it is to reach as far as 1 LY. Each individual laser might have a mirror only 1 meter diameter, which would let the beam spread before it reached Mars orbit. To avoid being so limited by diffraction, a bank of lasers must be locked together in phase to produce effectively one large beam. It is not sufficient simply to spread a multitude of small lasers over 100 km of space and operate them as one coherent beam. Thanks to the "thinned array curse" [proof reported in a-Forward 1984], the combined laser power interferes with itself and will not be concentrated in a narrow beam. They must be put at the focus of a 100 km mirror or lens, or possibly some other optical method will be developed for combining them into one beam. Lasers or other means to generate shorter wavelengths (hard ultraviolet, x-ray) might become feasible at high power and efficiency. Such a radiation beam could remain compact over tens of lightyears from the same aperture. This technology seems very unlikely and rather dangerous at present. Lasers might be more suitable to launch small probes, as discussed in Chapter 6. Lasers can solve many other interstellar systems problems as discussed throughout, perhaps better than they can solve propulsion.

Whether laser or just mirror, the lightbeam system has one advantage, it can be reused indefinitely (with some maintenance). It will be needed about 5 years per mission. One proposal [a-Forward 1984, 1986] is that the system be used for deceleration and even return of the mission, but an incredible 1000 km beam is needed. The starship would carry a series of reflecting sails. A large heavy one (1000 km diameter, good over 10 LY) would be detached near the destination to continue on as a very tight beam is focused and reflected from it to decelerate the smaller starship with sail. For the return trip, light is to be sent again from here, although leaving an automatic mirror there for a return sunbeam might be less risky. Another way to go is to lengthen the mission and use an electro-magnetic method to be described later to curve the starship path around the destination so that it is heading back toward us behind the destination [a-Norem]. Then laser light or sunbeam can decelerate it. Return is possible by the same means. All these sketches of missions require much continued commitment from home, as discussed later. Here just the possibilities with lightbeams are outlined.

Difficulties arise when relativistic speeds are attempted by pushing with light. As pointed out for high speeds [a-Redding], the intensity of light as seen from the starship decreases so that the lightbeam becomes ineffective beyond about 0.8c. A maximum efficiency was found to be about 0.3 for starship speed about 0.4c. The momentum and energy carried by photons depends inversely on their wavelength, and as this becomes longer by doppler shift, the intensity decreases rapidly (as the square of the doppler shift). No better ways to use lightbeams for propulsion seem possible, but the preceding are inefficient. For much of the acceleration period of the sail, most of the energy sent to it is reflected back at us rather than going into starship energy. The speed of photons is mismatched severely to the starship until the starship reaches nearly 0.5c.

A fresnel lens is an optical device which takes advantage of the diffraction of light. For use in space on a large scale, it is not a disk of plastic such as found in projectors but a series of rings of plastic, outer ones more narrow than inner ones according to a certain geometric requirement [b-Hecht; a-Sussman; b-Mauldin]. In the Rayleigh-Wood form each ring is thin and provides a delay for phase reversal in the light going through it. Between each flat ring is a calculated empty place, with outer ones again more narrow. The delays and the spacings are arranged so that parallel light entering the lens and diffracted by the rings is all in step in terms of waves at several points called foci. This lens works well only with light of nearly one color (monochromatic), such as lasers produce. It will not focus a good spot from the Sun. The lens is to be used in reverse, converting a small source of intense monochromatic light partly into a large, nearly parallel beam. For a focal length of 1 billion km used by placing the lens beyond Jupiter and the slightly-spreading laser source near the Sun, a 100 km lens requires about 10,000 rings. About half the energy goes into the main beam and the rest into secondary beams in various directions where diffracted waves are also in step [a-Kirz]. No lens can defeat the natural diffraction spreading of any beam, but efficiency can be increased with special variations in zone thickness or zone material. Because of its complexity, losses, and limitation to one wavelength, the fresnel lens does not seem like an improvement over large mirrors which also must be constructed to high precision.

A recent method of concentrating light has been developed called "nonimaging" [a-Winston]. This method has already been demonstrated to concentrate sunlight to intensity greater than the surface of the Sun and a laser might be energized with such intense radiation. For a direct beam on a large scale the method seems to require a very long reflecting cone and probably could not achieve diffraction-limited beam size. Further study is needed and likely once the interstellar research community ponders this idea.

Other radiation such as microwaves has been considered instead of light. Easily generated with technology at hand, microwave radiation such as radar pulses already greatly outshine the Sun in the radio part of the spectrum. But the amount of radiation needed is much much greater or else the mission will have low speed and long travel. Microwaves are typically 1 cm in wavelength and provide a new problem in diffraction. A 100 km transmitter "dish" can produce a beam only until about 0.0001 LY out. Possible improvements include: waves as short as 1 mm, and a dish 10 times larger. Larger dishes are somewhat easier to make than optical mirrors because one can be made of mesh with hole size about half the wavelength, and it need not be smoother than this size. The "sail" would also be mesh. If it is also 1000 km diameter, the beam can be effective to about 0.1 LY. With an ambitious acceleration of 0.1 gee from a microwave beam, only 0.15c would be achieved before the beam began to spread beyond the mesh sail. This is an upper estimate of future capability, and the use of microwaves does not seem an improvement.

A published mission design using microwave beam [a-Jones] to push a starship to 0.1c assumes a rather massive starship and finds almost 10,000 km diameter mesh "sail" needed for 3 cm microwaves. Because of diffraction, useful acceleration must be accomplished within about 0.05 LY even with the larger dish and "sail" used. A less massive mission would need smaller transmitter and mesh "sail" but then acceleration must be accomplished over a shorter distance. Microwaves appear more suited to massive missions provided that civilization can afford the (10)15 kw transmitter and find a way to stop the starship at the destination and cope with the extra strong cables required. At the other extreme, a probe mission [a-Mallove 1987] would have the 1 km mesh "sail" as the entire 100 kg vehicle, carrying detection and data transmission circuits on the mesh and boosted at more than 100 gees to 0.2c by a 10 Gw beam.

4.3 PHOTON DRIVE

Some decades ago it was asked, why not use photons as the exhaust [b-Sänger]. Thrust would be produced by their momentum according to the intensity of the light as discussed for solar sails. Nothing is faster, and this might be the way to the highest speeds. Photons are at no disadvantage compared to particles with mass. At ultra-relativistic speeds massive particles act in some ways more and more like high-energy photons. If there is a choice of frequency (wavelength), the only advantage of high-frequency photons is that less are needed to produce the same thrust, and the hypothetical conversion equipment is likely to be smaller. Radiowaves would make a photon drive but the mass-energy carried by the photons is so meager the size of the generator is almost inconceivable. Even the equipment for an optical photon generator is expected to be very massive and inefficient [a-Bond 1971].

A Rocket Equation has been worked out for photons [e.g. b-French; a-Pierce; a-Huth; a-Oliver] giving the mass ratio needed for the rocket when photons are exhausted to reach a given speed [see Appendix C for 3.6]. As for any rocket, some sort of fuel must be carried and converted to exhausted photons. The speed can be found from mass ratio as before except the fraction of mass converted to photons is involved. If large gamma is desired, the mass ratio needed is approximately twice the value of gamma wanted [a-von Hoerner]. A nice gamma of 5 requires mass ratio 10:1 for one leg, or 100:1 for a two-leg journey, or 10,000:1 to go and return. This seems barely feasible. Those cases where the entire mass of the Earth is needed to deliver one proton occur when matter is far from totally converted to photon exhaust.

Einstein's rule applies to converting ordinary matter to photons: the energy obtained is c-squared times the mass used. But this rule does not indicate how the conversion is to be done. Photons could be obtained from a small fraction of RE as occurs for fusion power. Heating a material to glow white hot like the Sun will not be sufficient. As seen with solar mirrors, that approach requires a hot area say 100 km in diameter to obtain enough photon flow. Without needing to count the enormous number of photons, the intensity and force from radiation is proportional to temperature with an exponent four [e.g. b-Hecht & Appendix C]. Suppose that the photon drive consists of some object heated to a suitable temperature for radiation. To obtain a typical 10 million newtons of thrust from an exhaust chamber of a reasonable size such as 100 m diameter, a perfectly radiating object at about 40,000°K would be needed. It would radiate about 2500 times more intensely than the Sun. The wavelength around which the radiation would be centered is about (10)-7 meter or harsh ultraviolet (UV) with appreciable soft x-rays. The source would likely radiate this in all directions. It is not too difficult to find reflectors to collimate UV radiation to form an exhaust beam (and absorbers for the modest x-radiation). But to shield the rest of the starship and protect itself, a reflector must not absorb more heat than could injure its surface, limiting it to a temperature of say 1/40 of the above source. The intensity it can handle is the exponent four of the ratio of temperatures or less than a millionth of the intensity of the drive source. Even with a major cooling system this is a severe requirement on reflectivity. If collimating the photons from a hot source is a serious problem, perhaps a very efficient and powerful laser of suitable wavelength can be developed. High-power lasers, too, have difficulties with overheating their mirrors and other components.

The radiation may be intense but unless most of the matter involved is converted to radiation, the starship will be too massive to accelerate near c. The photon drive is useless, and worse than other methods, if the energy comes from fusion which consumes only a tiny fraction of mass. As seen before, when only half the fuel mass is converted, a case that may well apply to the best photon drives, then the mass ratio is twice as worse. The factor of two for one application of the Rocket Equation results in a factor of 16 for a two-way trip. However, the incompletely consumed mass could be dumped into the hot exhaust to make it more effective. Any more shortfall in total conversion and a two-way mission approaches impossibility. Ways are scarce to break matter down totally to produce just photons. Two ways are explored in the next section, and they are far from ideal. Regardless of the fraction of mass converted, the overall efficiency of the photon drive is near zero at low speed and rises to 50% at best for a large mass ratio [a-Oliver & Appendix C, 3.6].

4.4 TOTAL CONVERSION AND ANTI-MATTER REACTION

An early proposal for total conversion of matter to energy (as radiation for a nearly ideal photon drive) was the use of electrons and positrons [b-Sänger]. Positrons are scarce in nature and are the anti-particles of electrons. The two have the same RE and annihilate when they approach to release two high-energy photons (sometimes called gamma "rays", no direct connection to the gamma of relativity). Each has energy about 0.5 MeV and they come out in opposite directions. Immediately the problem of reflecting these photons arises so that an exhaust in one direction can be formed. No ways are known to reflect high-energy photons on a large scale. If this was solved, then a method of carrying large quantities of electrons and positrons would be required. They would have to comprise most of the mass of the starship.

The common massive particles in nature are protons (and neutrons). The anti-particles to protons can exist but are virtually never found in nature. They were discovered in 1955 [a-Segre]. Like anti-electrons, anti-protons would soon disappear in a burst of radiation as they encountered ordinary matter. It is possible to make anti-protons, and they have been made on a small scale for many years in various particle accelerators for use in basic research. Noteworthy are the anti-proton storage ring at CERN in Europe and the superconducting ring at Fermilab near Chicago. The quantity that can be stored for days at CERN is about a trillion anti-protons with total mass about (10)-15 kg [b-Forward]. At Fermilab 200 billion anti-protons or about (10)-16 kg is stored a few hours [a-Lederman]. Anti-matter or anti-particles have also been called "mirror matter" because they are in some sense an opposite of normal matter, being opposite in charge and in magnetic and a few other properties. Many subtle rules of nature apply to the conversion or annihilation of particles. Protons have a property (baryon number) which can only be canceled out by anti-protons carrying the opposite of that property. Proton anti-proton annihilation is a complex process but results in a number of new lighter particles being formed. It should be mentioned that protons are composed of three "quarks" each, of different kinds, held together by a high-energy field of "gluons" (naturally), a counterpart of photons. The quarks are mentioned here because maybe someday a physicist-reader will find a way to use them in a starship drive. They have the peculiar property of never appearing alone. Thus annihilation does not result in a burst of quarks but rather new particles made of pairs of quarks, called pions and kaons.

Decades of experiments have found the statistics of proton annihilation [a-Forward 1982; a-Cassenti 1982; a-Vulpetti 1986]. Almost all of the complex details of this reaction are crucial to designing a reactor or drive that uses protons and anti-protons. A simplified description is provided here, adapted from (and sometimes correcting) many references. More attention must be paid to exact masses, and the particles of interest are first listed in Table 4.1.

Each time proton p and anti-proton $\bar{p}$ annihilate, a different mix of particles is produced. The average results are that about two π°, about 1.5 π^+, and about 1.5 π^- are produced. The three to seven or more pions (typically five) are about 99% of the products. The other 1% is comprised of kaons. Since net charge is zero to start with, the products must sum to zero charge. Actual products must consist of an even number of charged pions, two about as often as four. The numbers of particles and anti-particles of each kind must also sum to zero afterwards. The neutral pions decay so rapidly that each has become two high-energy photons before covering any significant distance. After annihilation these π^os emerge in any unpredictable directions. This π^o branch is virtually uncontrollable and represents a complete loss of reaction energy to hard radiation requiring shielding. The charged pions live much longer and travel in random directions. Then each decays to a muon and a muon type neutrino. The neutrinos almost never interact with anything and represent a complete loss of more energy. The π^+ goes to a μ^+ and a muon type neutrino (v_μ). The π^- goes to a μ^- and a muon type anti-neutrino ($\bar{v}_\mu$). The muons live about 100 times longer and decay to electrons and two neutrinos each. These four different kinds of neutrino need not be tracked, as they are uncontrollable.

Table 4.1
PARTICLES INVOLVED IN PROTON ANNIHILATION

Name	Symbol	Charge	Mass (mev)	Lifetime (s)
proton	p	+1	938	stable
anti-proton	$\bar{p}$	−1	938	stable
pion	π^o	0	135	(10)-16
	π^+	+1	140	3 (10)-8
(anti-pion)	π^-	−1	140	3 (10)-8
kaon	K^+	+1	499	(10)-8
	K^-	−1	499	(10)-8
	K^o & $\bar{K}^o$	0	498	complex
electron	e^-	−1	0.5	stable
anti-electron	e^+	+1	0.5	stable
muon	μ^-	−1	106	2 (10)-6
anti-muon	μ^+	+1	106	2 (10)-6
e neutrino	v_e	0	0?	stable?
μ neutrino	v_μ	0	0?	stable?

Notes: Numbers are rounded somewhat. For each neutrino there is an anti-neutrino.

Upon p $\bar{p}$ annihilation about 1880 MeV are available to form products. In most designed drives, no significant energy has been added to the incoming p and $\bar{p}$, so that this is all the energy. The typical 3 charged pions and 2 neutral ones take up 690 MeV of RE in being formed, leaving about 1190 MeV as KE. The π^os together carry away 270 MeV of RE and typically 430 MeV of KE. A total of about 700 MeV is lost to photons by that track, about 37% of the original energy. The typical three charged pions carry RE of about 420 MeV and share about 750 MeV of KE. Their gamma is about 2.7 so that their speed is about 0.93c. At this speed their lifetime is about 7(10)-8 second because of time dilation and they travel about 21 meters before decaying. This is a useful distance.

The pions decay to muons (RE 105 MeV each), losing about 250 MeV to KE of neutrinos and retaining about 570 MeV as KE of muons. Muon gamma is about 2.8, and they live almost 3 times longer or about 6(10)-6 second, traveling almost 2 km. The muons then decay to electrons or positrons and neutrinos. Since these neutrinos end up taking the majority of the energy at this point, whatever use a drive or reactor made of the products of annihilation should be completed before this point. Energy can be extracted before it is lost to any neutrinos. Although nothing will stop a decay, neutrinos need not be let to carry away KE and momentum before it is used.

Kaons are produced only 1% of the time, but they are more massive and in a large reactor must be accounted for. In the torrent of energy from an anti-matter drive, the 1% portion is too large to ignore, and the kaons must be controlled along with the pions and perhaps muons. Tracking the range of energies that go to each kind of particle after annihilation is a difficult task attempted in the literature and will be necessary for any real design [a-Vulpetti 1984].

The general result of proton annihilation is production of typically three short-lived charged particles (pions) possessing about 63% of the original p p̄ RE. But they carry only about 740 MeV as KE, or about 39% of the input RE. Their typical speed is close enough to light to be very useful if it can all be directed one way as exhaust. Their mass is valuable in that it represents momentum. Total conversion of mass to energy remains an elusive goal, since protons reacting with anti-protons does not result in just photons, immediately or later. A bunch of particles result which necessarily travel slightly less than c. It would not be fair to continue to call the annihilation of protons a total conversion (TC) drive. That term should be reserved for when some other yet unknown process is discovered that provides entirely photons and/or near lightspeed particles which all can be controlled. With proton annihilation we have the best of known processes, but it is not the ultimate. This drive determines the prospects for interstellar travel for the future as best known science can predict.

4.5 ANTI-MATTER DRIVE AND PRODUCTION

Given the proton annihilation reaction, a reactor or drive is to be designed that will make best use of its products. The goal is a momentum drive as in previous rockets, with fuel energy converted to exhausted reaction mass. No good ways are known yet to redirect momentum carried by neutral particles and high-energy photons, but charged particles like pions are easily controlled with electric and magnetic fields. The first approach to try is the magnetic "mirror" again [a-Forward 1982; a-Morgan]. Particle speeds are much higher than in the fusion reaction chamber, and adequate magnetic field will be more difficult. Assuming that a large chamber, or even a bank of large chambers, is practical, the Daedalus design is taken as a guide, but reduced a factor of 2 to about 50 m diameter for reasons to be shown. The magnetic field is to increase toward the front end, making exhaust escape behind [see Figure 4.2]. One must keep in mind that pions with the typical gamma of 2.7 decay in about 20 meters and have almost 3 times their rest mass, but many have higher or lower gamma. The momentum of those initially heading forward should interact with the mirror field before decay occurs. Those already heading backwards make no further contribution to thrust, their principle role having been to provide action-reaction to send the other particles forward after annihilation.

Particles emerge from annihilation with velocities with components parallel and perpendicular (transverse) to the prevailing magnetic field lines. Those few without transverse velocity go backward or forward without being affected by the field. The rest spiral around field lines with a radius of gyration (Larmor) proportional to their transverse velocity. Their spiraling is left- or right-handed depending on the sign of their charge. The exhaust is not straight-line particles but madly spiraling ones. This makes no difference in propulsion since there are equal numbers of each charge and the exhaust is naturally neutral. The background field B0 needed to make pions with maximum transverse velocity spiral around it in radius 1 meter is a feasible 1 tesla [method in Appendix B for 2.5]. The current needed is about 40 million amps. The field could be raised to 10 tesla to catch most of the higher-energy pions, since more field is needed to control more massive ones. Magnetic reflection must occur within 10 meters, and forward particles must be reversed in about 3(10)-8 second. The increase in field must be about 4 tesla. The end field must then be 5 tesla around a smaller section, and again about 40 million amps (A) of current is needed. This current may sound high, but many turns of superconducting wire can be used. The new "cool" superconductors have been made to carry at least 1 billion A/m^2 [a-Waldrop], and a total wire cross-section of 0.04 m^2 seems sufficient and reasonable. More improvements in superconductors are expected, at least in the magnitude of the field they can be immersed in without failure. Five tesla is surely feasible since Fermilab achieves this amount of field with cold superconductors.

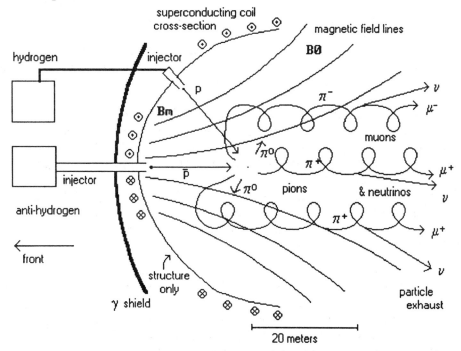

Figure 4.2 Simple hypothetical view in cross-section of an anti-matter rocket drive.
Proton and anti-proton annihilation produces charged particles which live just long enough
to interact with the strong magnetic field before decaying. Indicated theoretical particle
paths have exaggerated gyration radii. [inspired partly by e.g. b-Forward & Davis].

In this design the charged pions have been reflected backwards before many can decay to muons and neutrinos. No magnetic mirror can reflect all charged particles. Ones moving parallel to the axis of the system or with excessive transverse velocity cannot be controlled by a given magnetic field. The drive design must allow for some energetic particles going forward and capture them in shielding and dissipate the enormous heat. Electric fields can deflect them away from the axis so as not to interfere with anti-proton injection there. Another part of the design is to inject continuously or in pellets an unreacted mixture of protons (hydrogen) and anti-protons (anti-hydrogen). One design [b-Forward] sends a beam of anti-hydrogen backward down the axis and a beam of hydrogen in from the side so that they meet at the chosen center of the chamber. Both beams are neutral, and a bit of gamma radiation is produced when the electrons and positrons of these atoms also interact. The interacting beams pose another set of complex problems. The p and $\bar{p}$ must move slowly and become close before their mutual attraction pulls them together to annihilate. The necessary amount of dispersal and mixing must be determined. The rate of reaction is limited, or else particles from reactions may prevent other p and $\bar{p}$ from interacting.

The major obstacle to an anti-matter drive will probably be the production of anti-protons. Research accelerators produce them at best in quantities of trillionths of a kilogram, but research accelerators were not designed for large-scale anti-matter production. The process is to smash high-energy protons (about 100 GeV) into a block of light metal such as beryllium. Almost all known particles can be produced, but most common are pions and a few kaons. Anti-protons at best comprise about 1% of the products, depending on input energy [a-Baker]. Except for the scarcity of anti-protons (partly because they are seven times more massive than the easily-produced pions), the results resemble those from proton annihilation. Nature has anti-proton production low on the priority list of particles produced in collisions, although there may be yet undiscovered ways to set up a more ideal interaction that favors anti-protons. At least 5 GeV is needed for an incoming proton to produce one p back (to conserve the baryon count) and one more p and one $\bar{p}$, because only about 3 GeV is available in the frame of the collision. Using higher energy makes more energy available to produce more anti-protons from one collision and results in higher production rate, up to 1% of the pion production rate at 300 GeV input [a-Baker] when the anti-proton KE is low. High KE anti-protons are very undesirable, and it would be nice to produce them almost at rest. Energy efficiency is not good at high input energies because more input energy is lost in transition to the frame of the collision.

Another production method is to use two straight-line proton beams of about 100 GeV aimed at each other, but collisions are much less likely. Another way is to use two oppositely directed beams in two intersecting accelerator rings. Those that do not collide the first time get repeated chances at collision with negligible extra expenditure of energy. Accelerator production is feasible only because superconducting magnets have enabled operation with much less input of electric energy than formerly used [a-Lederman], but still much more than appears in the proton beam. At present for every unit of energy put in less than a millionth results in RE stored in an anti-proton. There are many ways to increase efficiency and production [b-Forward]. The energy lost to other particles could be captured as heat or nuclear energy and used to generate power to operate part of the accelerator. Using heavy ions such as uranium might also increase production [a-Chapline].

A quite different production method is to materialize pairs of particles and anti-particles from high electric fields. A (10)21 watt laser has been estimated to have the power needed to produce electric field greater than (10)18 volts per meter in a tiny spot, from whence electron-positron pairs will spew forth at an efficiency as high as the laser itself [a-Crowe]. Even in pulse mode this laser is millions of times more powerful than the state of the art in lasers. To obtain proton and anti-proton pairs by this method would require 2000 times more laser power and a way to prevent energy loss to other less massive particle pairs. But any sort of matter and anti-matter pairs is acceptable if they are storable. Again, if this method works it still cannot be hooked to Earthly power plants for lack of capacity to produce enough anti-matter for starship use.

One might think that anti-protons are too dangerous to handle, that storage in a bottle of regular matter is impossible, and that any accident results in a titanic explosion. Many ways have been suggested besides large particle storage rings to store and handle anti-protons [a-Forward 1982; a-Zito 1982, 1983; a-Morgan], probably as very cold or frozen anti-hydrogen. Anti-protons are likely to be produced at substantial KE (at least 1/10 their RE, say) and must be collected and cooled in a series of steps from their original randomly oriented paths. Much of this has been solved at CERN and other facilities. Once the anti-protons are moving slowly there may be ways to store them in more compact magnetic bottles or Penning traps using electric and magnetic fields or suspended in laser beams (but lasers, unlike magnets and charged plates, need energy to operate). The anti-protons, which carry negative charge, are converted into neutral atoms by supplying positrons (from another accelerator system) as soon as appreciable amounts are collected. Rather than a gas the anti-hydrogen might be collected as frozen clusters or tiny crystals of atoms or molecules (two bound atoms). For use in a reaction chamber the clusters must be carefully reduced to a stream of atoms or molecules, ionized, and gently accelerated into the chamber along its magnetic axis.

If cold enough (1°K), anti-hydrogen molecules would escape the solid form too rarely to cause much damage when they encounter normal matter. The clusters probably cannot be dumped into any ordinary container but must be held in magnetic traps or "bottles" continually. As for danger, a small amount of anti-hydrogen encountering normal matter produces hard radiation that goes readily through people and materials. Large amounts of anti-hydrogen should not be stored in one place without great care isolating it because once it starts sizzling on large scale no one would want to go near it. Violent explosion is unlikely, however, because the annihilation products penetrate ordinary matter too well to produce rapid heating.

4.6 MISSIONS WITH ANTI-MATTER

The bottom line for anti-proton propulsion is the question of how much is needed. There are several approaches to analyzing the capability of this propulsion.

As a first example, consider the fast small mission using a 1000 tonne starship. At 0.9c gamma is about 2.3. For any final speed from any given exhaust velocity, the Rocket Equation tells what ratio of fuel to payload mass is needed, regardless of other complexities. The mass ratio to reach 0.9c with exhaust near c is easily estimated at twice gamma or about 4.6 [Appendix C for 3.6]. However, although some part of the exhaust is at about 0.94c, not all the fuel mass is converted to exhaust mass or KE. Instead about 1/3 of the fuel RE appears as exhaust KE and about 1/3 is totally lost as neutral pions. An elaborate equation has been developed [a-Huth; Appendix C] for determining the effective exhaust velocity for general imperfect rocket drives. When no other material is added to dilute the exhaust, the exhaust velocity is effectively about 0.75c. The mass ratio can then be calculated to be about 7 for achieving a speed of 0.9c [Appendix C for 3.6]. 6000 tonnes of fuel would be needed, half anti-protons and a formidable amount of exotic substance to be carefully stored and handled. Stopping this mission would require an overall mass ratio about 50:1. Thus 1/3 total conversion of mass to propulsion energy is not a clear solution to interstellar travel over modest distances.

Consider a mission at 0.1c to determine how much smaller the anti-proton requirement would be. The optimum exhaust velocity should be about 0.06c. If anti-proton fuel is diluted about 1:200 with ordinary hydrogen or other common material, this exhaust velocity is achieved (according to the equation cited). The pions from the anti-matter reaction are somehow made to collide with and transfer their KE to a large amount of ordinary ionized matter in the reaction chamber. Another way to determine the dilution needed is to use the earlier relation of RE, KE, and speed, where 1/200 of RE is to be put into KE to achieve 0.1c. This dilution factor can be used with any size starship to determine the amount of anti-hydrogen needed. The mass ratio for a 1000 tonne starship with exhaust velocity 0.06c is found to be about 5:1, so that 20 tonnes of anti-protons are needed. Stopping would require a mass ratio of about 25:1 and 5 times more anti-hydrogen.

One hundred tonnes of anti-hydrogen sounds very costly. While most cost estimates are left to a later chapter, curiosity demands a brief answer here. If the Einstein rest-mass energy RE in it were purchased as electric power, its (10)22 joules would cost about 300 trillion dollars. (1 kg would cost about 3 billion dollars.) Maximum production efficiency appears to be somewhere between 0.001% and 0.1%, so the cost will be 1000 to 100,000 times greater [b-Forward]. Clearly a very rich civilization is needed to produce this most compact fuel for starship propulsion.

Anti-matter has more potential for smaller but important missions. If the mission is reduced to a one tonne flyby probe at 0.1c [b-Forward; a-Cassenti 1983], the material needed is 1/1000 of the preceding example, or 4 tonnes for reaction mass and 20 kg of anti-protons, costing perhaps 40 US GNPs and still out of reach of present human society. Some have studied optimizing anti-matter usage [a-Cassenti 1982, 1983] but no amount of optimization will noticeably reduce these costs. Clearly, fuel or energy for starships must be found ready-made in nature and not manufactured using present powerplants. Another remote possibility is solar-powered semi-automatic anti-proton production facilities in orbit operating for a long time. Anti-matter production could be a space industry, since it is a very valuable product that can be sold for many uses.

The flexibility of anti-matter for smaller-scale local missions (and other uses) has been pointed out [b-Forward], although many technical problems must be identified and solved for any sort of anti-matter reactor. It is a near ideal heat source for ordinary rockets using any kind of reaction mass (water, dust, or whatever). The mass ratio could usually be near 1. Rockets can make the best use of anti-proton heating by using that tiny amount which will heat the reaction mass to that exhaust velocity which is about 0.6 of the desired final speed of the rocket. Typical missions (in space, not Earth launch) would need the order of milligrams to grams of anti-matter. At ten trillion dollars per kilogram, anti-matter-energized rockets can compete well with the cost of lifting chemical fuels into space for use there. If over the long term a space anti-matter industry serving interplanetary transportation (and a few uses on Earth) develops, then the capacity for anti-matter interstellar probes might accumulate and the cost fall.

Another suggested approach to using anti-matter for interstellar travel is to employ it in a reactor decoupled from the drive [a-Nordley]. The reactor would be large and attempt to capture as heat about 80% of the results of annihilation (nearly all except the neutrinos). Then a 40% efficient thermal generator would provide electric power for whatever drive is desired. The overall efficiency does not seem much different from the direct drive described above. If the amount of anti-matter used is to be minimized, the exhaust velocity must and would be controllable to match it to starship speed. This may be a better method than diluting an annihilation reaction with hydrogen.

4.7 INTERSTELLAR GAS AND DUST

Interstellar space is not sufficiently empty for safe relativistic travel, yet not full enough for collecting much fuel along the way. The contents of space are not known accurately but estimated from slight effects on starlight and radiowaves and from light and radiowaves emitted by gas in it. On a large scale the various constituents of interstellar space have a complex physics [b-Spitzer 1978, 1982; b-Hollenbach; a-Ney; a-McKee]. The material is mostly hydrogen and some helium, sometimes neutral, sometimes ionized to a tenuous plasma of protons and electrons. The sparser regions have about 100,000 to 1 million atoms per cubic meter and the denser clouds up to a billion atoms per cubic meter [a-Martin 1978]. The overall average is about $1/cm^3$ or about $(10)-21$ kg/m^3. For better or worse we live in a sparse region with $0.1/cm^3$ and density about $(10)-22$ kg/m^3. The hydrogen seems to be moving about 20 km/s from approximately the direction of Alpha Centauri [a-Paresce]. The solar wind collides with this interstellar gas upstream and is pulled into a long tail downstream (about 0.01 LY). The shockfront is outside the orbit of Jupiter and shields most of the planets from interstellar gas clouds unless they are unusually dense. Atoms more massive than hydrogen such as helium and surely deuterium are pulled into a thin tail downstream from the Sun and also are ionized within 10 million km of the Sun.

Brighter stars than our Sun ionize the hydrogen gas around them to produce what are called HII regions. Far from stars the HI regions are about 1% to 10% ionized, probably due to UV starlight, occasional high-energy photons, and other cosmic particles passing through. The gas is about 1/10 to 1/2 molecular, the rest individual atoms [a-Blitz]. Helium-4 comprises about 1/10 the gas (1/5 within the Solar System), deuterium about (10)-4 to (10)-5, and helium-3 about (10)-5 [a-Pasachoff]. The latter two are fusion fuels and therefore of great interest. Carbon, nitrogen, and neon are around (10)-4. Lithium-7 is way down around (10)-9. Elements beyond helium comprise less than 1% of interstellar matter, and dust is another 1% by mass. Somewhat common molecules (about (10)-4) are water, carbon monoxide, ammonia, and formaldehyde. In some places vast clouds with small concentrations of more complex gases exist, with up to 100 different kinds of organic molecules [a-Blitz; a-Lada]. The closest are much farther than 10 LY and may seem only of academic interest, but other civilizations may live in and travel through them. Their densities are about 100 to 1000 times average and their size hundreds of lightyears, with lightyear size "lumps" up to 100,000 times average density. Recent photographs of nebula such as in Orion show very complex structure with wisps and tendrils too. The gases are cold but regions of gas are moving at many kilometers per second. Magnetic field ten times stronger than Earth's have been observed in gas clouds.

Our Sun happens to be just a few thousand light years outside a denser ring of dust and gas (mapped by carbon monoxide, a better approach than mapping hydrogen) that encircles the galaxy in its plane and constitutes the principle star-forming region [a-Gordon]. The ratio of matter in stars to interstellar matter is estimated at 25 to 1 there. This may be help or hindrance for local interstellar travel but has implications for the existence of other civilizations farther inward. Our Sun is relatively isolated and was formed in a region with just enough material to make good planets, nearly sweeping the vicinity clean. Assigning to each nearby star a proportionate volume of space (300 cubic lightyears), the interstellar matter remaining in space has mass less than about 10% of that in the star. If our Sun were closer to galactic center, the Solar System might be plagued with passage through dust clouds that absorb some sunlight and life here might have been prevented.

Interstellar dust grains are very small and scarce, about 1% of all matter in space, but still of concern. Their presence is known because they block much starlight, reddening and polarizing what does come through and reemitting starlight as infrared. Typically several thousand lightyears of dust are sufficient to reduce light intensity by half. If the Solar System passed through a good dust cloud, no stars would be visible at night. The weak magnetic fields out there partially orient the dust so that it causes large-scale polarization of light. The dust seems to consist of carbon, nitrogen, and oxygen compounds of silicon, magnesium, and iron covered with water, methane, ammonia and organic ices. They seem to have grown from dispersed atoms and molecules collecting around a silicate seed [a-Greenberg 1967, 1984]. Scattering of light shows sizes not far from the wavelength of light, with most grains between (10)-8 and (10)-7 meter. The larger ones are as rare as one per cubic kilometer. The densities cannot be much more than twice water and the masses must be in the range of (10)-18 to (10)-16 kg. These sound low but are about 10 billion times the mass of hydrogen atoms, indicating that grains contain about a billion atoms of the various kinds named. It is helpful to know that one average grain can be found in each million cubic meters, almost a trillion times rarer than hydrogen atoms by count but 100 times rarer by mass.

The low-intensity UV and more energetic photons are sufficient to break up all molecules and to reassemble the icy surfaces of dust grains in hundreds of years. Because of accumulation of very reactive molecules in the cold ice, the surfaces of grains are expected to explode periodically, reducing the grain and preventing larger grains than about $(10)-7$ meter. The observed nearly uniform attenuation of starlight in all directions indicates that dust grains are uniformly small everywhere. If pebbles and rocks exist in interstellar space they are much rarer, probably rare escapees from planetary systems or comet clouds. There is no known mechanism to make pebbles in space from grains. Comets seem to be made from interstellar grains when a star and planets form (condense) from the interstellar medium.

In interstellar travel the gas produces slight drag and erosion on the forward surfaces of the starship and the dust may produce severe erosion [a-Martin 1972]. The gas density is about $(10)-21$ of Earth's atmosphere. But at 0.5c it rushes by about 100 million times faster than an Earth breeze. Since windpower increases with the cube of speed, this relative gas speed more than makes up for its low density. Various studies have various conclusions about the severity of erosion [a-Benedikt; a-Powell; a-Langton 1973, 1977; a-Martin 1978]. Debate centers around whether high-speed grains remove much starship material upon collision. Some believe that vaporized metal will mostly cool and recrystallize in place. Like most other interstellar phenomena, this one has not yet been subject to experimental test. The best experiments to date have been with relativistic uranium atoms, not grains of dust. High-speed tests with dust seem feasible. Atomic ions at high speed seem to penetrate a millimeter or so into metals and will pass through thin foils without leaving much energy as heat [a-Forward ltr 1986]. The rarer dust grains will leave holes or pits. Nuclear tests are appropriate because some small number of collisions with gas and dust will involve nuclear reactions and production of radioactive isotopes. This is a separate large area of study since nuclear reaction rates (cross-sections) depend very much on which nuclei are involved and their energies. An estimate at about 0.1c shows about $(10)-7$ of the nuclei in matter would be so affected per year.

If the interstellar material encountered adhered permanently to the starship, the mass acquired would be a negligible one kilogram per year at high speed. Eroding material requires energy to gasify it and break all bonds, about 10 million J/kg for a light metal. If, as found below, at least 100 J is deposited in a square meter per second at 0.9c by dust, then about $(10)-5$ kg of material could be removed per second. With surface material of density at least 1000 kg/m^3, the top $(10)-8$ meter would be removed each second. A thin sail would be eroded away in minutes to hours. A 1 meter thick aluminum shield would last about 3 years. However, the problem may not be so serious as there is good possibility of cooling the sail or shield. More methods of protection from interstellar matter are discussed in the next section and Chapters 5 and 8.

The typical KE of a dust grain relative to a starship moving at about 0.9c with gamma 2.3 is about 1 joule. The density of matter ahead appears to be compressed by gamma. This sort of impact can be absorbed unless there are too many. For a skinny starship with frontal area 10 meters diameter, there are about 60,000 hits per second, each depositing a substantial fraction of 1 joule in the forward surfaces [see Appendix C]. In the unlikely case that all KE is absorbed, 600 w/m^2 must be dissipated if heating is not to melt materials. By radiation alone a surface can dissipate this heat only if its temperature rises to at least 300°K, just lukewarm. The energy deposited when hitting the hydrogen gas is about 100 times greater since it is about 100 times denser. Each atom carries about (10)-10 joule relative to the starship and the (10)16 hits per second deposit about 1 million joules. Ten kw/m^2 must be dissipated, requiring at least 700°K to radiate. The radiation itself retards the starship, but too little to matter. The problem can be reduced to manageable levels by modest reduction in speed (say 0.5c) and streamlining so that little KE is absorbed. Erosion of smooth forward surfaces by dust grains may then be the most serious part of the starship interaction with interstellar matter.

The drag force (friction) can be estimated from Newton's law [Appendix C] for a starship in our region of space with frontal cross-section area 1000 m^2 to be less than 0.0001 newton at 0.1c. This is negligible compared to planned thrusts. Drag force is proportional to speed squared, to frontal area, and to the density of interstellar gas. Beyond 0.1c drag increases more rapidly as gamma becomes significant. At a gamma of ten the drag would be about 0.1 newton, still negligible. For practical braking some means would be needed to increase greatly the frontal area, using thin material such as a sail, but not so thin as to be eroded away prematurely. A 100 km drag sail presents 10 million times more area and would provide braking of 1000 newtons at 0.1c, still not enough, and far too little at lower speeds. Three years would be required to remove 10 km/s from the speed. Interstellar gas is so sparse that obtaining a substantial braking force of a million newtons would require very large technology at high speed and would not be feasible at low speed. The increased gas density near a destination star would provide too little help too late.

The temperature of interstellar space is never less than about 3°K when no stars are nearby and often is much greater. The density of gas is so low that at higher temperatures it still has no significant heating effect on a stationary vehicle (an unlikely situation). The sparse gas molecules are moving randomly and sufficiently fast to represent up to 100°K in HI regions and 10,000°K in HII regions. Similarly, the material in space has negligible cooling effect and all cooling (dissipating unwanted heat) must be done by radiation. This works very well against a background of 3°K radiation left over from the cosmic "big bang". Dust grains are typically at about 10°K in HI regions, somewhat more in HII. The background radiation itself has been shown to have negligible effect on starship motion [a-Bolie].

4.8 INTERSTELLAR ELECTRIC AND MAGNETIC FIELDS AND THEIR USE

Space is necessarily electrically neutral although typically 1% to 10% of the gas consists of unattached ions and electrons moving about randomly in equal numbers to form a plasma. Any large-scale separation of these would result in large electric fields that bring them back together. The possibility of using artificial electric fields to work on interstellar material leads to many imaginative ideas related to propulsion although no useful thrust. Where ions are scarce, laser and other methods to produce ions are worth considering. Electric fields can be used to repel or attract charged particles (moving at high speed relative to the starship) and to cause drag intentionally for a braking effect. Decelerating without carrying and expending fuel on the second leg is very important for the feasibility of interstellar missions. Electric shielding for radiation protection is discussed more in Chapters 5 and 8.

Ways of ionizing the hydrogen ahead of the starship have been proposed, either by collision with foil or dust thrown ahead or by shining UV laser light tuned to knock electrons from hydrogen atoms [a-Matloff 1975, a-Whitmire 1975]. Because of doppler shift the laser must be finely tunable so that apparently approaching hydrogen receives the wavelength needed. Hydrogen has a cross-section to tuned UV light larger than its atomic size, and about 0.001 LY of interstellar gas has enough hydrogen to absorb almost all light shone through it. A first estimate shows that at 0.1c about a day's worth of ions can be produced by a forward pulse of about one joule, very reasonable but differing from other estimates. Laser ionization may cause the ions to disperse before they can be collected, since it leaves them hot too. Foil ionization requires a high gamma to work well.

The starship may unintentionally become charged and therefore carry an electric field. Electrons in space are more likely to strike and adhere to its surfaces, and its consequent negative charge will attract more ions to it than it would normally encounter [a-Martin]. The result is enhanced drag on the starship. Uncharging a starship or its intentionally charged equipment is a challenge perhaps solved by carrying a selection of radioisotopes which emit electrons and positrons. Kept in conducting boxes they remain neutral. If a long charged wire is not needed or is replaceable, it can simply be let go. Detecting and measuring the starship's charged state can probably be done by observing what ions come in from the side to attach themselves.

For braking, drag is intentionally encouraged, and use of electric or magnetic fields can enhance it by increasing the effective area and volume of interstellar matter encountered. A charged sail ahead, or better a network of charged wires or conducting mesh, a "drag screen" [see also a-Matloff 1985], would interact strongly with the ionized component of interstellar matter while being of low mass and suffering little erosion. Consider this example with values chosen by hindsight to get a ballpark estimate of the possibilities: If a 100 km screen has a zappy charge of + 1000 coulombs and an area of (10)10 m^2, then its electric field of (10)4 V/m is effective about 10 km ahead. The ions in front which "feel" the field comprise perhaps 10% of the interstellar medium and have typical density of (10)-15 coulomb/m^3. The ions are not fixed in place but very mobile, and the literature seems to overlook or over-estimate the possible braking effect for the small total mass of ions involved. Regardless of the reach of the field ahead, each ion effectively interacts with the starship once and is scattered forward, taking very little of starship momentum and KE. The screen charge and field strength are chosen to reach far enough ahead to give affected ions time (about 0.001 s) to be kicked away near lightspeed from the starship. Else it would catch up with those ions, and braking effect would be lost. A drag calculation [Appendix C for 4.7] shows effective braking force is about 100 newtons, and much worse at lower speeds. If braking could be increased to 1000 newtons on a 1000 tonne starship, deceleration would be (10)-4 gee. In 100 years speed would be reduced by only 0.01c, not much effect.

The amount of charge needed on the screen is large and may be difficult to get, although it is spread out very thinly. Charge is obtained with an accelerator or from a radio-isotope. One pure kilogram of an electron emitter releases about (10)6 coulombs. For a practical half-life greater than 1000 years, a year would be needed to accumulate 1000 coulombs. Charging up is slow unless more of this dangerous substance is carried. Useful braking may be obtained with larger screen and more charge, especially if interstellar matter can be made to act more like a plasma and less electrically neutral. The screen attracts electrons, but they can be separated as below. The screen must also be made one-sided so that it does not repel ions equally from behind (an electric drive probably being impossible). At relativistic speeds the electric force is reduced by gamma-squared [b-French] and the apparent charge density is greater by gamma. The net effect is less electric braking.

The drag screen attracts electrons strongly and would soon become neutralized, but ways to keep electrons away have been devised [a-Matloff 1976, 1977]. A long negatively-charged wire pointing ahead of the starship provides electric field which repels electrons sufficiently that they pass by the main starship screen without interaction. The wire must be 100,000 km long (longer for speeds over 0.1c) and charged to about -1 coulomb by a radioisotope source that emits positrons. At 0.1c electrons need about 3 seconds to pass by. The sideways field [Appendix C] can push them aside as much as 100,000 km in this time to miss the main field. A special small propulsion unit is needed to pull the wire gently taut in front. This electron repeller can be used for many purposes, including preventing electron interaction (therefore drag on) negatively-charged structures and aiding operation of ramscoops in the next section.

Another kind of drag screen also could have multiple uses. A small positive charge is put on a large grid outside a very negatively-charged starship [a-Matloff 1976]. Incoming protons attracted (at a distance) to the overall negative starship are halted at the positive grid and build up a space charge in front of the grid to repel further protons. Both protons and electrons would be repelled by this system if it stabilizes. While this screen could provide braking, it would not be a good way to shield from dust because a large mass of trapped protons would be needed and must be moved forward by the drive. To make a shield appear solid, a tonne or so per square meter must be there, whether as a thin gas or as bulk material. Protons could also not be made to pile up thickly because of their mutual electric repulsion.

Magnetic fields in space are complexly distributed on an immense scale, but also very weak. The estimated average field is between (10)-10 and (10)-9 tesla, about (10)-6 to (10)-5 of Earth's (surface) magnetic field [a-Berge; a-Parker]. The fields are probably residual from stars. Our Sun's field may extend lightyears into space, intermingling with fields of other stars. Motion of stars and ionized matter through space stir up the weak fields and generate more and stronger ones. Weak though interstellar fields are, they bend the paths of high-speed ions over many lightyears, so that no charged particle can reach Earth in a straight line traceable to its source. Local slow ions follow paths hundreds of kilometers in radius and electrons spiral around with radii of the order of kilometers in the prevailing fields [a-Sagdeev]. The strength and orientation of fields can be deduced from polarization of starlight, but over vast distances, not the nearest few lightyears. The field direction is primarily in the plane of the galaxy, tangential (azimuthal) to galactic radii, but there are many irregular loops of field extending above and below the plane. Using the fields for travel might best be done for destinations "north" or "south" of the galactic plane, but local field direction should be determined first. It probably changes very slowly.

Methods of using weak interstellar magnetic fields have been proposed. Since magnetic force is always at right angles to the motion of any charged object, no direct thrust can be obtained. For the earlier laser-propelled mission, the mission path is to curve around behind the destination star so that the laser can be used (if it can reach that far) to decelerate and reaccelerate the sail-carrying starship. Or a curved path may work only for a flyby. The curved path can be obtained by establishing a large charged structure around the starship that interacts with the prevailing magnetic field to force it into a curved path [see Figure 4.3; a-Norem; a-Forward 1964]. One method is to use many charged wires stretched in all directions radially from the starship. The plane of the curved path must be perpendicular to the magnetic field lines, and therefore the field in the region of the destination must be known before the mission begins, so as to know where to aim the laser. The main questions are how much charge is needed to obtain enough magnetic (Lorentz) force and how large is the turning radius. The analyses reported are complex yet seem inconclusive, indicating for a small probe that the wires should be about 1000 km long and charged to about 1 million volts. Sometimes overlooked is that the wires must be strong and nearly rigid from repulsion so that they can pull the starship to the side.

A simple estimate of the problem can be made by assuming a turn of radius 1 LY at speed 0.3c. The inward radial force needed to do this for a 1000 tonne starship is 1 million newtons, and this must be supplied approximately by a charge of 10 million coulombs moving in the stated magnetic field. However, wires or almost any large conducting structure of size 1000 km at 1 million volts can hold only about 100 coulombs [Appendix C]. Therefore a large starship cannot be maneuvered by this means, but a 10 kg probe could. Lowering the speed would help reach feasibility too. The interstellar field appears to be too weak for most applications. It may also vary too much in direction for a regular predictable turn. This problem illustrates the difficulty of making major course changes in space at high speed. Changing direction enroute can require as much energy as slowing, reorienting, and reaccelerating. But if a force is available always transverse to the direction of travel, no energy need be used by the starship while its momentum is turned around. The mission is lengthened by the extra lightyears of the turn, and increasing the radius would make this aspect of the mission worse.

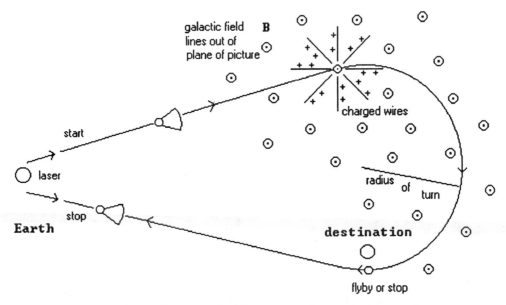

Figure 4.3   Interstellar magnetic field maneuver for round trip around star
[as described by a-Norem].

Charging the long wires, but not the spacecraft, can be done by insulating the wires from it with high-strength insulator fiber and putting a radioactive element on each wire. The isotope emits, say, positrons at high speed, leaving the wires negative, and 1 million volts is the practical limit. Stretching the wires out happens automatically as they charge up and repel each other. The wires will attract a flow of ions from space to neutralize them unless the local ion density is low. The estimated drag on the wires and spacecraft is low. As the spacecraft nears a star, the increased ion density will both neutralize and add to the mass. A way is needed to disconnect the radioactive sources from the conducting wires and to discharge them. Perhaps sources that emit electrons can be connected.

An intriguing new idea that apparently obtains negative thrust with a magnetic field is the "magsail" [a-Andrews]. A large loop carrying current is in a plane perpendicular to the starship motion and dragging behind the starship on cables. It produces a modest magnetic field around and behind the starship. Any charged particles encountering this field react with it. The Lorentz force on the particles deflects them to the side and the reaction force imparts momentum to the field and therefore to the starship ahead of it. Both ions and electrons are deflected, but in opposite ways. The ions will dominate the interaction and cause rotation of the field and loop. Some charges would funnel through the loop, but the starship is ahead of that position. This method has been called a "sail" because if pushed by a wind of ions (solar wind) it should be accelerated until it follows the wind (too slow for interstellar travel).

More importantly, after the starship is at high speed and deceleration is wanted, the fast "wind" of interstellar ions coming at it provides braking force. A large loop is best because more ions are deflected for a given size and current. A 100 km design needs 1 million amps and produces about $(10)-5$ tesla within 100 km around it. This field can deflect 0.1c ions away with typical force $(10)-17$ newton per ion. The whole field was thought to develop $(10)4$ or more newtons of useful braking force in interstellar plasma. However, the mass of ions encountered per second is again much less than starship mass, scattered ions take little of starship momentum, and action-reaction leaves an effective braking of about 100 newtons [again use Appendix C for 4.7]. This is too weak to stop in 1000 years. Denser plasma could be found by traveling along a star's plasma tail, but usually only 1/10 ionized interstellar gas is available. A larger loop might be necessary, but for a 100 km loop of superconducting wire with 1 cm^2 cross-section, the mass is already about 100 tonnes. Like all magnetic coils, the loop expands and circularizes itself when released into space, and tension in the wire must be of concern. Force on the loop is proportional to its speed, a good feature. This "magsail" system deserves further study on the force for a given size and current in typical interstellar plasma and on the severity of the side problems it seems to have. Of concern should be fastening numerous thin cables to the fragile and undesirably rotating loop.

Force can be produced on an artificial magnet due to variations (gradient) in the magnetic field. The latter has been studied for propulsion near Earth [a-Engelberger] but the force is too small and variable. Using the variation in a field is making use of a second order (i.e., small) effect. Propulsive force would be at right angles to the direction in which field lines grow farther apart, making tricky sailing. A low-mass 65 kg loop carrying 4000 amps and 10 km in diameter applies superconductor to the present limit. A force on it of 50 newtons was estimated in the upper Earth field. The gradients of Earth's field are at best $(10)-10$ tesla per meter. The current needed to provide a force of 1 million newtons when the loop is restricted to 30 m diameter (the body of a small starship) is about 15 trillion amps [see Appendix C]. This unachievably high current provides a similarly incredible local field of about 500,000 tesla. The interstellar field is $(10)-5$ weaker than Earth's, and its gradients are expected to be spread over vast regions, possibly being of order $(10)-23$ tesla per meter. A propulsive coil millions of kilometers in diameter would need a similarly high current. Moreover, magnetic propulsion is not free and energy must be put in to keep current flowing in the coil in this case, despite its zero resistance.

Other ways to interact with interstellar magnetic fields include using the current induced in a conductor moving through the field (Faraday effect). If a 100 km loop is used to attempt 0.1 gee of side force, the loop resistance must be less than a millionth ohm at 0.1c. Much superconductor would be needed. Using loops to generate power for operations is a subject for Chapters 5 and 8. Very little power can be produced for within ship use without unfeasibly large structures, and even less for propulsion. More use of magnetic fields for propulsion is examined in the next section.

4.9 INTERSTELLAR RAMJETS

All principal topics needed to discuss ramjets have now been introduced. The ramjet need not go at relativistic speed, but some think that it could be the best way for relativistic travel. This method, conceived about 1960 [a-Pierce; a-Bussard; a-Marx] and called the Bussard ramjet, collects material from space (method then unspecified), such as hydrogen and/or heavier nuclei, and uses it for reaction mass, fusion, or possibly TC. Because hydrogen is very tenuous, a collection apparatus covering hundreds of square kilometers or more would be needed. Such large sizes appear for nearly every starship propulsion system proposed. The interstellar gas will cause more drag on the starship as collection area increases. The amount collected must be small at low speeds but can be substantial at relativistic speeds. The ramjet turns out to have many more problems making it a much less likely propulsion method, nor does it seem to belong among relativistic topics after all. The widely quoted classic Bussard paper provides mainly analysis of motion and very little technical detail.

Extracting fusion energy from hydrogen raises problems far beyond the fusion power plants discussed earlier. Protons fuse at least $(10)20$ times less readily than deuterium because the nuclear force between two protons cannot overcome the electric repulsion. Stars succeed with protons because they can hold a much higher density of plasma for the millions of years needed for two protons to encounter just as one turns into a neutron by a slow process. Reactions that convert protons to helium also give off unpleasant gamma rays and some neutrons. The steps from protons to helium are several, implying that the reactor must hold the particles awhile [see Table 2.1 for reactions and energies]. The first and slowest step is production of deuterium from protons. The proton-deuterium reaction occurs much more readily, but still about $(10)-4$ as fast as deuterium-deuterium, the latter giving a mixture of helium-3 and tritium (hydrogen-3) plus neutrons. The last principle step is to fuse deuterium with helium-3 and tritium nuclei to obtain helium-4 (and neutron or proton) and more energy, rather than let the mass-3 nuclei exhaust unreacted. A way should be found to make protons fuse on the fly, so that their high speed relative to the starship need not be removed first, but this is incompatible with a slow multi-step reaction. Because deuterium and helium-3 are also present in space, they might be collected along with protons, or in preference to protons. They can be fused much more readily but still not easily.

Design of a ramjet might begin with considering how much scoop area is needed, and that is determined by how much propulsion is desired. Suppose a starship of 1000 tonnes moving at 0.1 c needs an acceleration of 0.1 gee. The thrust needed is about 1 million newtons. Providing this at 0.1c requires a power of $3(10)13$ watts and rising. Allowing for an optimistic 33% efficiency, let this requirement round to $(10)14$ watts or J/s. A modest fusion fuel provides $(10)14$ J/kg, so that 1 kg of fuel must be scooped per second to feed this ramjet. Suppose that it is hydrogen only, present at $(10)-22$ kg/m^3. A volume of $(10)22$ m^3 of space must go through the scoop each second. At 0.1c the starship moves through 3(10)7 m/s, so that it needs a cross-section of about 3(10)14 m², or over 10,000 km diameter. This is not an encouraging result. The thrust requirement must be lowered. Lower speed will reduce the amount of material that can be scooped per second but also the power requirement.

A simple (nonrelativistic) relation can be found between scoop area and acceleration [Appendix C] showing area proportional to starship mass and acceleration and finding a limiting acceleration independent of speed. References [a-Bussard; a-Marx] show that acceleration is approximately proportional to speed up to the limit, and that when refinements such as relativity and various losses are included, starship speed reaches a limit less than c. The fusion energy content of nearby space can be reduced to a simple number, about (10)-8 J/m³. For the above 1000 tonne starship, and a possibly manageable scoop of 100 km diameter, the acceleration is limited to 0.0001 m/s², not a good result. There are not enough fusible protons in space for this relatively low-mass starship. (If there were, fusing them is another severe problem discussed later). In this simple case exhaust velocity can be shown to be inversely proportional to starship speed [Appendix C] and is about 0.01c for a 0.1c starship.

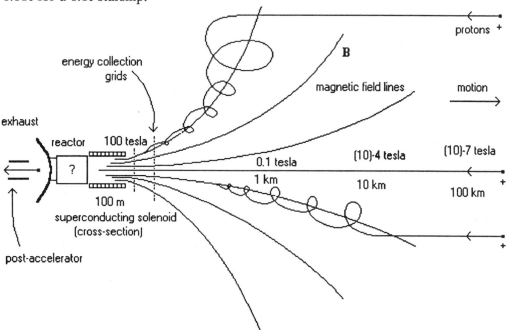

Figure 4.4   Magnetic field ion scoop for ramjet, shown in cross-section.
Scale varies widely. Radii of gyration of indicated theoretical particle paths decrease from
1000 km to 1 mm as protons enter solenoid.

The next aspect to examine is the scoop itself. Intensive study has occurred on the magnetic method without arriving at a clearly feasible system for high speed [a-Matloff 1974; a-Fishback; a-Martin 1971, 1972, 1973]. Mechanical systems 100 km or larger are unlikely. Electric and magnetic fields have no intrinsic mass and can reach long distances. Ingenious arrangements have been designed. The magnetic field from a solenoid [Figure 4.4; Appendix B] can reach out long distances if the current in its coil is sufficiently large. Incoming particles feel a Lorentz force that causes them to spiral around magnetic field lines that they cross. Far out, most cross nearly transverse to lines, feel very weak force, and spiral in large orbits. As they near the starship the field grows stronger and they spiral in tighter gyrations. Their frequency of gyration also increases in proportion to field strength and for protons is about $(10)7$ multiplied by field strength. Comparing the effect far out and close in, the Larmor radius of gyration for a 0.03c proton in a distant field of $(10)-8$ tesla is about 10,000 km, larger than the supposed scoop. Such protons are unlikely to come into position for capture before the starship passes. At $(10)-6$ tesla the gyrations are about 100 km and may be captured. At $(10)-4$ tesla the gyrations are about 1 km. At this position frequency of gyration is about 1000 per second. At 0.01 tesla gyrations are about 10 meters in radius and clearly going to fit into the solenoid part of the scoop.

Magnetic field far from a loop or coil of wire diminishes approximately with the cube of distance, so that physical law is against a good cheap scoop. The field curves around the solenoid bluntly and not in the long funnel shape often drawn. Scoop designers have usually considered the scoop to be the size the field has when it has diminished to prevailing background field B0 of $(10)-9$ tesla. This is optimistic. Consider a solenoid producing field Bm of 100 tesla in its center and radius 100 meters. This field is three times larger than the highest steady field produced at the MIT magnet lab [a-Hamilton]. There is doubt that superconductors, cold or cool, can be used in and for fields as high as 100 tesla. Nevertheless, assuming this is technically feasible, field Bm can decrease 11 orders of magnitude before lost in space. This occurs approximately in a distance 4000 times larger than the solenoid, or 400 km away. A scoop of size 100 km seems possible. Mapping the field [see Figure 4.4], $(10)-7$ tesla occurs about 100 km out there, $(10)-4$ tesla at about 10 km out, and 0.1 tesla is about 1 km out. Comparing the above radii of gyration with these field sizes, it appears that, close enough in, gyration becomes smaller than scoop size and protons will be captured. Inside the solenoid those protons captured with large transverse velocities now gyrate in tight spirals about 1 millimeter in diameter! There will be no problem controlling them in a field of 100 tesla. Overall this field was effective out to about 10 km to 100 km, well before it fell to background level. The bad news is that at higher speed the effective size is smaller because the protons gyrate in larger radii.

Approximate travel time into the scoop at 0.03c is in the range of 0.01 to 0.001 seconds. Depending on reactor structure, the solenoid with high field Bm might be 100 m to 1 km long. At 1 km away frequency of gyration is about 1 MHz and in Bm the frequency is about 1 GHz, like radar. Protons coming in near the axis of this scoop do not cross sufficient field lines to spiral around them. However they will be guided through the solenoid and therefore have been scooped. Only protons farther than about 100 km from the axis will miss being scooped. Electrons are also captured and much more easily. They spiral 2000 times more tightly (with opposite gyration) and will be captured from farther out. Thus a net negative charge will be flowing through the scoop. Spiraling electrons radiate energy intensely.

Of greater concern is proton radiation as protons approach the solenoid. "Synchrotron radiation" is a range of frequencies of electro-magnetic radiation given off in a beam by a charged particle when it is forced to follow a curved or circular path [Appendix C; b-Jackson]. It is inversely proportional to the radius of curve and depends strongly on speed and gamma. The peak radiation frequency is in a band just below the gyration frequency. An incoming proton carries about (10)-12 J of KE at 0.1c. Inside the solenoid it gyrates with 1 mm radius about 1 billion times per second and emits about (10)-26 J per revolution. Thus it would not lose all its KE until after about 100,000 seconds. Negligible energy is lost in the scoop and reactor. However, the loss worsens near c and with high gamma. If the starship has gamma 10, the radiation in the scoop is 1 million times larger, and noticeable proton energy is lost this way. In accelerators such as needed to make anti-protons, synchrotron radiation is a serious energy loss because of the high gammas. In a hypothetical fusion reactor following the scoop, synchrotron radiation is thought to be large enough to prevent energy gain for proton reactions and perhaps for deuterium [a-Martin 1973]. If not that, other radiation losses cast doubt upon the efficiency of ramjet fusion [a-Heppenheimer].

Since 100 tesla is a very high magnetic field, the current and conditions to produce it should be examined. If the solenoid is 100 meters long, a sheet of current of ten billion amps is needed. This can be divided up among many turns of wire, say 1 million turns carrying 10,000 amps. Superconductor can carry limited current per unit area, (10)9 to (10)10 A/m^2 in present practice, before losing superconduction [a-Waldrop]. At best, a cross-section of 1 m^2 of super-conductor is needed, or 1 cm thick over the length of the solenoid. The mass of the magnet wire (diameter 100 m, typical density 3 tonnes/m^3) can be estimated at about 1000 tonnes, a serious problem for a 1000 tonne starship. A 10 meter diameter solenoid reduces the material need by 100 but also reduces the effective scoop diameter by 10 and area by 100. Since the superconducting wire would be spread out as a thin cylindrical sheet of turns, holding its form under the stress of high magnetic field becomes a problem. Most of the stress is expansive (also called "hoop stress"). The literature has addressed this subproblem [a-Matloff 1974]. The ability of materials to contain the magnetic field pressure alone limits the size of a ramscoop field to the 1000 km range. A solenoid containing 100 tesla is like a thin-walled tank trying to hold 5 billion pascals of pressure or about 50,000 Earth atmospheres. Another serious problem is that the starship is immersed in the high magnetic field.

The ramscoop has been proposed as a brake. Indeed it already is a brake above some speed. More energy is used to compress the rare gas into the throat of the scoop for fusion than is recovered from fusion [a-Heppenheimer]. At speed 0.1c, protons come in with KE about 5 MeV. In most scenarios they must be essentially halted in a fusion reactor. When four of them emerge later as a 4 MeV helium nucleus, there is a net loss even if the system is perfectly efficient. The incoming protons react with the magnetic field to slow the starship, and the loss is not made up by released fusion energy above about 0.05c. One solution [a-Whitmire 1975] is stop the protons with an electric field in such a way as to store the resulting electric energy change. After fusion this electric energy is sent to a post-accelerator [Figure 4.4] and used to boost the exhaust. Even when this energy is recovered and all losses are ignored, analysis [Appendix C] shows that little extra speed can be gained after the starship reaches 0.1c.

A decelerator with no feed-through results if magnetic field strength is made larger in a small central region. A magnetic mirror is formed, and most incoming protons are reflected back out. Again a problem of sufficient field and/or size occurs. As with electric drag braking before, a very large area of incoming protons must be repelled in order to achieve say 0.1 gee of braking. Obtaining an effective 1000 km scoop size has been shown to be nearly impossible. Perhaps the ramscoop will be more useful for local space travel where ion densities are higher and required speed is much lower [a-Fennelly 1976]. The same 100 km scoop would provide about 0.1 gee. Or put in orbit the scoop (without the ram) could collect solar wind to provide fuel for other purposes.

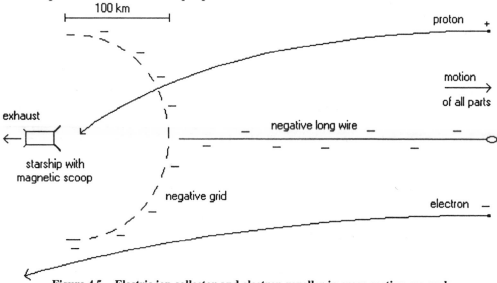

Figure 4.5   Electric ion collector and electron repeller in cross-section, no scale [inspired partly by e.g. Matloff & Fennelly 1977].

As difficulties with the magnetic ramscoop were found, attention turned to electric methods of gathering protons from space [a-Matloff 1977]. Again, the only material captured will be ions. Since most interstellar gas is not ionized, effective magnetic and electric scoops are severely constrained unless ions are created artificially as described earlier. The principle of electric ion collection or "scooping" is to charge negative a grid at the front of the starship (and a very long forward wire as before) to collect ions from surrounding space while repelling electrons, as in Figure 4.5 [a-Matloff 1977]. Other designs are possible [a-Matloff 1976]. Minus 20 coulombs on the large 100 km grid is expected to attract ions from up to 1 million km away, and the electron repeller is expected to push most electrons away from being drawn in. The maximum range of an electric field is estimated by comparing the field between distant ions (no more than 0.1 m apart) to the field from the collector charge. The effective range is expected to be less because of the shielding and neutrality inherent in interstellar plasma. The mass of ions collected can be calculated as in the reference [a-Matloff 1977]. Electric collection has the advantage of providing nonspiraling protons. But the ions pass through the grid into a neutral area where a magnetic solenoid does gyrate them and direct them into a still hypothetical reactor.

Despite the large collection area, study shows that electric collection also seems limited to providing fuel for very low thrust (less than 0.0001 gee) and seems worse than a working magnetic scoop. Speeds are of the order of 0.01c rather than 0.1c, although electric collection provides thrust independent of speed. These estimates include selection of rare deuterium and helium-3 ions instead of protons for fusion fuel. There are many more problems in designing a workable system. One is that protons arrive with extra speed at least 0.05c from being attracted. The electron repelling wire is not a trivial system to add on. And the charged screen must be propelled some large distance in front of the starship. For deceleration the electric grid can be charged positive but using it to deflect ions away and shield the starship seems less feasible.

Most efforts at ion collection come to naught if fusion energy cannot be obtained from collected protons [a-Heppenheimer]. Although a working fusion power reactor is yet to be demonstrated on Earth, considerable speculation has occurred on details of ram-fed fusion reactors in the various references. Fields as above could be arranged to collect deuterium and helium-3 (if they were ionized), but their densities in space are only 0.0001 to 0.00001 of hydrogen, requiring at least 10,000 times larger scoop area for similar energy gain. Even with the thrust requirement relaxed to compensate, ramjet travel time becomes so long that other methods compete well for interest and resources. Extraordinary effort is needed to use proton fuel, and a bold idea is catalytic fusion [a-Whitmire 1975]. Here carbon is used as a catalyst to speed p-p reactions up to nearly "normal" as in the CNO cycle of hot stars. Unfortunately a very high (1000 tesla) magnetic field seems required. If this medium ion fusion reactor could be made, it would be valuable on Earth, and on starships it also would work with hydrogen fuel that is carried rather than scooped.

A variant on the ramjet is the "ram augmented interstellar rocket" (RAIR) [a-Bond 1974; Powell 1975, 1976]. Here the scoop gathers reaction mass for use in a rocket energized by other means, such as anti-protons or fusion fuel carried and diluted with extra reaction mass, or a laser focused from afar to heat it [a-Whitmire 1977]. According to one study, RAIR is several times more efficient than fusion, the two rocket modes cannot be mixed, and there would be two simultaneous exhausts, one from ramjet and one from fusion reactor. If electric collection is used, it would be for gathering protons. It could provide a proton beam coming in at about 1 MeV that could initiate fusion against deuterium, lithium, or even boron. Then only deuterium or lithium fuel need be carried.

Another way to save the ramjet concept is to place fuel artificially in its path. For a mission carried out over millennia, a stream of pellets of fusion fuel could be placed between here and at the destination and a scoop used to collect them. Simply freezing the fuel might not keep it intact for long times. The scoop would still need a large area to find all fuel pellets that would have dispersed due to errors of aim and placement. Still another procedure is to use the scoop to collect natural fuel for deceleration while the starship accelerates with carried fuel and then coasts for decades through space [a-Roberts]. Fuel must be collected over a long time with minimum drag or the starship slows too much during collection. Probably the best approach here is to separate the high-quality fuels deuterium and helium-3 from the protons, but any procedure should avoid decelerating the much more abundant incoming protons.

4.10 SUMMARY OF PROPULSION, AND BEYOND

Despite all the ideas for propulsion presented here [see Table 4.2], the great barrier to fast interstellar travel remains getting enough energy and finding the low-mass technology that handles that energy efficiently and safely. The many important considerations are too complex to pick out the one best method to try from this table. Most promising but still very hypothetical are: a large low-mass solar sail reaching 0.001c to 0.01c or a sail driven by mirror-focused sunbeam up to 0.1c (fuel and reaction mass not carried), a pulsed fusion drive reaching possibly 0.1c (difficult fuel acquisition and use), or an anti-proton drive to better than 0.1c (incredibly expensive fuel carried in minimum mass, used at no better than 30% TC). Speed is likely to be less than 0.1c, perhaps lower than 0.01c, requiring voyages 100 to 1000 years. Required acceleration is in the range 0.01 gee to 0.0001 gee. Power requirements vary widely, but a minimum for fusion drive for 1000 tonnes payload reaching 0.1c is (10)11 kw. Handled by 100 tonnes of drive equipment, this requires an incredible million kw/kg.

Table 4.2
SUMMARY OF INTERSTELLAR TRAVEL METHODS

Method	Speed	Power (kw/kg)
chemical rocket	<0.0001c	1000
gravity boosts from planets	<0.001c	high?
fission powered electric ion rocket	0.01c	0.1
* fusion powered direct rocket (probably pulsed)	0.1c	100
* fusion reactor powered electric ion rocket	0.1c	1?
* raw sunlight pushed sail	>0.001c	100,000?
* focused sunbeam pushed sail	0.1c	1,000,000
laser beam pushed sail	0.01c	1,000,000
anti-matter direct rocket	0.1c	10,000?
anti-matter powered diluted rocket	0.1c	1000?
fusion rocket using fuel collected from space (ramjet)	0.01c	100?
* braking by sail	<0.01c	
* braking by electric deflection of interstellar plasma	0.1c?	
* braking by magnetic deflection of interstellar plasma	0.1c?	

Notes: * marks all that seem better bets for further study. Not all conceivable methods are listed here, nor can these be ordered simply. Judgment of good and not-good is rather subjective. All of these and more deserve continued consideration, and scientific opinion has been diverse. Estimated maximum speed is chosen as one of many selection criteria, rounded and listed. Data are distilled from many studies and are more conservative than detailed estimates earlier in the text. Specific power in kilowatts handled per kilogram (by power conversion equipment only, not including fuel mass) is another criterion for which uncertain but optimistic capabilities are listed here (not to be confused with requirements). Only moving mass is considered.

Combinations are important, especially for one-way voyages where one of these propulsion methods starts the voyage and a low-mass low-energy braking method ends it. Possible brakes have included a solar sail, an electric interstellar plasma deflector, and the magnetic deflection of interstellar plasma ("magsail"). Considering many practical factors, the most likely mode at this point seems to be one where fuel is not carried for the main propulsion but just for auxiliary help. Speed is acquired by sail and braking done by one or more interactive methods. The more recent literature has many summaries of propulsion prospects [e.g., a-Ehricke; a-Andrews; a-Nordley; a-Bond 1971, 1986; a-Matloff 1976, 1983, 1985; a-Norem; a-Cassenti 1982; a-Mallove 1987; a-Forward 1976, 1980, 1986; a-Vulpetti 1979; a-Powell 1974; a-Massier 1982; a-von Hoerner; a-Dyson; a-Jones & Finney; a-Kooy; a-Marx; a-Archer; b-Forward; b-Mallove; b-Friedman; b-Nicolson].

Just how much more energy would be needed than anti-proton annihilation provides, if we could dream a moment? It would be nice to have the mass involved be a small fraction of the starship so that the number of legs in the voyage do not matter. And to be able to refuel at any destination. If a 1000 tonne starship needs only 100 tonnes devoted to whatever mass provides its propulsive energy and wants to reach a very helpful and optimistic gamma of 10 (0.995c), then it must have the means to acquire a KE of gamma times its RE (gamma minus one, actually), or (10)24 J. This amount is 1000 times the human world energy consumption and not much less than the solar energy impinging on Earth in one year. Carrying this in 50 tonnes (reserving another 50 to stop), the energy density of this hypothetical source is over (10)19 J/kg, or about 100 times TC. Very ultra-relativistic travel would require more. But even if the source of energy is "massless", energy itself acts as if it has mass. This starship would effectively mass 20,000 tonnes, and the proposed solution is impossible. General relativity is relevant here. Any onboard energy source will act as if it has inertia and be difficult to accelerate. A basic rule re-emerges: do not carry the fuel or energy for propulsion aboard. The starship cannot easily carry it no matter what its form. The search for external sources must continue. Faint hope of other exotic sources is offered in the final chapter.

Efficiency is assumed perfect here, and if it were not, the smallest leakage or waste would be of cataclysmic size. The exhaust jet itself would not quite cut a moon in two but would certainly blast its surface and be an unmentionable weapon. The technology to control such energies must be precise beyond present comprehension, and very fail-safe. A sizable part of a star's output is to be funneled through a relatively small delicate mechanism of some sort. This is not to say that an ultimate starship drive would use ordinary materials directly. The alternative might be a drive thousands of kilometers in size, but then how can it be low-mass so that it can move itself rapidly? The hypothetical example above required power handling density of the order of a billion watts per kilogram. If other civilizations have developed and are using such ultimate drives, we have seen no sign of their use. We can imagine that this technology is everyday stuff to "somebody" somewhere, part of the regular run to star colonies, never to return to the original society (thanks to the high gamma). Considering the planet-scorching energies involved, even "they" will use such drives carefully, perhaps not cranking them up fully until far from the orbital plane of their planets. Large-scale use of feasible fusion and anti-matter drives would increase the intensity of hazardous particles in local space.

Chapter 5

Starships as Systems

The purpose of this chapter is not to design a working starship, but to recognize it as a complex system. A general description of its parts is given as an interacting system. We might like to know whether human technology is capable of engineering a starship at present or in the foreseeable future. The basic constraints provided by the physics of motion, the physics of energy conversion and control in assorted propulsion methods, and the properties of space determine most of the boundaries of designing an interstellar mission. The question of propulsion is put aside temporarily here, and it is assumed where necessary that propulsion capable of 0.01c to 0.1c will be developed. It is also assumed that human social systems will support mission design, construction, and long-term operation. Some technical, biological, psychological, and social factors are surveyed here and covered in more detail in later chapters. They are crucial to success with complex technology and further constrain mission design. Two underlying questions occur: Is there anything about the starship as a complex system that makes interstellar travel very difficult. Could other civilizations do this, and how would they?

5.1 ENGINEERING RULES

A number of empirical rules for designing complex systems have emerged from centuries of practice with simpler systems. In many ways late 20th Century technology has been the most demanding of human abilities. A question is whether modern high-tech society is approaching the competence needed for a starship system. One of the largest present human technical systems, the "skyscraper" building, massing at several hundred thousand tonnes, is comparable to some possible starships. The largest transportation on Earth is a loaded oil tanker at 300,000 tonnes or more. The largest complex systems for space thus far were the Apollo missions (about 3000 tonnes launched), although the hardware, size, and population are dwarfed by the preceding and by a starship. Some of the empirical rules found before and during Apollo have served the US space program well and therefore have set some basis for starship design [a-Kleinknecht; a-Dawe]. (The wayward compromises by NASA and Congress on the shuttle are a later development.)

. It is an unanswerable question whether other civilizations would "engineer" the way we do, although the rules for systems seem largely dictated by nature and therefore universal. Whether a starship program will follow the usual course of engineering projects is uncertain. From experience, the stages might be approximately these: idea and motive, feasibility studies, increasing commitment outside the technical community, experimental tests of uncertain science and technology, design of likely system, building working prototypes, tests of subsystems and whole systems, simulated missions, real missions on smallest useful scale starting with probes, larger mission based on preceding achievements, series of more ambitious missions. Depending on mission duration and speed, interstellar travel presents at a later stage an instance where human designers cannot obtain feedback on their work. Up to a point messages can be radioed back on what works and does not work, but eventually the starship is out of range in space or time. If it is never heard from again, results are indeterminate.

In an approximate order from the general to the specific, some engineering rules from experience are given here [see also b-Krick]. Their value to starship design should be contemplated. Possibly the only way to find out if our engineering ability is sufficient is to begin trying.

Rule 0: Engineering is setting sizes and making systems that work.

Rule 1: If there is a way for a system to fail, it will happen (Murphy's law). Failure analysis can never be complete, either. For example, some nuclear reactor accidents have involved a chain of 4 improbable failures.

Rule 1a: Design fail-safe. That is, if something fails, try to prevent anything else going down with it, or possibly arrange for a beneficial failure or a self-fixing one.

Rule 2: Humans, at least, tend to build wholes from parts. Since a whole system cannot be deduced entirely from its parts, induction is part of the design process.

Rule 2a: The whole can be greater than the parts. That is, the function of the whole system can be qualitatively, not just quantitatively, different from the functions of its parts. (Note latter comments on systems.)

Rule 3: Define a problem such as propulsion as broadly as possible. Even more broadly, ask: What is the purpose of interstellar travel? Identify unnecessary assumptions and biases which limit the problem artificially and prevent many kinds of solutions. Do not assume the solution (e.g. a rocket), find it. Avoid vested interest until the final design is very clear.

Rule 4: Once primary and subsidiary problems are well-defined in detail, the solutions may be half accomplished.

Rule 5: Keep everything as simple as possible.

Rule 6: Know the limits set by known science. Engineering cannot exceed known science but must wait (and ask for) further science. This does not mean that engineering cannot find new solutions to problems. The realm of engineering creativity, a result of intelligence working against entropy, is probably far larger than the realm of scientific models and laws.

Rule 6a: Be conservative with technology. This contradicts the wide-eyed technical ideas in most interstellar studies, but this rule is for using established technology at the working hardware stage, not the exploratory stage.

Rule 6b: Work from the known to the unknown. Start with experience, but take early risks to expand that experience.

Rule 6c: Models do not scale up linearly (e.g. strength-mass ratio, fluid flows).

Rule 7: Watch out for interactions. Every functioning part can foul up other needed functions by many other parts.

Rule 8: Most systems involve the need to optimize conflicting functions, and trade-offs are needed where each function is less than ideal.

Rule 9: If a system of mass m needs parts, each of which masses 0.1 m (or even 1 m as often occurs in propulsion), stop and redesign until each part does not call for more total mass than intended.

Rule 9a: If other considerations are equal, the masses of parts of a system (and some other characteristics) might be equal in an optimum design. This rule derives from impedance matching.

Rule 10: A practical system must be reliable (to an extreme and well-defined degree for interstellar travel), operable (reliably controllable), and repairable (or better, easily maintained so breakdown is unlikely).

Rule 11: Use multiple backups for subsystems (3 or 4 for interstellar travel). Each backup must be designed as a main system, not an inferior secondary version. Use a different proven method for each system rather than achieve redundancy by making the same system three times. Hint: use several different propulsion methods. (The Voyager probes did not have backups for many components because Voyager was expected to function less than 50 years. Some of the few backups did need to be used.)

Rule 11a: Where substantial failure rate is unavoidable, use tens or hundreds of identical copies of a subsystem in parallel. (Big example: why rely on one huge tricky drive when banks of identical small independent ones are better. If a few fail, 90% of drive capability should remain.)

Rule 11b: In counterpoint to multiple backups is multiple use of the same system to save mass. This is gambling. (It is better to have 3 sails, each of which can also be an antenna.)

Rule 12: Temperature is a major confounding factor. Beyond "room" temperature (20°C) the failure rate generally doubles with each 10°C increase in temperature, the old Arrhenius rule. The size of materials changes at different rates with temperature, causing misfits and stress.

Rule 13: Systems are rarely self-stabilizing and usually have numerous ways to destabilize themselves. Know the system so well that stability can be made inherent.

Rule 14: Use self-repairing subsystems and hope that this repairs the whole system. (This rule spins off from the space age.)

Rule 15: Use monitoring sensors more reliable than system components, else they are worse than worthless, they mislead during breakdowns. (This rule spins off the nuclear age.)

Rule 16: Avoid designs by committee and layers of bureaucratic review. Find the top experts who will live for the project, give them unlimited support, and leave them alone.

Rule 17: Use the same parts in as many different places as possible. (Reduce the number of different parts needed.)

Rule 18 (especially for starships): No repair services are available enroute, and no radio or rescue help after leaving the Solar System.

5.2 STARSHIP AS SYSTEM

Study of starships is unavoidably colored by fictional representations of them as large, easily relativistic, marvelous machines with a little drive stuck on the back. That image is replaced here with a view of the starship as a whole complex system of technology. Systems can fail where every component is apparently successful and human oversight is optimal. Systems are composed of subsystems, sometimes with flaws in how the boundaries are determined or connected. A number of subsystems are described in this chapter without much technical detail. The main purpose is to have an overview of how things must fit together. The weaknesses in many components are explored later.

The starship system is basically hardware, although explicit programs and plans (software) and implicit human ways of doing things are needed for its operation. The system includes all hardware needed to begin its journey including any external components that provide propulsion. The system can be as "simple" as a low-mass probe without humans aboard, or as complex as a large colony of humans. It is very unlikely that colonies will be sent out until inanimate probes have tested most subsystems and provided sufficient data to determine human mission size and purpose. For discussion in this chapter a "modest" human expedition is usually assumed. It might have to become a colony. If design appears formidable, probes as discussed in the next chapter become the focus. If a human mission is workable, then grander plans might be laid.

A system is studied differently depending on whether it is closed or open. Most human technical systems are assumed open, with unlimited energy, material, and human resources available for construction, operation, and correction. An open system presumably has no physical boundaries but there are bounds on human activity. Earth is a closed system for most activities, a fact not widely appreciated yet. Solar energy inflow and heat radiation outflow are the only significant flows through its upper atmosphere boundary. A starship has more stringent boundaries and is in most ways a closed system, both as a technical system and as an ecology if humans are aboard. Its shell or envelope cannot be perfect and leakages of material and energy through its envelope are to be considered. Some propulsion systems bring in energy from outside, others release it from fuel aboard. In either case the subsequent energy and exhaust material flows are away and so large that they must be accounted for precisely to avoid overheating.

Systems have interacting parts called subsystems. Interaction occurs via flow of energy (and information, closely related and sometimes called a "signal"). When part A affects part B and part B affects A there is a closed cycle involving feedback. Real systems have many such cycles or loops, many unintentional. For a stable system the feedback must be negative (out of phase or delayed); any misbehavior of part B results in a signal to A that then sends a change to B to reduce the unwanted behavior. Such a system must be linear. When the signal or energy flow is positive feedback (in phase), it causes amplification of behavior and vibration (oscillation) or runaway of the system. Runaway always stops somewhere, either by reaching the limit of system function without harming it, or by breaking up the system so that recovery is impossible. Dissipation (damping) of system energy reduces response and oscillation. A new discovery is the possibility of chaos (not to be confused with truly random behavior) where the system never does the same thing twice but varies unpredictably and perhaps uncontrollably. Most systems are either inherently "nonlinear" or become so if pushed very far.

An example could occur with a solar sail. Such a large structure has many ways it can vibrate (modes). A very small disturbance in its configuration can be amplified by the system, which has continual solar energy coming in as a source of energy for vibration. The sail might shake itself apart by exceeding the strength of cables somewhere. Or it might stabilize because the vibration energy is dissipated as heat during flexing of cables. Analysis and testing are needed to determine what might happen. With some caution computers can be used to intervene in systems and calculate what small compensating action would prevent a given instability. A fusion reactor drive is another example surely fraught with complex instabilities that can build up from small fluctuations in magnetic field and plasma flow. Such a high-power system must have all remotely possible behaviors well-understood to avoid runaway or extinction of operation.

Starship design has been considered most carefully for a robotic probe mission just beyond the solar system [a-Jaffe] and for the Daedalus robotic flyby of Barnard's Star 6 LY away [a-Bond 1978]. Such missions are mainly scientific and exploratory and carry no people. Many subsystems have been identified but much work remains to be done on their interaction, and on problems within subsystems. Table 5.1 lists several categories of subsystems (themselves higher kinds of subsystems), including some needed for long-term human comfort on more ambitious missions. The list is surely not complete. The extra components needed for reproducing probes are discussed later. A subsystem list should be considered as a matrix. Every subsystem can interact with (and interfere with) every other, during design and in operation. For example, the drive might produce radiation that interferes with sensors, computers, and food growing. Without estimating subsystem masses, it should be clear that a human mission adds much wider variety of needs (as well as capabilities) and involves many more massive items.

Each subsystem has sub-subsystems, and each of those is likely to have sub-sub-subsystems and so on down for three or five levels. Any part of a building (heating system) or car (engine) is a subassembly or subsystem. Cars have 3 levels or so, down to a pile of nuts, bolts, and formed metal parts. Buildings might have 4 or 5 levels. Note that a car is a subsystem of the national transportation system (which is a subsystem of a modern society), and a building is a subsystem of a city, which is sub to a society which is sub to the world. Starships will be more like small worlds with seven or more levels down to the level of screws and resistors. The number of different parts could be millions unless effort is made to use identical parts everywhere possible. The total number of parts could be billions. (Voyager, mass about 1 tonne, had 65,000 parts). Most of these parts are crucial, else they would not be used. While the distinctions are elusive and possibly useless, the lowest level of parts is generally replaceable rather than repairable and subsystems at all levels are where repairs and changes are made. On the upper end, the higher the subsystem, the less it may be understood.

Table 5.1
A LIST OF SUBSYSTEMS OF A GENERIC STARSHIP

Big heavy duty subsystems:

- o   Structure
- o   Propulsion or drive
- o   Fuel or energy source for propulsion
- o   Shield system (including ionizing gas and destroying large dust and rocks)
- o   Ecosystem (see human needs, below)
- o   Electric power supply, multiple, with storage for emergency power.

Fine subsystems (robotic or human missions):

- o   Sensing, external
- o   Sensing, onboard, of all possible functions
- o   High-level control (computers, humans)
- o   Activation devices (electronic controlling of all possible subsystems)
- o   Communication, internal and external (to Earth)
- o   Possible mobile repair units
- o   Instruments for science and exploration enroute and upon arrival
- o   Extra: subprobes with propulsion, instruments, computing, communication.

Subsystems for human needs:

- o   Plans and information about starship and its mission
- o   Storage of extra water, oxygen, most chemical elements and compounds
- o   Food storage, growing, processing, manufacturing
- o   Human washing, grooming, waste cubicles
- o   Sanitary processing, recycling of wet and dry waste
- o   Atmosphere processing
- o   Private rooms for sleeping, fun, meditation, illness
- o   Work areas for domestic, scientific, and all other tasks
- o   Heating for humans, agriculture, some equipment
- o   Lighting for work, recreation, agriculture
- o   Artificial gee for humans, some work, some agriculture, some manufacturing
- o   Shops with tools and parts for fabrication and repair of all technologies aboard
- o   Specialized tools, parts, supplies for all starship subsystems
- o   Spacesuit storage, maintenance
- o   Complete medical clinic-hospital
- o   Education or training facility
- o   Cultural studios
- o   Exercise facilities
- o   Meeting rooms
- o   Archives of science, technology, arts, human studies, much more.

Extra equipment:

- o   Auxiliary space and landing craft
- o   Exploring and/or colonizing equipment (ground cars; camping; solar energy, water, and air collection, purification, and storage; mining; more).

The design of subsystems is where serious engineering begins after the whole concept is proven feasible. Some subsystems are discussed further in this chapter, where they are called systems for short. A general survey of factors and problems is given here and more technical detail is given in a later chapters for various categories of subsystems.

5.3 STARSHIP SIZE AND BIG SUBSYSTEMS

Starship size depends very much on mission. Mission planning depends on possible starships. Which comes first, mission (software) or starship (hardware)? Unless the mission is a probe, living-ware in the form of people is considered first here. A minimum mission size might need the skills of 100 people. (Selection and specialization are covered in Chapter 10.) Each needs a certain amount of supporting mass to live and to work. High-tech lifestyle on Earth involves enormous tonnages of materials, equipment, and waste per year. Even if each person requires only one tonne of personal equipment for comfortable and productive survival, the habitat and structural requirements could be much more. An attempt is made to estimate the mass needed per person. After all essentials are summed, then the required propulsion size could be determined, including an estimate to move the propulsion itself.

An estimated mass for a small self-sufficient space colony was 500,000 tonnes to support 5,000 people, or 100 tonnes per person [a-O'Neill, a-Matloff 1976]. When discount is made for not needing the large volume (1 km long by 100 m diameter) and additions are made for long-term survival and mission equipment, the requirement is somewhere between 10 and 100 tonnes per person. A NASA study of space colonies estimated a low 10,000 tonnes for 1000 people or 10 tonnes per person [b-Johnson; a-Matloff 1981]. A starship at a minimum could mass 1000 tonnes (plus fuel and drive). This was the "fast" model mission, now downgraded to moderately fast (0.1c) in view of difficulty obtaining antimatter propulsion. If such 0.1c propulsion is not feasible, then the moderately slow 0.01c model mission of at least 10,000 tonnes would be required (plus fuel).

The volume of starship needed begins with the population but is mainly determined by material storage and workshop space, also by fuel and large equipment such as subprobes and landers (probably strapped into an open frame somewhere). One hundred people need at least 300 m^3 per person (100 m^2 with 3 m overhead) in the closed ecosystem habitat for enough room to enjoy being with each other for decades or centuries. The population density in Rome is 40 m^2 per person with sky overhead. NASA studies estimated a squeezy 35 m^2 to 47 m^2 area (not counting work area) in a very spacious toroidal geometry with overhead providing about 800 m^3 per. The 30,000 m^3 habitat volume need not be mostly empty space but could be partly filled with equipment and work stations, provided that open space is available somewhere (growing areas, meeting rooms, game rooms). If this living and working volume is 30 m in diameter, with floors about 700 m^2, then a stack of 14 levels about 40 meters tall is needed. It is not surprising that the habitat could be like a high-rise apartment building with 4 apartments on each of 14 floors. This is only the habitat and work space and is perhaps less than half the payload volume. It is necessary later to know its total envelope area, which is about 5000 m^2. If there is never significant gee, then "floors" are not defined, all habitat walls are usable, and all of the starship interior could be accessed by open jumpways, with cables for massive loads.

Terrestrial transportation equipment for human use (vans, jetplanes) typically has a volume to mass ratio of about 10 m^3 per tonne, 1/10 the density of water and not very roomy for living in. The habitat portion of the starship would seem to require 3000 tonnes. Apparently, designing a 1000 tonne starship for 100 people is too difficult and cramped. A 10,000 tonne starship would have about 7000 tonnes left for propulsion (no fuel) and big equipment and supplies, also packed at about 10 m^3 per tonne for human access and use. Its volume would be 100,000 m^3. In cylindrical geometry the diameter could be 30 m and length about 140 m. The proportions could be a thin 10 m diameter and about 1300 m long or a disk 100 m diameter and about 13 m thick. A sphere about 60 m in diameter also has this volume. Fuel tanks would be extra. As to be seen, 30 m is too small for rotating the whole starship for gee, and 100 m is rather large to shield.

A variety of propulsion methods to move mass of 1000 to 10,000 tonnes have been discussed, but without clear conclusion as to the best method. Light-pushed sails and fusion drives, both slow, could be the most likely methods. Sails have cable mass problems for the larger starship mass. A 0.1c mission could use about 10 years to accelerate, needing about a billion kw at constant power for 1000 tonnes. Propulsion which needs 1 kg of equipment to supply 1 kw is hopelessly massive. Fusion would need to provide an unlikely 1000 kw/kg to be feasible, and more when the mass of fuel is considered. A sail can handle a million kw/kg, but that is irrelevant compared to the absolute need for about 100 km diameter sail massing 1000 tonnes or less. This hinders astrogation and space studies. If the sail cannot stop the starship, then extra mass is needed for whatever will. Preliminary designs generally showed that propulsion required about 10 times more mass than payload. A 2-leg mission which stops would require mass ratio at least 100:1 unless sailing and/or electric/magnetic braking are used. Good engineering design dictates that no single drive be relied upon. Either there should be three propulsion methods, all independently working, or one method should be used in multiple form of perhaps 10 to 100 units.

The strongest effect on mission mass is the elimination of propulsion and fuel mass needed to stop the starship. Except for the rare case of a roundtrip back to Earth by humans without any stops (a great circle also needing continuous propulsion), human missions must make at least one stop. If mission speed is no higher than 0.01c, three different braking methods should be considered and perhaps all used: sail, ion drag, and magnetic plasma drag (magsail). None needs equipment larger than 100 km if plenty of time is allotted. On a 1000 year mission, 100 years or more can be used for braking, a deceleration of 0.0001 gee. As was shown, a sail will develop negligible drag force from interstellar gas. The sail can be used at the end when starlight is intense and speed-dependent brakes are inadequate.

If a human mission is too massive for a first mission, a robotic probe can be sent. But if a 10,000 tonne starship is far too large to send, then cutting by factor of 10,000 to a 1 tonne probe (plus propulsion) might not be sufficient, especially if speed is raised by 10. Indeed 1 kg to 100 kg probes have been proposed. One tonne seems about enough for a variety of precision durable microinstruments, one telescope, nuclear power supply, transmitter, and antenna, like Voyager but needing much more transmitter power. Redundancy would be minimal. Launch would be by external means. Continuous lightbeam is desirable to provide a better power supply since it may not be possible to make a reliable nuclear power source to last 1000 years. More probe options are considered in the next chapter.

Others have thought grandly and proposed "world ships" up to 100 billion tonnes carrying hundreds of thousands of people for millennia [a-Martin 1984; a-Bond 1984]. Possibly designed in reaction to the enormous expense and difficulty of going fast, these colony- or city-sized vessels seem limited to leisurely travel (0.005c for 1000 years or more). Sizes as large as 15 km radius and 90 km length would be rotating and certainly could provide comfortable living for centuries, with plenty of people to work and play with. They illustrate the economic theory that if the money is available, it is better to spend money on size rather than speed, since fuel expense increases rapidly (squared) with speed. Chapter 6 covers more of the "world ship" concept.

Although a human habitat requires many levels and rooms, the mass for structural support need not be great if no substantial gee forces ever occur. Structure could be of aluminum, steel, or the difficult-to-form titanium. Even long-lived wood could be used. Materials that outgas or store charge should be avoided (green and glued wood, plastics). If the habitat is spun, then the problem is much different, and the mass would larger. Structural materials are not a difficult problem unless the starship requires enormous structures, either large sails attached with cables, or long wires, or a large magnetic coil, or an enormous solar energy collection system. The materials must be used near theoretical limits of strength. Many involve carbon, which must be found in asteroids and comets.

Heavy shielding may be needed in front to stop radiation and damage from impinging gas and dust. It is needed behind if there is a drive emitting radiation. Lithium hydride is a low-mass material which absorbs neutrons, and the waste heat can be useful, but enough to catch most neutrons would be too massive. Noncritical supplies and fuel can serve as partial shielding. The sides of the starship need shielding from cosmic particles, and some supplies and storage can be on the interior of outside walls to help shield the habitat. Humans require radiation levels below about 0.5 rem long-term (explained further in Chapter 9). With some methods of propulsion, the starship is to be turned over to use the drive as a brake, pointing it forward. Then shielding interferes with braking. Perhaps the drive would be adequate shielding, but the exhaust is likely to be fast rather than dense.

A starship might be a cylinder divided into 7 (unequal) layers [Figure 5.1a]: front shield, big equipment, supplies, living quarters, fuel, rear shield, and drive. The front end should not be blunt because tapering ("streamlining") the front shield would reduce drag and damage. This general form is reminiscent of an ancient skyrocket. (Some purpose might even require a "stick" behind.) Since the shield is massive, the starship should be slender to put as much behind a narrow shield as possible. An alternative arrangement is to put massive amounts of frozen fusion fuel in front, or a piece of comet ice. A disk shape or wheel [Figure 5.1b] can be shielded by fuel and a massive ring to protect only the outer part of a large toroidal rotating habitat. A compact design [Figure 5.1c] would have the starship spherical, with a conical front shield. Humans would live in the center, surrounded by equipment, supplies, and fuel as shielding behind the front shield. Active electric and magnetic shielding is discussed later. Front shielding aside, more compact regular round shapes give more strength from less structural mass.

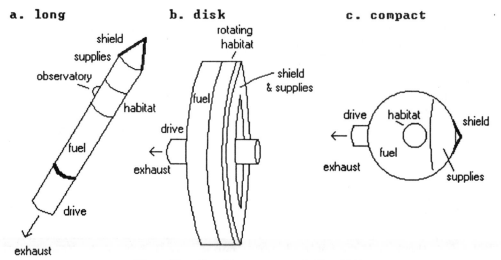

Figure 5.1   Starship arrangements (fanciful).
Fuel volume may be much greater than other components.

Last but not least in the list of big subsystems are the power supplies, which must come from energy stored in the starship or collected somehow from space. The main concern is reliable electric power to operate all the subsystems covered in this chapter. If heat is also needed, it is best obtained from the waste heat of a drive or reactor. Some have suggested the old standby, nuclear fission power, for example, on the Daedalus mission [a-Bond 1978]. But this is likely to be heavier than projected and has been rather problematic on Earth. Nuclear waste may not be a problem, but reliability, corrosion, radiation, refueling, and many other concerns must be reconsidered. A simple low-mass fail-safe fission reactor which works for centuries from one fuel loading is needed. If controlled fusion becomes available, then a couple of fusion reactors would be preferred since their mass may be low and fusion fuel is inherently low-mass. Fusion reactors seem repairable, whereas fission reactors require a large, specialized, and toxic industry to prepare and could be unrepairable. Onboard power is closely related to the choice of propulsion. Any energy source for propulsion can provide abundant power on the side, even one receiving lightbeams or pellets from afar. If a successful electric or magnetic ion scoop is developed, then scarce fusion fuel ions could be collected from space sufficient for onboard power via a fusion reactor [a-Matloff 1977]. Only 1 Mw is needed here whereas a drive would have required over a million times more ion collection.

There should be several major sources of daily and emergency power, distributed throughout the starship so that no more than a few percent of capacity can be lost in one accident. Power distribution should be multiple, with separate wiring so that there is no way that key components can lose all operating power. The size of the power supplies cannot be estimated from the people alone. Humans could get by on 100 w each for personal needs despite present terrestrial extravagance, or 10 kw total. Others have estimated 100 times more. For all activities power needs may be about 1 kw per person. Lights for food growing and power for recycling pumps raise the total to near a megawatt. Starship control, cable pulling, wheel or gyro spinning, propulsion restart, charge ejection, and other big starship needs could at times require high power from nonpropulsion sources, and at least two power supplies should have high temporary capability, possibly a gigawatt. Still higher power could be provided only through designs that permit drawing power pulses from capacitor storage charged from smaller power sources.

Whether humans are present or not, temperature control is needed. Present space technology is used almost entirely at normal temperatures ("room", about 300°K), even aboard Pioneers and Voyagers. Heat is retained by low-mass high-efficiency insulators with many reflective layers. Excess heat, of which there should be none except from a drive, can be radiated into space with difficulty. Large hot surface area is needed to get rid of heat. A metal or other heat conducting surface with mass 1 kg at 500° can radiate at most 3.5 kw/m^2 and therefore 3.5 kw/kg at most. It may be difficult to distribute heat for radiation if less metal were used. For environmental comfort waste heat may be dumped at half this temperature, leading to 1/16 as much radiation. Some have proposed molten metal droplet and other systems which can dissipate about 60 kw/kg [a-Aston]. A molten system requires high temperature to operate also. Electronics, motors, valves, many parts, even structural materials give problems at colder-than-Earth temperatures. Progress is slow in developing low-temperature versions of needed components.

5.4 FINE SUBSYSTEMS

In another unfortunate interaction between systems, the need to look directly ahead is inhibited by shielding. Indeed at higher speeds, the hail of radiation would prohibit anyone or most instruments from peering around the shield to look ahead. At 0.1c hydrogen gas arrives like 5 MeV radiation. Navigation could be done without looking ahead, but detection of objects and changes in gas/ion density requires lasers and radar transmitters which can look ahead without damage. Another interaction can occur with the immersion of the starship in a lightbeam from behind, which is reflected from ahead by a sail. This would also interfere with navigation and space studies and would be blinding to look at.

All parts of the starship need sensors to monitor temperature, dust impacts, radiation, stress, air flow through leaks, and many more variables. All sensors should be as reliable as the information needed is critical. An information buss or network is needed all over the starship to carry information to avoid the mass, complexity, and unreliability of a million wires and connectors. The network should be optical fiber carrying digital code. Information is put on and taken off at junctions in an orderly way. Each sensor needs its own low-power input and converts its findings into a standard digital code. Periodic checking of all sensors by computer should show malfunction of any point in the network.

Two more similar networks could carry communication as voice and picture among human stations and instructions for various systems. The drive may have its own communication system. It is often forgotten that computers are only part of a control system. Actual operations are done by the control of electric power to actuators (motors, solenoids, thermal elements, etc.) Digital instructions must be converted to analog signals or power pulses by means of reliable solid state power devices. On a robotic probe human communication is irrelevant but computers must be smarter, to adjust to unforeseen problems. The computers may not be able to repair every fault by reconnection of systems. Some sort of onboard moving repair robot may be necessary [as found for Daedalus, a-Bond 1978]. It needs passageways, radio computer links, sensors, places to obtain parts, and a wide variety of precision tools to carry. Several may be needed to specialize in large or micro-structures. They need backup power storage, frequent places to plug in, and ways to obtain power when a main power system is down for repair. Open questions are what repairs the long-lived robots (themselves?), and how many repair robots are sufficient to guarantee mission success.

Vital to the value of the mission are scientific instruments, both automatic and human-monitored (if any). Telescopes and intensity-measuring instruments throughout the spectrum provide valuable data to send to Earth or to keep resident scientists busy. They must be protected from damage at high speed and may only look to the sides and behind. Observing stations would be on the sides of the starship. Sensors for magnetic and electric fields, atoms, molecules, and dust, cosmic high-energy particles, and many more to be listed are to be located compatibly with starship operation. Telescopes must accurately measure positions and motions of stars for determination of starship course and better stellar data. Large-scale mass measurements are possible if a radio link to Earth can be kept. But most antennas would be on the sides of the starship out of the path of fast gas and dust. The Daedalus study supposed about 100 tonnes of instrumentation, plus over 200 tonnes of subprobes. Local missions need less than 1 tonne for less instrumentation.

Communication to the Earth requires large parabolic dishes and high-power transmitters. Voyager uses 20 watts radio transmission and a mission to about 0.01 LY can get by on 40 w [a-Jaffe]. 40 megawatts would be required from 10 LY away unless larger starship and even larger Earth antennas are used (as shown in Chapter 7). Multiple use of large systems can be considered for transmission. Some forms of solar sail could be shaped to work as radio antenna and even as shield. Daedalus had the framework of the fusion chamber shaped to be a radio dish, already pointing at Earth. Unfortunately Earth is directly behind and some kinds of drives would interfere with radio communication. Perhaps the drive can be turned off periodically for transmission and reception of data. Even so, the antenna would be damaged if kept protruding beyond the shield for long times.

Information storage must be vast, especially for a one-way human colony mission. At least the equivalent of a major university library is needed (10 million volumes including journals, 10 terabytes or 10 trillion bytes where one byte encodes one character), plus a wide sample of Earth arts and other material. A large store of moving pictures as video would require much more storage. Videotape is presently our densest storage but not a long-term medium and easily erased or damaged. Optical digital recording is most promising, now storing about 1 gigabyte per disk. Its permanence needs more study. Density is still low since one hour of mediocre color video would require several disks. Important information should be carried in triplicate and most other in duplicate. Perhaps ten tonnes must be devoted to archives plus the readers. Even so, scientists must be judicious about data stored. Instruments could collect a terabyte in an hour and this must be processed for its meaning (sometimes just a few bytes) rather than stored in permanent media.

5.5 SUBSYSTEMS FOR BASIC HUMAN NEEDS

The main problem with feeding humans is that food energy is obtained very inefficiently. Plants are the most efficient source of food by far but photosynthesis traps and stores only about 1% of the light energy impinging on plants. Human bodies require about 100 w of energy, and if 100 people are living aboard, 1 Mw may be needed just to illuminate plants. This 10 kw per person for food growing is in the range found by others [a-Matloff 1985, but see Chapter 9]. If food could be manufactured palatably, considerable power could be saved. The mass of agricultural tanks and growing areas is estimated at about 1/10 of starship mass. The water mass may be as high as 100 tonnes for all needs, and its storage and distribution system should be non-corrodible. Very limited meat production may be possible, perhaps fish in tanks, conceivably poultry. Humans need about 1 kg of quality food per day, 3/4 of it being water and indigestible fiber. If food were all carried in storage, the total need for 100 years would be about 100 tonnes, a significant part of starship mass, with possible long-term deterioration. To cover emergencies several years of adequate cold or freeze-dried food should be accumulate, and growing areas should be larger than minimal. Growing areas should be several in number and isolated so that no ecological problem can kill all at once.

An artificial gee (not to be confused with gravity, and not always identical in effect) may be needed for survival. There has been no hope of finding a starship drive which could operate continually even at 1/2 gee. Therefore a rotating structure, either for continual habitation or as a place to spend a significant fraction of each "day", should be incorporated into the starship design. This "wheel" will be massive and must be strong. Larger rotating space colony designs require materials working near ultimate tensile strength [b-Johnson]. Living quarters would be arranged quite differently in a rotating wheel than in a stack of 14 levels. Because of its mass, the wheel's angular momentum is also a problem. If its axis is transverse to the starship travel axis, any change in starship course will be complicated by interaction with the wheel. Therefore the wheel axis must be along the starship axis. Friction in the bearings will tend to rotate the starship oppositely, and jets are needed to stop starship rotation frequently. A partial solution is to have two counter-rotating wheels, requiring additional mass and complexity.

A small wheel must rotate rapidly to provide at least 1/2 gee and human orientation is upset by this [b-Johnson; more in Chapter 9]. Even 2 revolutions per minute is unpleasant, yet requires about 100 m radius wheel. Radius is strongly dependent on rotation period [see Appendix D]. Transfer from rotating wheel to other parts of the starship must be done at the axis. During acceleration (linear gee), movement among levels would require heavy elevators, ramps, or stairs. Rotating the starship seems out of the question unless it is a huge "world" ship. A large wheel implies a large starship with a large front cross-section. An extremely good electric and magnetic radiation shield might be feasible for protecting this wheel [b-Johnson]. Any material shielding should be added to the starship envelope, not to the already massive wheel. The need for an artificial gee wheel is a very serious problem. Some other method of keeping people fit while living in zero gee is needed.

Heat could come mainly from propulsion waste. If there is none, or the drive is off, then another source is needed. It turns out that insulation can keep losses so low that ordinary power usage (perhaps agricultural lights alone) could provide sufficient heat despite radiation losses to outside at 3°K. Space missions thus far have taken an Earth-like temperature with them. The Voyagers heading for the stars are using onboard heat to hold about 20°C on all electronics and many other systems. Future probes may be designed for nonheated environment (as cold as 3°K), but little of the technology is now ready. With humans aboard, much of the starship must have an Earth-like environment, and most other equipment need not be exposed to interstellar "cold". For an environment with interesting variation, the temperature level (and possibly also humidity, light, and air circulation) should fluctuate daily and randomly to simulate our adaption to weather. Lighting probably cannot use present technology but more efficient methods that give spectra suitable for plants or for people, last for decades, and do not release dangerous metals like mercury if broken. Lighting for human work can be 100 times dimmer than plants need. Daily exposure to brighter light is needed to set the 23-25 hour human clock to a chosen day length. Most lights may be switched off where not in use, with dim background lighting in all areas as emergency lighting and to avoid gloominess.

Air quality control is crucial. Air pressure can be about 1/2 atm with 1/3 of this pressure for oxygen as if at high-altitude, and 2/3 for nitrogen [b-Johnson]. Helium and/or argon might replace part of the nitrogen. Growing plants may extract approximately the amount of CO_2 that human respiration puts into the atmosphere, but CO_2 must be kept within certain low limits. Plants also interact with oxygen and nitrogen. Other gases must be removed by efficient filters, some of which involve chemicals which must be replenished or refreshed. Methane and ammonia come from decay and from people and animals. Other more dangerous gases might be emitted in labs and shops. Water vapor can be a serious problem. Plants emit it copiously and it must be removed rapidly or corrosion and mold growth will occur in many unseen areas. Air should be as dry as remains comfortable to reduce mold and avoid large amounts of condensation at cold spots. Most present plastics should be avoided for two reasons, the vapors they emit, and the accumulation of charge. Most partitions and structures should be metal to avoid charge accumulation as well as to help absorb radiation. The living area may need several airlocks so that sudden leaks cannot affect all areas at once. Storage areas would have airlocks. No matter how tight the starship is built, small leaks over decades result in serious loss of atmosphere if not replenished.

Waste is anathema on a starship. The rocket exhaust, if any, is no place to dump paper towels, food and human waste, or any precious organics. Dirt is also not tolerable and regular cleaning of body and habitat must be done in new ways with greater care. Closed cycle sewage and resource recovery systems seem possible. With a plant growing area, recycling waste into the food supply is made more esthetically palatable, but proper chemical systems can make any waste quite safe. Another way to look at it is that the human body rebuilds itself several times in a lifetime. Thus 100 people at 70 kg each constitute 7 tonnes of vital organic material recycled every decade or so. Carbon and other elements in the form of gases will inevitably leak away. If every material on board is a vital resource, how much loss can be tolerated? If the total vital material is 1000 tonnes and the voyage is 1000 years and 1% loss is acceptable, then only 10 kg can be permitted to "disappear" per year. That is a very tight system. A loss of 300 micrograms per second is about $(10)19$ atoms (of light elements).

5.6 SUBSYSTEMS FOR ADVANCED HUMAN NEEDS

Extra gases for air, extra water, and extra soil head the long list of materials, supplies, and parts needed to restore inevitable deterioration in the living system and in other components. One book cannot begin to list all the supplies and cultural resources that should be taken on a star voyage. The list depends in part on the duration and purpose of the mission, also on the number, needs, and skills of the voyagers and the type of propulsion. Because of the long time and distance, a microcosm of the most advanced human society must be carried, plus the means to do further research enroute and produce technology to upgrade the starship. The number and mass of supplies needed could be a serious problem. A few thousand tonnes might not be enough for massive stocks of materials and parts needed to repair everything on board even if voyagers spend much time recycling all broken parts. There cannot be stores of fresh toilet paper and towels, of shoes and pants, maybe not even bolts of cloth. All equipment and supplies that deteriorate must be avoided or made from more basic material, perhaps fiber spun from plants or extruded from bulk plastic, itself remade from scrap. The emphasis on minimum deterioration and waste will have more effect on "lifestyle" and social structure than often realized. Consider the benefits of shipboard near-nudity.

NASA has had to solve many problems associated with human needs for the short term, just weeks in space [b-McAleer]. NASA may have been overzealous in some areas such as toilets, orange drinks, and what color to paint walls [a-Sc 10 Nov 89] while neglecting to work toward closed ecosystems. The USSR in its space program has dealt with human needs over 1/2 year in space with reported hardships and deterioration of health. Successful survival in space for weeks is far from success for 2 years on a mission to Mars with only radio help available and centuries in a starship with zero outside help. Just the medical aspects are formidable even when space (and terrestrial) medicine is well understood. Humans cannot have all the diseases they carry removed, but voyagers may be selected as much for their good health and genes as for their intelligence and resourcefulness. New diseases may appear. Reliance on many medicines seems impossible, given the scale of that industry and the variety of new drugs. A 100-person mission cannot have many medical experts and full hospital resources. Death from accident, illness, and old age will occur, and disposal of bodies is another difficult problem. In 1000 years bodies alone amount to about 100 tonnes or 1% of starship.

Human work areas replace limited robotic repair with full-scale ongoing refurbishing of the starship. Shops must be able to handle all components, small and large, as virtually all may deteriorate in 100 to 1000 years. Like the people, the parts that arrive are not the same ones that departed. Shops may be patterned after present technical specialties, plus new ones for space equipment and propulsion. Thus there might be mechanical, metallurgical, optical, electronic, chemical, organic, biological, even nuclear-particle-radiation shops. Some space technology work might be done outside if high radiation can be avoided. Vacuum (space) and radiation protection suits would be handled in a shop and needed for outside work and repair. Shops need adjoining labs for basic and supporting study and measurement. Computer repair would be a steady occupation, but would new chips be made? There would also be shops for personal and domestic supplies (clothing, paper, pens, cooking tools, furniture, plumbing, cables, etc). It may be possible to upgrade most of the starship, even its propulsion (although additional energy is unlikely to be obtained enroute). It should be possible to alter another shop to do a particular task if an accident destroys both a component and the shop to repair it. The starship should be designed so that any part can be removed, repaired, replaced during travel with no loss of life or any vital supplies or equipment. This kind of transportation has never been attempted before.

Very important and symbolic is the control room. Surely the starship will not run itself, and perpetual human hands-on monitoring, and monitoring of the monitoring, must go on. Humans will want to know they are in charge. The modest control room would have the main instrument monitors and controls (big levers for the titanic drive, if possible), and all starship and mission plans in assorted media (on line, on disk, on paper). For science as well as for the thrill of deep space several observing bubbles or other facilities should be placed around the sides of the habitat volume. Humans also need room for large and small meetings, exercise and sports, and studios to practice various arts. The main meeting hall should be versatile for large dinners, artistic performance, and computer-aided conferences. Humans need private rooms for sleeping, meditation, interpersonal fun, intimate communication, counseling, and illness. Research work not requiring special equipment can be done in personal rooms, shops, or labs. Beyond growing food, there must be facilities for cleaning, processing, preparing, storing, and manufacturing food.

The human subsystem is treated more in later chapters too. In a small crowded starship the voyagers will want more facilities than can be provided, because they have many needs and interests. Instead of getting to run across countryside, they must exercise in machines or wheels. Some might want impractical clothes, natural materials for hobbies (wood, stone), old Earth books, and more. Vanities will be carried to the stars along with human culture. And there must be room and safe places for children to play.

5.7 STAGING AND LANDING

A starship will not be launched from Earth. It will be assembled in space, either near Earth, near the Moon as material source, at a Lagrange point (L4, L5) on lunar orbit equidistant from Earth and Moon, near some asteroids, near Jupiter, or in a solar orbit. Difficulties of bringing much material from Earth are due to the deep gravity well. Chemical rockets are barely able to do the launch, and study of interstellar propulsion has revealed no other method with such high thrust. If super-strong cables are developed for a starship, they may also solve the Earth launch problem by serving as elevator cables from ground to synchronous orbit. A site not in the plane of the planets or near the Sun requires extra chemical or nuclear propulsion to reach. A site at Jupiter and beyond requires long times to transport supplies or faster interplanetary travel.

The Moon has good supplies of (in approximately this order) silicon, oxygen, iron, aluminum, calcium, magnesium, titanium, and some manganese, chromium, potassium, phosphorus, all very useful [b-Johnson]. The Moon also has small amounts of lithium, boron, and most other elements but is deficient in hydrogen (therefore water), nitrogen, and carbon. A comet, or an icy asteroid would be needed for massive quantities of these life-supporting elements. Structure and cables depending on polymer plastics would also require carbon and hydrogen. The Moon has the material needed to make solar cells or mirrors. The Moon's escape velocity of about 2 km/s is about 30 times easier to deal with than Earth's but still a barrier for large projects. A continuous mass driver has been suggested for throwing minerals into space. Fusion fuel might be obtained with difficulty locally from the solar wind or at great effort from Jupiter's atmosphere [a-Parkinson].

Some proposals for propulsion have humans working all over the solar system— million kilometer long launchers, giant mirrors or lasers near the Sun and giant lenses way out there, dropping a starship from Saturn to go round the Sun, solar collectors all around the Sun, and so forth. None of these plans can be settled here for lack of knowledge of best methods, but some must be kept in mind as the starship system is studied.

At the other end of staging is landing. For exploration or for colonization the starship needs an assortment of big equipment and small supplies. These might include subprobes for space or surface exploration, spacesuits, landers for people to use in vacuum, gliders for slight atmospheres, ablation capsules, shuttles for some more massive planets with air, and the means to depart planets again. Some very difficult problems of landing are yet to be discussed. On the ground, after a long thousand years, are needed cars, water, air, mineral, and solar energy collection and storage systems, camping equipment initially pressurized or sealed, and a multitude of other items, some transported from Earth, some manufactured enroute.

5.8 OPTIMIZATION AND MORE ENGINEERING

Optimization has become a fine human and engineering skill. Nevertheless, many technical studies for interstellar travel have attempted to squeeze the last decimal place from idealized methods, using fuels at their maximum energy content, using materials at their ultimate strength, assuming infinite amount of money, and other extravagances. No room is left for error, for overlooked factors. Starship design is so demanding that the temptation is to push every aspect to the limit to reach feasibility. Some concluding remarks on optimization and general design seem in order here.

Studies attempting to optimize propulsion systems, or combinations of them, have been cited, such as how to vary exhaust velocity for most efficient use of fuel, or how to minimize the travel time or mass ratio, given various constraints [a-Powell 1974; a-Bond 1978]. Another line of planning has been to use a series of large expensive local propulsion methods to gain a few more 100 km/s as a starship leaves the solar system and shave a hundred years from a thousand year voyage. A form of optimization called cost-benefit analysis would bring into question a starship or mission with a complex series of operations which separately or together bring little gain. The costs and benefits can be measured in dollars, energy, time, or other parameters. For the purposes of this book, these are fine tuning methods, rarely yielding a major new result which will tell the best way to design a first starship. The most effort should go into hardware research for and practice with fusion, light sailing, and closed ecosystems, and let optimization come later with experience.

A simple example of optimizing of a different kind is given here. It has been shown that a faster voyage permits a smaller mass of starship but requires much more propulsion mass. Voyage time is inversely proportional to effective speed, and fuel mass is proportional to speed squared. A slow voyage requires more people and extra supplies so that again a larger starship mass is required, perhaps a mass proportional to the duration of voyage [see Figure 5.2]. When a parameter such as mass becomes large for small time and large for large time, there might be an optimum minimum mass between and a corresponding optimum travel time. Enough studies of propulsion and starship size have been done to make a guess at the functions and find the minimum [see Appendix D]. The meta-analysis can be done for each effect without worrying about the interaction of propulsion and size. Suppose that the minimum size starship is zero, that a 100 year voyage requires just 1000 tonnes of people and supplies, and that a 100 year voyage requires 1 million tonnes in terms of propulsion. (The model missions have been altered in light of succeeding studies of propulsion.) Then the optimum mass is about 10,000 tonnes and voyage time 1000 years. Further, it can be seen that the major contributor to mass is the long duration.

Optimization is expected over time too. It is often said that whatever starship is sent out, in a hundred years faster ones are developed and sent out to pass the first one by and reach the destination first. This does not mean we can wait for better technology, or it will never come. Accepting later improved starships will be a necessary part of a starship program, although in a political climate like the present the public will wait forever for what will then never happen.

A questionable habit in many published mission plans is the jettisoning (abandoning) of unused massive components. Items have included "engines", tanks, structure, sails, cables, wire, and more. Starship design should be such that all possible parts are carried to the end. The materials may be needed. Of course having to decelerate unwanted mass is a problem. If available means are tight, then last minute decisions can be made as to what to release or cut away. Perhaps in hundreds of years of coasting scientists and engineers aboard will find a way to use a dead drive as a brake, such as rigging a mass driver and chopping up the fuel tanks for throwing mass (if energy is available in some form), or deploying many kilometers of charged wires.

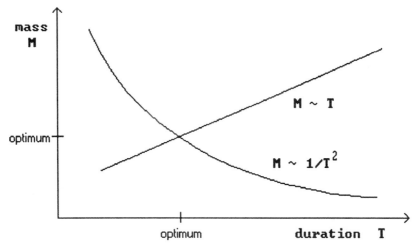

Figure 5.2 Optimization of starship payload mass and trip duration (on sketched, not exact graph). Very approximate results are 10,000 tonne payload and 1000 years duration.

In early considerations of propulsion, each component of propulsion tended to mass as much as 1 to 10 basic starship (payload) masses. This is design trouble and usually indicates need to start over. If the starship masses 1000 tonnes and the minimum drive (all moving parts and fuel) masses 10 times more, this is about the limit of feasibility. For a 2 leg mission (one way and stop) 100:1 mass ratio might be needed to get the 1000 tonne payload there. The same dilemma can occur with smaller parts if there are too many of them. If each of 1000 vital components masses 1 tonne each, the whole 1000 tonne starship is already filled before any big parts have been designed.

The ultimate in design redundancy is multiple starships. There is no reason not to build a fleet of ten smaller ships except for the extra cost and mass to provide envelope and structure for each. All other parts should be duplicated anyway: propulsion, fuel tanks, ecosystems, human experts, computers, archives, tools, everything. If the starships travel within tens of kilometers of each other, with means of crossing between, then each can serve as backup for the others. Sizing should be generous so that one starship might absorb one half of the people and most valuable supplies from another, in case of mishap. Still, the saving of mass from having less backups and carrying less spare parts and supplies will not make a fleet less massive than one starship with the same people and mission. During cruise, an abandoned starship will travel with the fleet and is always available for parts. If its braking system is operative, all of its parts can be had at the destination. Fleets do not seem to have been proposed elsewhere except for the case of sail-ships [a-Matloff 1986], where cable mass grows faster in design than starship mass and forces rethinking.

Chapter 6

Missions

This chapter concerns where an interstellar voyage might go and general discussion of who or what might go. The mission plan rather than the hardware system is the emphasis here. Various mission studies mentioned earlier are examined further, along with new suggestions for probes and expeditions. While some crucial matters need more investigation, it is possible at this point to begin to see what missions foreseeable technology permits. The matter of who plans and controls a mission is left to Chapter 10. Again the underlying question should be considered: Can and would "others" do it this way if they were in our situation with our local density and selection of stars and our developing technology. In such discussions about "others", "we" refers to humanity in general, with scientists and mission planners representing us.

6.1 DESTINATIONS

Any probes or human missions to another star would take full advantage of the scientific opportunities near the outer Solar System. Early missions are likely to be local and are sometimes called "precursor missions" because they explore local interstellar space not far from the Sun while providing opportunity to test methods of traveling farther. Space beyond Pluto might be thought of as empty and dull, maybe a little gas and dust and weak fields to measure, nothing much to see or photograph. Human missions this far are unlikely because there is little to literally see. If exploration were not so expensive, some humans might wish to travel for years out beyond Pluto just to do it, but the size and complexity of the spacecraft would be much larger than a one tonne probe. Such a human mission might establish knowledge about how humans would survive and behave prior to a human mission to a near star.

Beyond Pluto and its big moon Charon, the next interesting place out there is a region where the solar wind interacts with interstellar matter [a-Cesarone; a-Paresce; a-Van Allen]. The distance to this shock region and heliopause is uncertain, about 10 to 30 billion km (0.001 to 0.003 LY or about 70 to 200 AU) and a goal of any mission. [Many reports use a common "astronomical unit" AU, the average distance of Sun to Earth, to discuss distances in and near the Solar System.] The interstellar gas flow (called a "wind", although zephyr would be more appropriate) is thought to come from the direction of Ophiuchus. It has relative motion about 22 km/s, due to the Solar System moving in it. Where this neutral gas meets the 400 km/s solar plasma wind and magnetic field, a collisionless shock occurs (so-named because electro-magnetic fields accompanying the plasma do the interacting, not the particles). Solar magnetic field lines spiraling out at the same rotation rate as the Sun are disrupted and carried downwind along with solar plasma to form the heliopause, a blimp-shaped boundary with the Sun in the blunt end and a long tail behind. *Pioneer 10* is now heading down the tail. *Voyager 1* is now heading almost directly upwind and will reach the shock front about year 2015 when its power supply is expected to become too weak for data transmission. Where the heliopause occurs depends on the density of interstellar gas ahead, which must vary as the Solar System pushes into further regions of it. The Solar System is moving toward a somewhat different point about 50° "North" of the interstellar flow apex [Figure 6.1].

Much further beyond is the conjectured Oort cloud of primordial comets 0.1LY to 1 LY away and of major interest for studies of cometary collisions with Earth in the past and future [a-Jones 1983; a-Marochnik]. While the number of comets could be trillions, each a disaster if it hits anything, the volume for them is so large that they will be difficult to spot even with radar. They would be moving very slowly, less than 1 km/s, and of course would not have glowing tails. Scientific probes to the more moderate distance of about 0.01 LY are within our technical capability now, given major commitment. Such a mission is not usually counted as interstellar travel but it would be important preparation. Nearly empty space itself is of interest for another reason, accurate measurements of star positions and speeds. Comparison of positions as seen from the Solar System and as seen from a lightyear out provides a very long baseline for vital measurements.

Since many stars have dim companions, the Sun could have a very dim one farther out. Thus far no direct evidence has been found for one, but the hypothetical companion is called Nemesis and might orbit between 0.1 LY and 1 LY out. A mission is unlikely to be able to detect or visit such a burned out or failed star (brown or black dwarf up to 1/10 mass of Sun), although they may be common. The nearest star system (Alpha Centauri, a double star at 4.4 LY, with a third component Proxima) is not necessarily the best destination for a first true interstellar mission. One star of the main pair is as bright as our Sun, an unusual coincidence in our neighborhood. But it is not expected to have planets because most studies of planetary formation indicate that condensing material ends up either as a pair of rapidly orbiting stars or as a single star with more distantly orbiting planets and debris. A probe to Alpha is needed sometime, if only to determine whether there is useful material orbiting these stars and to test planetary theory.

The next star out, Barnard's Star at about 6 LY, was chosen for the Daedalus mission study because it was the closest one for which slight evidence has been found of one or more large planets. Barnard's is a small dim star and one of the fastest local ones (108 km/s), probably wandered here from the interior of our galaxy, old and poor in heavier elements, and unlikely to have habitable planets. It does not seem like an exciting destination, but Daedalus was undertaken for general proof of feasibility more than fascination with Barnard's. A robotic mission to any star at all is certain to learn much, but some destinations will impress those who fund the mission more than others.

The nearest brighter star is Sirius (about 9 LY), but it is accompanied by a white dwarf (period 50 years and occupying the habitable zone). A white dwarf indicates an aged system that has gone through catastrophe with loose material blown away or swept up. In general, bright stars are not good destinations for any missions seeking life as they live too briefly. The white dwarf is a very interesting object for scientific study. The next interesting choices that could have planets are Epsilon Eridani, 61 Cygni (double), Tau Ceti, Epsilon Indi, and BD + 50 1725 at 11 to 15 LY. Such missions would be longer than desirable for a first mission. They are worth considering because four of these stars are single, not much dimmer than our Sun, and likely to have planets in a habitable zone. Epsilon Eridani seems to be too young for life [a-Croswell]. 61 Cygni's two stars are farther apart than Pluto from our Sun, giving room for hospitable planets. Almost all of these stars have been checked for obvious radio signals. Figure 6.1 places some of the near stars on a starmap using Earth-based coordinates. That some stars are far from the plane of Earth orbit ("ecliptic") is not a problem for interstellar travel since any useful propulsion method must produce high speed in any direction.

Some ten pairs of near stars of interest happen to appear close to one another in direction, although one is several lightyears beyond the other [a-Heppenheimer; despite title, covers this subject with map of pairs]. A single probe could visit 2 and sometimes 3 or 4 stars in one mission. At 0.1c up to two centuries would pass before data is received. Beyond Barnard's Star is a medium bright double star 70 Ophiuchi (16.7 LY) after a turn of only about 5°. Beyond Luyten 726-8, a dim star, is the medium star Tau Ceti after a turn of about 7°. Beyond Epsilon Eridani is another medium but triplet star 40 (Omicron) Eridani after a large turn of 30°. Significant changes in course require substantial propulsion energy as discussed in Chapter 7; small turn jets will not have noticeable effect at 0.1c. The first star's gravity is not much help in making the turn because of the high speed.

Destinations for interstellar travel depend on the purpose. The likely purpose in the next few hundred years is exploration to determine what places there are to visit, what the resources are there, and what humans might do there. Thus far no definite evidence of planets has been found at other near stars. Although smaller and more difficult to detect, a habitable planet with water and oxygen is likely to have life. What to do upon arrival is a complex question requiring future study. [Some procedures are suggested in Chapter 11.] Aside from the appeal of distant stars and objects, there is little need to study farther stars as destinations because it is rather clear that interstellar exploration and colonization will occur in small steps. If humans reach one or two near stars and start viable civilizations there, later humans will be just as able to step to a few more stars. Robotic probes are likely to be the first attempts to explore, although we cannot be certain yet that these will be more successful than larger-scale human missions. Enough stars of interest occur within 20 LY to use ten or more probes for exploration, but few likely destinations for people are within 10 LY. More on destinations is given in the next section and Chapter 11. [see also a-Archer; a-Martin 1978]

6.2 NEARBY STARS AND POSSIBLE PLANETS

The detailed properties of stars are very important both to interstellar missions and to estimating the probability of life elsewhere. Much more needs to be learned than is presently known, and this book can provide only a small portion of the large quantity of pertinent observations found and theories established by astronomers. To discuss stars statistically, some descriptors of stars are needed. Spectral types have been defined for observed stars and most normal stars are categorized spectrally with letters B, A, F, G, K, and M. These tell approximately their temperature and are listed in decreasing temperature from about 25,000°K "blue-white" very bright stars to the 5000°K range for our Sun a white-yellow G star, to less than 3500°K for red M stars [b-Abell; b-Baker; a-Mattinson].

In accord with the radiation law, emitted intensity or luminosity varies very much more than temperature varies. The empirical relation found for most observed stars is shown on a Hertsprung-Russell diagram, a graph of absolute brightness versus temperature. Most stars fall on a "main sequence" curve where luminosity is approximately proportional to temperature. They behave this way for billions of years as they fuse hydrogen to helium and release nuclear energy stably, doing naturally what technologists face great difficulties to achieve. They are reasonably well understood (in theory). Less understood are the rarer temporary stages of stars such as giant red helium-burning stars, bigger and more luminous than their temperature merits, and white dwarfs, small and dimmer and very hot for their size as they take billions of years to cool down after having run out of nuclear fuel.

Luminosity also depends strongly on star mass, with an exponent of at least 4, in the main sequence. A factor of 2 change in mass causes approximately a factor of 10 change in luminosity. F stars (6000 to 7500°K range) are only about 1.5 times more massive and up to 5 times more luminous than G stars, and K stars (3500 to 5000°K range) are about half as massive and 1/10 as luminous. Luminosity is an intrinsic property of stars and is to be distinguished from the brightness that is observed for stars at a distance. Table 6.1 summarizes these characteristics for the common types of local stars.

Table 6.1
GENERAL TYPES OF COMMON STARS (FOUND NEARBY)

Spectral Type	Temperature (°K)	Mass (Sun = 1)	Number Nearby	Luminosity (Sun = 1)
B	10000-25000	3-15	0	100-5000
A	7500-10000	2	1	5-100
F	6000-7500	1.5	1	2-5
G	5000-6000	1	3	0.3-2
K	3500-5000	0.5	7	0.03-0.3
M	2500-3500	0.05-0.3	35	0.0003-0.03

Notes: Ranges or values are typical and approximate only; nearby count for 52 stars within 16 LY (remainder 5 white dwarfs).

Although our Sun is average in luminosity, it is a rare star, since most stars are smaller and dimmer. Only F, G, and K stars have luminosity suitable for habitable planets. The abundant M stars are very small and dim. A planet can get warm very near one but then it is strongly affected by the star's gravity. Dim M stars with as little mass as 1/10 of our Sun's are still formidable objects. Several white dwarf (WD) stars are also found near us, but no O or B stars are nearby. Large hot stars radiate their energy away copiously and do not last long enough for life to evolve on planets around them. Hot O, B, and A stars are to be avoided if planets are sought that have been there more than a few billion years. Type F may last 4 to 10 billion years, long enough for life to evolve. G stars can last 10 to 30 billion years, and K much longer. The amount of heat available from any star decreases as the inverse square of distance from it and therefore within some range of distances there is a habitable zone or biozone where temperatures on a planet with suitable atmosphere permit life as we know it. [Biozones are discussed further in Chapter 11.]

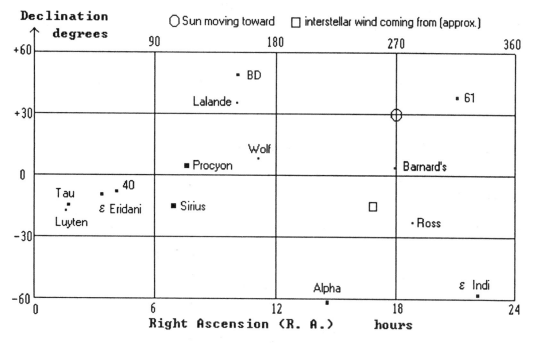

Figure 6.1 Starmap of near stars in Earth-based coordinates. All stars to 10 LY are shown, and selected stars to 16 LY. Several more pairs of near stars than shown are close in line of sight and permit possible dual missions. Dot size indicates luminosity (absolute magnitude). Shortened names are from Table 6.2.

Stars are named for the constellation they are found in (preceded by a Greek letter to designate the order of visual brightness) [e.g. b-Moore; b-Eicher], or by their Arabic name, or for their coordinates, or for their discoverer and a number, or by other methods, or by several names/numbers, with no general rules. Perhaps the most beautiful-sounding star system by name nearby is Omicron Eridani. Many nearest stars are dim and known by complex numbers only. To describe approximate directions in the sky the regions of the traditional constellations are sometimes used by name. For astrogation and other purposes, star coordinates are very important. One common old system is based on Earth's orientation and rotation. Star positions are measured in "right ascension" (RA) in hour-minute coordinates from the spring equinox, and in declination in degrees from the equatorial plane as projected in the sky. Thus Earth's tilt is included. Figure 6.1 shows a flat map of the stars within 10 LY and other near ones of interest in these coordinates. RA can also be expressed in degrees (at 15° per "hour"). The third vital coordinate is the distance from us. These spherical coordinates can be converted to any other 3-dimensional set convenient for astrogation. Other planes can serve as the basis for coordinates, such as the plane of Earth's orbit and the Solar System, tilted about 23.5° from Earth's equatorial plane (for celestial coordinates), and the plane of our flat spiral galaxy (for galactic coordinates, but still centered on our Sun).

Table 6.2
NEAR STARS (WITHIN 10 LY AND OTHER BRIGHT ONES)

Name	Components	Distance LY	Coordinates RA	Dec.	Type	Mass	Lumin.	Magnitude (Visual)
Sun		0.0			G	1.0	1.0	−27.0
Alpha Centauri	A	4.4	14h36	−60°38	G	1.1	1.3	0.0
	B				K	0.9	0.4	+1.3
(Proxima)	C	4.3			M	0.1	0.00006	+11.0
Barnard's Star		5.9	17h55	+4°33	M	0.15	0.0004	+9.5
Wolf 359		7.6	10h54	+7°19	M	0.2	0.00002	+14.0
Lalande 21185		8.1	10h01	+36°18	M	0.35	0.005	+7.5
Sirius	A	8.7	6h43	−16°39	A	2.3	23.0	−1.5
	B				WD	1.0	0.003	+8.7
Luyten 726-8	A	8.9	1h36	−18°13	M	0.15	0.00006	+12.0
	B				M	0.15	0.00004	+13.0
Ross 154		9.5	18h47	−23°53	M	0.3	0.0004	+11.0
Epsilon Eridani		10.7	3h31	−9°38	K	0.8	0.3	+3.7
61 Cygni	A	11.2	21h05	+38°30	K	0.6	0.08	+5.2
	B				K	0.5	0.04	+6.0
Epsilon Indi		11.2	22h00	−57°00	K	0.7	0.1	+4.7
Procyon	A	11.4	7h37	+5°21	F	1.8	7.6	+0.4
	B				WD	0.6	0.0005	+10.7
Tau Ceti		11.9	1h42	−16°12	G	0.8	0.4	+3.5
BD +50 1725 (Groombridge 1618)		15.0	10h08	+49°42	K	0.6	0.04	+6.6
Omicron (40) Eridani	A	15.9	4h13	−7°44	K	0.1	0.3	+4.4
	B				WD	0.4	0.003	+9.5
	C				M	0.2	0.0006	+11.0

Notes: Mass and luminosity referenced to our Sun as unit mass and luminosity. Components of a star system are labeled A, B, C in decreasing brightness. Right ascension RA is in hours and minutes, declination "dec." in degrees and minutes. Magnitude is visual, as seen from here. Alpha has magnitude approx. 0, brighter stars negative, Sun very bright. Magnitude is a log scale with an unusual base. Different sources vary on values for data. [b-Abell; a-Mattinson; b-Baker; a-Webb]

The 12 stars within 10 LY of us are in 9 systems and consist of 7 M stars, 1 A, 2 G, 1 K, and 1 white dwarf. Table 6.2 gives basic data on these and other near ones of special interest. Fifty two stars including our Sun (or 53, depending on the source of data) are presently within 16 LY of us [b-Abell, a-Mattinson; a-Webb; b-Baker; a-van de Kamp]. There are 39 (or 40) separate star systems in this 17,000 cubic LY volume, with average separation about 7.5 LY. Some analyses find less separation because all 52 known stars within 16 LY are counted, instead of star systems. This misrepresents the distance one must travel and the number of habitable sites since multiple stars are less likely to have planets. With only 4.4 LY to Alpha, we are less isolated than most potential civilizations in this part of the galaxy. Of these stars, 35 are M stars, 1 is A, 1 is F, 3 are G, 7 are K, and 5 are white dwarfs, frankly a mediocre collection with few life-supporting ones. As single stars there are only 2 of type G and 3 of K within 16 LY, for an average separation of 15 LY between stars more likely to have habitable planets.

Near us there are no nebula, giant stars, new stars, neutron stars, suspected black holes, or other interesting objects. Active regions with new stars, and black holes and neutron stars would be very hazardous to visit anyway. All white dwarfs but one closely accompany another bright star. Also of interest to astronomers is that many of these stars have some variability in their light, unlike our Sun. Our Sun is among the brighter of the local stars but not special in the context of a wider population of stars. Very hot stars are very rare. There are 9 double star systems and 2 triple within 16 LY. The single G and K stars have already been named as possible destinations: Epsilon Eridani, Tau Ceti, Epsilon Indi, and the obscurely-named BD + 50 1725. Forty nearest star systems (including Sun and Altair, a bright star at 16.6 LY) are shown in Figure 6.2 in perspective from a viewpoint outside the group, with depicted "size" encoding brightness.

The formation of stars as multiple or having planets depends in still unknown ways on the distribution of rotational ("angular") momentum in the condensing gas cloud that forms a disk of material [a-Stahle]. If one star results from further contraction, it must end up with less than $(10)-5$ of the typical cloud angular momentum or fly apart before it can form. Therefore planets should form from the disk also. Single stars are less common than binary sets and are most expected to have planets. Large hot stars are known to spin rapidly as they incorporate as much angular momentum as possible. Two medium stars take up most of the angular momentum if one large rapidly spinning proto-star breaks into two similar parts [b-Ridpath]. Different characteristics are observed for binary stars, depending on whether they are close and similar, co-orbiting in less than 100 years, or distant and dissimilar. For the latter, some models show that a collapsing gas cloud tends to break up into two similar parts long before stars form [a-Boss] or that two adjacent but gravitationally-bound clouds form two often unequal stars [a-Abt]. Planets may be a common spinoff in that case.

Often two stars revolve far enough apart that some inner planets could have reasonably stable orbits. The stars must be about 4 times farther apart than the planetary orbits. A Kepler law states that an orbit of a given size has a certain period, regardless of the stars and objects involved [see Appendix B for 1.4]. The distance-period relation observed for our planets and Sun holds everywhere. Typical binary stars with multi-hundred year orbits are sufficiently far apart (beyond the orbit of Uranus) that some planets could be in the biozone. Far beyond a double star more stable orbits are possible, but planets there would be cold and of less interest (except to scientists or as sources of material). Very distant companion stars to single and double stars (e.g., Proxima) sometimes occur but it is not yet understood whether they are captured stars or part of an unusual gas cloud collapse.

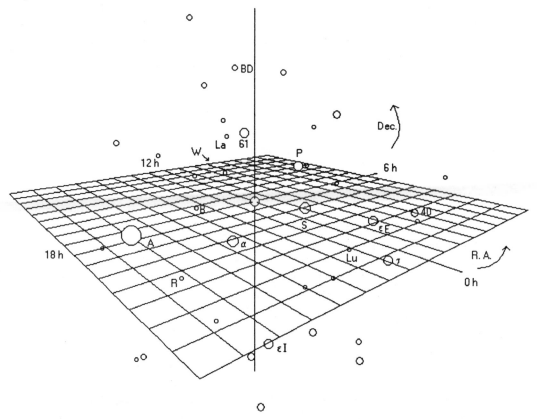

Figure 6.2   Forty nearest stars within 16 LY, shown in perspective as seen from 38 LY away, from 20° declination above equatorial plane, and from -50° azimuth or 20 h 40 m right ascension. Sun is at center. Grid unit size is 2 LY. Only the most luminous star of each system is shown, with depicted size indicating luminosity after distance is accounted for. Key to closer or brighter stars: α = Alpha Centauri, B = Barnard's Star, W = Wolf 359, La = Lalande 21185, S = Sirius, Lu = Luyten 726-8, R = Ross 154, εE = Epsilon Eridani, 61 = 61 Cygni, εI = Epsilon Indi, P = Procyon, τ = Tau Ceti, BD = BD 50°1725, 40 = 40 Eridani, A = Altair.

Collapsing interstellar gas and dust to form stars and planets tends to start spherically symmetric but ends as a thin, not necessarily circular, disk, confining all planets to a narrow plane [a-Boss; a-Cameron]. Present infra-red observations find disks of condensing dust (larger than interstellar grains) around new stars or in star-forming regions [a-Black 1991; a-Gatley]. These sites, also called T Tauri stars after the first one found, would be wonderful to visit, but none are near enough for an early expedition. Infrared in excess of what normal stars should emit has been found for as many as 20% of a sample of normal main sequence stars [a-Aumann]. Better infra-red telescopes might be able to see planets next to their bright suns.

The other method of detecting planets has been long-term observations of a star's motion across the sky. Planets larger than Jupiter-size are needed for wobbles to be observable, and such planets tend to have long period orbits 10 years up, as seems to be observed. No planets have definitely been found yet, although there is inconclusive evidence for large ones at Barnard's and at several stars more distant than the nearby set covered here [a-Black 1973, 1991]. Better infra-red and visible light observatories with much finer resolution are needed in space or on the Moon. They would be crucial to advancing the search for destinations for travel, as well as for many other scientific questions. More discussion of planets at other stars is provided in Chapter 11 in regard to the conditions for finding life.

Two components of stellar motion can be observed, radial velocity toward and away from us as measured by doppler shift, and transverse velocity as measured by long-term comparisons of star positions. These are relative to the Sun's own motion. Almost all local stars are in motion with speeds from zero to several hundred km/s with respect to our Sun [both components listed in a-van de Kamp]. Typical speeds are from 0 to 20 km/s, some toward us, some away. They have a general motion around the galaxy and random motions superimposed, probably due to their many close encounters over the eons. Over periods greater than a million years, stars that are our neighbors now will not remain so. This has implications for long-lived civilizations. For almost any mission, stars move fast enough that accurate astrogation requires knowing and accounting for their motion. As mentioned earlier, the present Voyager and Pioneer missions will first pass some stars that will have approached closer to the Solar System at that time [a-Cesarone]. However, no stars are moving so fast as to provide special close (less than 1 LY) interstellar mission opportunity in the next hundred thousand years. Humans must wait about 815,000 years for star DM + 61 366, type K and now at 33 LY, to pass about 0.3 LY from the Solar System.

The local stars are part of a wave of star formation that started about 5 billion years ago and still goes on. We share their history, unexciting as it is now. The material came from previous stars which blew up as supernovas and spread material out in spherical clouds that collide [b-Spitzer]. Two ancient relatively nearby supernova remnants have been identified [a-Cesarone]. Our Sun and most near stars seem to be third generation, having the enriched full range of elements that supernovas produce. The previous generation, also Population I stars, condensed from less rich material that came from the first generation called Population II stars (the illogical naming is from the first and closer population to be described). A few local stars have spectra showing only the lightest elements and are probably primordial stars. They can live tens of billions of years, depending on mass. Stars also emit material (light and heavier elements) continuously as" winds" in small amounts or in periodic minor bursts and explosions. The majestic motions of stellar winds and supernova remnants constitute the ongoing evolution of our galaxy which will produce new waves of gas cloud condensation and new formation of stars near our Sun in the next billion years or so [b-Gilmore].

6.3 SCIENTIFIC OPPORTUNITIES

A survey of scientific studies to be done by interstellar probe or starship is provided here, more to identify problem areas than to design a mission. Many of these can be done in near space and their value will justify "precursor" missions that gain experience for later missions. While travel among the stars sounds as if it would bring astronomers vividly in contact with their subject, the difficulties of travel to large and dangerous objects make most observation more attractive in local space. But if the past is any guide, scientific work in interstellar space will be important for science in many fields besides astronomy and space travel. Ambitious instrument programs for the Moon and space stations will also help, not only to develop more sensitive and spaceworthy ones, but also to gain more information on possible planets at other stars. General scientific focus would be on observations of the types of stars and their planets, on the composition of interstellar material [b-Spitzer 1982], on fields and high-energy particles in space, and on precise locations of stars. If humans are aboard, they (and computers) will study all data. If the mission is robotic, data are stored and transmitted to Earth according to a complex schedule. Interstellar missions also provide the time, vacuum, and near-zero gee for a variety of science experiments in and near the starship. Relativity observations are discussed in Chapter 7 and study of other life is discussed in Chapter 11.

Instrumentation for close observation of exotic objects such as white dwarfs, neutron stars, star-forming clouds, pulsars, and black holes is not discussed here because these are much farther away and unlikely to be visited in the near future. A visit to the Crab Nebula about 6000 LY away has been described to show what places of tantalizing beauty, science, and danger are available [a-Seward]. This expanding multi-colored gas cloud with a rotating pulsating neutron star at its center came from a supernova observed here in year 1054. The x- and gamma-radiation from it is about 1000 times more intense than the visible light. Heavy shielding would be required to approach the nebula, but with a lightyear size it might be best viewed from afar. Fortunately there are closer but less spectacular old supernova shock waves to observe.

Preparing scientific instruments for interstellar travel is a major task, since present space technology is not proven beyond several decades. Extrapolating from present experience to 100 year tests and then to 1000 year operation will be difficult tasks. Extensive onboard repair and upgrading by robots or humans might be needed for major and critical instruments. The technology for science during travel and at the destination is so interwoven with the technology for starship monitoring, operation, and navigation that the scientific part of the mission should not be considered separate. For example, the same telescopes observing stars are also used to establish precise position and speed of the starship. The same instruments measuring interstellar gas and dust are monitoring the safety of the starship and possibly detecting unexpected changes in the medium. The instruments identified for a precursor mission [a-Jaffe] serve as a guide for longer missions. Calibration of instruments has been a problem for short missions and becomes more difficult for long missions as instruments undergo changes not exactly known beforehand. All corrections must be determined and made by onboard computers (and human crew if available).

The present Hubble Space Telescope is 2.4 meters diameter, rather massive, and very expensive [a-Bahcall]. Its 1 tonne mirror can collect far UV (100 nm) through far IR (1 mm), an extended wavelength range useful only in space. It is a robotic system which can switch the image to six different instruments (electronic "cameras", spectrographs, and photometer). Resolution of 0.1 "arc was to have been 10 times better than Earth telescopes can achieve but a manufacturing error prevented that. The image sensing uses about 2 million pixels. All this is state of the art for space science as of about 1980. Two or more small precision telescopes can be used as an interferometer for much finer resolution close to bright stars. The first optical interferometer is being built at Mauna Kea (Hawaii). For the Moon 42 of 1.5 m telescopes are proposed in a 10 km array with computed-aperture. Called LOUISA, this Lunar Optical Ultraviolet Infrared Synthesis Array would be capable of studying Earth-size planets at near stars [a-Burns]. An Earth-size planet at 10 LY would have angular size about (10)-10 radian or about (10)-5 "arc, requiring an optical instrument of size at least 10 km to study. For a starship, telescopes up to 5 meters are desirable, about the size of the Mt. Palomar 200 inch instrument, but making them very low in mass yet rigid seems difficult. A 4 meter mirror is expected to be reduced to 500 kg and its structure in low gee to about 1 tonne [a-Burke], far too much for a low-mass probe. Optical telescopes require structural accuracy to about 0.1 micrometer.

Modern telescope images are often put on solid-state image sensors or sent to spectrometers to determine the spectral lines of optical sources from UV to IR. With film replaced by electronic sensors with a smaller number of pixels in a smaller area, a telescope cannot observe the finest detail (to the diffraction limit of its lens or mirror) and make fine measurements over large areas. Comparison of star positions requires having several adequately bright stars in one field of view or "picture", but bright stars are not that dense. In small pictures only the star of interest might show, rendering the measurement useless. Various complex solutions have been proposed [a-Jaffe], and like most instrument problems this one does not seem fatal. Accurate star positions could be one of the most important scientific measurements because the long travel of the starship permits very long baselines. Determining accurate distances to Cepheid variable stars thousands of lightyears away would help establish much better most of the scale of the Universe because these stars have a known relation between their luminosity and their period of variation.

Studying planets at other stars is needed not only to find destinations for peopled starships but also to establish better statistics on the occurrence of planets of all types. Finding planets requires finding what plane they orbit in. Once a planet is found in motion, the plane is partly defined and can be searched carefully for more. Often the rotation axis of the parent star can be determined and is a clue to the direction of planetary orbits [a-Doyle]. Observation must start at a long distance to permit time for many measurements. Planets are usually found by finding dim objects with noticeable motions against the background stars. Planets can be distinguished from numerous dim stars by their different light, since planets reflect light from the parent star and emit infra-red. All possible objects near the star must be measured and data stored. Starship motion also will cause planet positions to change against background stars, in the short term. Determination of the star's distance and luminosity can tell which planets are in the habitable zone, and spectral measurements at closer range can determine some gases in their atmospheres (if any). Estimated size indicates planetary mass and gravity.

Radio telescopes are rather easier to make, requiring low-mass non-precision wire mesh dishes. Some nearby stars and possibly their planets are interesting radio sources. Very low-frequency radio observations have also been suggested and require a very long pole antenna. A pair of radio telescopes can make precise position measurements. General microwave background intensity is also to be measured over a broad frequency range from radio up to infra-red with radiometers. Variations in the interstellar medium can be observed from variations in radio and lightwaves coming through it as the starship advances. On the doubtful list is use of radar or other means to search for comets (sans tails) at about 0.1 LY from stars. On the unlikely list is the detection of signals from other civilizations. A starship can have the time and better conditions for listening for intelligent signals but may not carry large dishes.

Measuring the atomic and molecular gases in space involve collection, identification, and counting. Mass spectrometers, which work naturally in a vacuum and use magnetic and electric fields to distinguish ions of different masses, may not work well with less than 1 atom per cm^3. High starship speed will permit collecting more material faster, to be stored for analysis, but high speed also would produce severe collisions in collection apparatus and result in contamination. A magnetic ion catcher is needed, like a miniature ramjet scoop. Distinguishing neutral gases from ions before collection seems relatively simple. A separate electric ion collector can determine the quantity of natural ions. Collecting the rarer interstellar dust grains requires a soft catcher. At high speeds this seems to be a difficult problem if the grains are to be recovered intact. The alternative is to demolish the grains by impact and analyze the debris by mass spectrometer. Near the Solar System the starship or probe may still be moving slowly as it accelerates. This is the place to measure interstellar gas and dust motion just before it impinges on Solar System magnetic fields and solar wind.

The measurement of interstellar medium properties is needed to know better what starship designs are possible. Just as gas and dust affect erosion, drag, and braking, magnetic fields need to be known for possible electro-magnetic applications such as turning, braking, and material scooping. Magnetic fields down to interstellar strengths can easily be measured with Josephson junction magnetometers on a long extending rod (boom). Direction is obtained simply by using at least three such devices in three orthogonal directions. Net charge on the starship can be measured by an insulated electric field meter or voltmeter located outside the conducting shell. Plasma wave detectors measure the slowly varying electro-magnetic fields of collisionless shock waves where interstellar medium encounters the heliosphere. Many other of the preceding measurements are wanted at this heliopause, and most must be measured far from the starship to avoid interference from fields and shocks around the starship structure.

Cosmic particles over a wide range of energies are of interest. Some travel too slowly to enter the solar magnetic field for local observation. Elaborate apparatus is needed to measure high energies (some protons carry over 1 joule!). Interstellar magnetic fields scramble ion directions too much to identify any sources. Gamma- and x-rays, actually photons, are also of interest and can be traced to their sources. Neutrinos are of interest from stellar and artificial sources (such as antimatter drives in use by "others"), but detectors have required a hundred tonnes of special material to detect one per day. Space is also permeated with slow neutrinos, undetectable thus far.

Other evidence of other civilizations is worth watching for in the low-noise interstellar environment, given the time and instruments, and this work is very appropriate for a mission on its way to find such a civilization. Besides radio signals, intentional and leakage, there might be laser signals, unusual doppler shifts in radio and optical point sources, unusual infrared radiation, spectral lines of elements not ordinarily found in nature (artificially radioactive elements from fission or fusion power plants), spectral evidence of high concentrations of medium and heavy elements in space where they might have been thrown (e.g. from ion drives), a wide variety of specific x- and gamma-radiation energies from various radioisotopes, and evidence of high magnetic fields [some suggested in a-Freitas 1985].

Large-scale observation of mass is possible by tracking radio signals from the departing probe or starship. The mass of the Solar System can be measured on as large a scale as signals can be tracked in direction. Any drive must be off so that residual acceleration of the starship is detectable, and this requires good control and designated off periods. Most kinds of drive would also interfere strongly with ion, electric, magnetic, and radiowave measurements, especially if high electric power is involved. Any nuclear drive or auxiliary power supply emits radiation (or its neutrons cause nearby structure to emit radiation) to produce false readings in the relevant parts of the x-ray and gamma spectrum. Neutron shields are thick and massive. Sometimes it is sufficient to shield just the instrument. And neutron-sensitive materials should be avoided in starship construction.

6.4 HUMAN MISSIONS

A human expedition, as distinguished from a colony voyage, would expect to return to Earth for the simple reason that there is no other known place for a hundred people to go. A suicide mission, where a small group of people travel until they run out of resources and choose some painless end to their voyage, is very unlikely. A mission that studies several likely stars might be in space thousands of years and return after people on Earth had lost interest. Earth society must maintain continuous or scheduled radio listening for that period of time to benefit. Some interest would continue if regular and fascinating radio transmissions are received. When a starship returns, interest would increase. Expeditions are thought to depend on Earth support, but after departure support diminishes to zero because of increasing communication delay time or inability to receive communication. If the mission depends on Earth for beamed propulsion, then the dependency is long-lasting and fragile. Although the mission plan may originate on Earth, the responsibility and control thereafter is in the hands of the voyagers. They may prefer to control propulsion themselves, a preference at odds with Solar System-based laser or sunbeams. If the expedition can choose to return to Earth, they will understand that they return to a foreign world.

A primary question is the prospects for a human mission to a near star. One hundred people has been taken as the minimum population to travel. With great difficulty and a drive more powerful than seems feasible, the size could be reduced to as low as ten hardy determined individuals. But the mass and cost of the starship may increase, due to the greater energy needed. Splitting 100 or more voyagers among 5 to 10 smaller starships may be worth the extra cost because of the much higher chances of success. Given 100 people and a 10,000 tonne starship (plus fuel) [est. also in a-Matloff 1980], the questions become when and how do they get there, and do they return. Because of size human expeditions are likely to travel as slow as 0.01c. A slow mission reduces the need for shielding and may permit a disk-shaped vehicle wide enough to accommodate a rotating habitat. Final data return to Earth would occur about 1000 years later. The demonstration of a 0.1c drive will much improve this, but 0.1c may raise the cost so much as to force a slower mission (as shown in earlier discussion of optimization). Even at 0.1c, a hundred year voyage must involve 4 generations, many of whom will not witness the end. As shown later [Chapter 10], the population must at times rise as high as double that planned.

Before a human expedition, one or more probes should successfully reach other stars and return data showing a place humans want to go. Or signals from another civilization have described a place humans can visit, preferably with an invitation. Else the mission would leave with much uncertainty and may lack flexibility in energy, propulsion, supplies, and equipment. The probes would also provide proof of methods of propulsion. Carrying fuel for several stops is out of the question. Even the difficulty of refueling at an unknown star system seems great and extra equipment might be needed. If a fuel station could be ascertained beforehand, then a starship with a stable small society committed to interstellar travel could continue to the next likely star. If the expedition depends on outside equipment such as mirrors for propulsion, then the task of exploring farther becomes very difficult because such equipment must be carried or built there. An advance robotic mission might carry a mirror system but the mass would be as large as for a human mission.

It may turn out that reliable probes cannot be built, or that none are heard from. Then Earth and local space society faces the decision of giving up or sending humans on a first mission on the basis of weak evidence of interesting planets obtained by large telescope arrays. Once a propulsion method is shown workable locally, there will be interest in a human expedition however risky. Humans carry the ability to monitor a starship carefully, process and study much relevant science along the way, repair many kinds of failures, overcome many kinds of emergencies, and generally may justify the expense of sending them to obtain a successful mission. Because of the long travel time, a major mission goal for most of the mission is survival, as enjoyably as possible.

Any mission capable of visiting more than one star is capable of returning to Earth. Returning an expedition to Earth is more difficult if propulsion energy must be carried the whole way. The mass ratio for a one-stop two-way voyage is between perhaps 10 thousand and 100 million to one. The best plan for return is a propulsion method which supplies energy from outside or acquires return fuel at the destination. Even less likely than returning is changing the mission plan as successive generations come into control. Accumulated momentum and scarce propulsion energy prohibit this.

No gain comes from using propulsion to go around a star without stopping. Rocket energy needed to travel in a loop that returns home is about as much as that needed to stop at the destination, turn around there, and accelerate toward home. (A reversal of outgoing momentum requires passage through a state of zero momentum.) Moreover, the sideways (radial) gee needed to turn in a partial circle of Earth orbit size at 0.01c, close enough to study a planet, is an unachievable and unendurable 10 gees. The difficulty is much worse at higher speeds and not much better for a larger turning circle. Generally there is a problem with making any turns. The star cannot help much. At 0.01c the starship must pass within about 1 million km of the center of a star of Sun-size mass to be deflected a mere 1°. A star with a close companion permits substantial gravity braking, although such a star system is unlikely to have planets of interest and would not be a good destination. Were such a system visited, the two stars could be departed with almost double their orbital velocity.

An unlikely option for a return mission is to use the charged-wires Lorentz force method and travel in an ice-cream-cone-shaped path from Earth around the star and back. The method is uncertain because the magnetic fields are weak and uncertain, the wires are very long, the trip is several lightyears longer, and at slow speed the Lorentz force is weak. A mission departing by sail could be halted by sail on its return. If the lightbeam extends many lightyears, the mission could possibly use it to stop at the destination and to reaccelerate to return [recall Chapter 4]. If the mission cannot stop, then centuries of travel are needed for a few days of close observation, a low-gain situation even if two stars could be visited on the same loop.

An interesting change in attitude toward a long mission would occur if biotechnology develops to the extent that people can hibernate as long as a thousand years or simply live that long. Given the choice most would choose deep sleep, thus reducing the mass, size, and resources the starship must have also. Some would take turns monitoring the long journey and hibernating. On Earth people will have long forgotten the mission, especially if exciting weekly reports are not received. Those who monitor the mission might want to sleep it out too, and be available at the end.

A one-way colony voyage cannot be sent until a destination is known, or until a starship can conduct searches for planets at several likely stars to find a suitable one. The mission will be slow and involve many generations. Thus the starship is sometimes called a "generation" ship. A 100 person expedition is a small colonizing mission, but high costs may require that colonies be started by a small number of people. Automated machinery can help with the work, and a wide variety of frozen human germ cells from Earth can help the gene pool. In a very serious colonization program, increased success would occur if several starships are sent together or in a series. A series of starships could bring people and supplies needed at later stages of colonization at a slower (cheaper) speed [a-Hodges].

Detailed planning of what colonists must take to colonize is better left until many facts about the destination planet are known. Room is not likely to be available to send bulldozers and other bigger equipment, and colonists would need to start nearly from scratch, camping out for decades using native materials until manufacturing ability is slowly built up. Dissembling the starship and bringing it to ground requires more landing capacity than can be provided. A colony mission could be prepared to build bigger space colonies from raw material if no habitable or no terrestrial planet is found. Advance probes may need to prove the quality and quantity of material in space there. Perhaps voyagers have adjusted genetically to zero gee over many generations and cannot or will not attempt planetary living.

The effects of relativity on missions are fun to contemplate although it is not clear that starships will ever reach high gamma. The advantage of slow time passage aboard ship has been much discussed. A related effect is the planning of stops. Most time is used starting and stopping. If 6 proper years are needed to travel 20 LY at 1 gee and stop, followed by another 6 years to reach a star another 20 LY away, the 12 proper years aboard ship could be used to cover 572 LY in one jump (at 1 gee). However, the energy cost is much greater because much higher gamma is reached. It is also possible to go some distance to a star and discover that a later expedition from Earth arrived there somewhat sooner, a result of higher speed, not of higher gamma. Regardless of speed, a faster starship is unlikely to have enough spare power to slow down and assist an earlier, slower, or disabled starship.

6.5 SIMPLE PROBES

Probes are likely to be one-way, probably one-leg missions. They will "fly by" their destinations or "fly through" expected planetary systems. At 0.1c a probe will spend only a few days in a large planetary system, a few hours in a small one. Slower probes would be more productive if mission sponsors could wait that long, or find low-mass effective brakes for fast ones. The probe needs large, high-resolution optical telescopes if any planets are to be photographed (electronically), a requirement in conflict with very small probes yet a principal goal of a mission. Perhaps radar would be used, which can scan from a low-mass array of small transmitters--no dish, no lenses, no mirrors. Even if the probe halts, no sample return mission is likely, since sending massive analysis equipment is faster and cheaper.

Probes may need to be small for feasible travel and realistic cost. One tonne is nominal, 1 kg has been suggested, and the 450 tonnes of the Daedalus is surely excessive unless people want to wait centuries for the first interstellar mission. Some Daedalus estimates of 100 billion dollars seem far less than the likely costs. To obtain results as soon as possible, speed should be 0.8c to 0.9c. Going faster so that gamma becomes substantial is of no benefit to us on Earth. There is little doubt that the probe must be a flyby, although it could visit several stars if small course changes can be made. Just how much mass could be sent to a near star at 0.9c at a cost in the hundred billion dollar range, and what propulsion would be used? The answer must begin with the energy requirement.

If 10 Gw of power (10 electric power plants here or one solar power plant in space) were dedicated to storing or sending propulsion energy for 10 years, the total energy available would be 3(10)18 joules. Purchased from existing power plants this energy would cost about 100 billion dollars. A program to build in space 10 Gw solar power satellites (SPS) would surely cost as much for the first station but less for additional ones. This subject has been thoroughly explored by NASA and others [e.g. b-DOE & NASA]. Clearly the rest of the probe program would need additional money, perhaps a similar amount.

Given about 3(10)18 J and the need for speed 0.9c (gamma about 2), the mass of the probe would be about 30 kg. If 30 kg is workable, then 10 kg is, and 10 kg could be chosen as design baseline for a starprobe program that could be started tomorrow. Major reduction in size of instruments, computing, and data storage would be required. The probe would be useless unless we can receive its data, and a large low-mass antenna and radio transmitter are necessary. As to be shown in Chapter 7, a megawatt of transmitter power from a 100 m dish may be needed to be detected here from 10 LY in a similar dish. Small optical and radio telescopes are needed to make observations and would also help the probe keep track of where the Sun is to aim its antenna precisely. Onboard energy supply for decades is a serious problem. Perhaps after being boosted to full speed by outside means and carefully aimed, the probe could fly dead without power until it reached the next star, then revive on solar power there (which would be strong only near any inner planets). Methods of sending power by lightbeams to it should be considered. Then the antenna can also be a sail and power receiver with none of these functions interfering with each other. If 1 kg were devoted to sail/antenna, it could be 100 m diameter. To gather 1 Mw of local solar power with this size at 10% efficiency (thermal generator or very thin solar cells), it would need to enter regions where the star is as bright as the Sun is to us here, a condition which would start too late and hold too briefly. Microwave power from here is out of the question because of diffraction over 10 LY (but see later). A laser (deep ultra-violet for less diffraction) or solar lightbeam may be feasible, although no such laser is now known for the power level required.

. The probe could be accelerated at high gee by lightbeam, but probably not by magnetic driver because of the millions of kilometers of track needed. However, the light source will be expensive and must be as small as possible, so that acceleration should be extended as long as possible. Acceleration could be done at constant power for a year or more, requiring 30 Gw. A laser installation would be very big, and a sunbeam should be attempted instead. A mirror just 1/2 kilometer in diameter could provide a sunbeam at this power from a 15 million km position from the Sun (as discussed in Chapter 4). Onboard power remains a serious problem since the transmitter may need a megawatt. At 10 LY away the small portion of sunbeam still hitting the collector would provide sufficient power if efficiently converted into electric power. A very low-mass power converter at about 1 Mw/kg is needed. Thermionic devices at the focus of the mirror would be far too massive. This probe could also use about 20 kg of antimatter as fuel in a miniature efficient drive of yet unknown form. The launch mass would obviously be larger but feasible. The fuel cost is horrendous compared to other methods for the same 10 kg probe, perhaps 200 trillion dollars.

No optimum probe design is offered here. Several other designs have been suggested [see laser-pushed ones in Chapter 4; a-Forward 1984]. The Daedalus 450 tonne probe (plus fuel and drive) has been studied in depth and shows feasibility in part but at enormous cost. The Daedalus study settled on fusion pulse power and projected a speed just over 0.1c, with a trip time of 50 years [general mission description a-Bond 1978]. This probe is planned to carry over 200 tonnes of subprobes in an attempt to gather more data despite the speed of flyby. With breakthroughs in fusion drives, a lower-mass system can surely be designed, and one that would use fuel gathered on or near Earth. A one tonne probe at 0.1c might be a better design focus than the 10 kg version above, requiring the same energy as the 0.9c probe but having more possibility of carrying the instruments, antennas, and means of collecting power. A medium small (63 kg) probe has been proposed to consist only of a flat 1 km solar sail carrying circuitry on its back for scientific observation and data return [a-Matloff 1981]. It would be started near the Sun at some 400 gees and reach 0.01c. How it maintains stability and steers does not seem to have been worked out.

The lowest mass proposal is "Starwisp" [a-Forward 1985], 0.016 kg of wire mesh circuitry boosted to 0.2c by 10 Gw microwave beam (about 100 gees) and sent to Alpha Centauri. The power source is intended to be a feasible solar power satellite. Microwave wavelength of 3 cm would push mesh with hole size 0.3 cm, and most materials would be superconductors. Microcircuits at each mesh position, presumably operating at the 3°K "temperature" of space or at higher temperatures near stars, would handle both optical and radio signals so that visual information at the destination could be converted to radio signals sent to Earth. The probe would not decelerate or maneuver. Upon arrival all operating power (just 2 watts) would come from reception of a small part of the diffuse microwave beam that reaches the mesh across 4 LY. A 50,000 km microwave fresnel lens must be used at our end to provide the beam, and to collect the faint returning signal. The mesh serves as reflector, receiver, and phased-array transmitter by actions of the microcircuits. It needs to be about 1 km diameter, and the number of microchips is about 100 billion with total mass 0.004 kg. These are very optimistic projections of our present technology in electronics, distributed data processing, and large space structures. Scaling up to 30 km sail with mass 14 kg pushed by 10 Tw was also studied for a

0.5c probe to Epsilon Eridani (about 11 LY). Earlier a 100 m mesh sphere with mass tens of kilograms was suggested with similar electronics [a-Forward 1976].

If fusion and light- or microwave-powered probes prove unfeasible in the near term, what power source and drive would be suitable for a small probe? Ion drive has low thrust and can be adjusted in exhaust velocity for more efficiency, but its mass is high. The only remaining nuclear energy source is a fission reactor which would provide electric power through a thermal cycle or by thermionic devices. These are known technologies, and a very optimistic design has a probe reaching 0.01c for a 400 year fly-by voyage to Alpha [a-Aston]. Low acceleration would continue for 65 years, seemingly the upper limit of possible reactor and drive reliability. Probe payload mass would be 5 tonnes and initial mass over 400 tonnes, about half mercury propellant, the other half reactor and drive. The 1 Gw ion drive has been discussed [Chapter 2] and represents a major up-sizing of present ion accelerators. The reactor performance of about 100 kw/kg pushes nuclear fission reactor technology rather far, but fission is theoretically capable of providing at least 0.01c. Reducing the payload and therefore the vital radio transmission antenna will not help because the reactor mass will not shrink much. Most of the reactor is devoted to moving itself, a hint to try redesigning. It is not clear whether a reactor can be revived after 400 years to provide enough power for data return.

6.6 ADVANCED PROBES

Launching probes has been discussed as an alternative to transmitting messages to selected or random stars [a-Bracewell; a-Freitas 1980, 1983; a-Oliver]. This book can only sample the large subject of searching for and communicating with other hypothesized civilizations (SETI & CETI) [see also Chapters 7 & 11], but a comparison of sending radio signals with sending hardware is instructive. Only some of the arguments can be summarized here. Since radio communication is aimed at finding intelligent technical civilizations, comparison is made with probes for the same purpose. (Probes can have many other purposes.) An ambitious long-term probe program is assumed, with extensive radio monitoring on or near Earth to receive results. The discussion also assumes that probes have or can obtain abundant energy if not extra material for their missions. Because of the enormous rocket mass ratio, the possibility of a probe bringing physical samples back to us is not considered further.

Some reasons favoring probes are the following:

a. A radio signal not replied to tells us nothing about "who" lives at the star system, but a probe visiting it can almost certainly determine if beings with technology live there and provide many details on "them".

b. Probes can carry strong transmitters directly to the star system of interest and operate them in immediate automatic (if limited) response to local conditions and behaviors.

c. Smart probes on site can carry on complex conversations in real time.

d. Probes can choose how much effort and time to expend at one star before moving on, depending on what is found (assuming a powerful drive is available).

e.   Radio is faster but the wait between replies may be as long as the time for a probe to travel there.

f.   A probe protects us from dangerous situations, since it can disguise its purpose and origin while warning us.

g.   Probes can become data stores about us (or monitor and meddle, "2001 monolith" style) if they simply wait in protected places at star systems until detected life evolves to civilization.

h.   If everyone out there is just listening and waiting, then the initiative goes to those who send something that gets attention and at least tells where and how to transmit messages back to us.

i.   Probes require less monitoring by us than listening for unpredictable radio signals. We would know about where they are and what frequencies they use.

Some reasons disfavoring probes (and some rebuttals) are:

a.   Probes are a long-term expensive program, and up to a point their hardware and launch costs more than carefully chosen megawatt radio transmissions.

b.   Probes cannot reach high gamma but this does not matter since time aboard a probe is irrelevant (except for preservation of equipment that needs to work nearly perfectly nearly forever).

c.   Probes may be too difficult to make long-term reliable. If none from elsewhere have made themselves known here, this is further evidence (but incomplete and inconclusive).

d.   Practical probes may be 10 to 100 times slower than radio signals. But if probes take as long to get there as the time for new civilizations to develop, say thousands of years between contacts, then radio can do no better while probes are already at the scene starting communication without delay.

e.   Probes may be so costly only advanced civilizations can do them. Let them do that work and we wait, watch, listen, and transmit prudently.

In the compromise position is the fact that radio and probe programs go together. Messages should be traveling (on tight beams) between "others" and their probes. We might intercept such a signal. And both probes and radio searches require similar monitoring equipment. Some further arguments presume self-perpetuating and self-reproducing probes, as covered below. These might, in the long term, lose track of where Earth is, or their efforts are wasted because Earth no longer is interested or exists. A very long-term and complete probe program (millions of years, millions of probes) would be needed to learn all that can be found on the incidence of life and intelligence in the galaxy. All this assumes very rich and patient civilizations.

Calculations that compare the energy cost per unit of information sent find probes at no disadvantage, but assume that every atom of material in the probe conveys useful information to the presumed recipient/finder [a-Freitas 1980, 1983]. The theoretical maximum information transmission rate has been estimated at $(10)33$ bits per second per joule. This is very optimistic since it is very near the inverse of Planck's constant, the quantum limit of about $(10)-33$ J s. At the least, the radio vs. probe argument shows that a society committed to SETI should invest a similar amount into probes. The question of whether we should search for probes from elsewhere coming into our Solar System is partly a topic beyond this book (as a contact from "them"), but finding one would have enormous effect on our plans for interstellar travel [see Chapter 11]. In designing our probes, we must consider whether and how we expect others to find them.

A bizarre sort of probe is the biological one, sending "seeds" of life across the galaxy or to selected star systems, hence the imaginative program name "panspermia". Some believe that "spores" of life travel among the stars perhaps without anyone's help, being driven high into the atmosphere, having developed radiation and vacuum protection, then drifting by light pressure until another habitable world is reached [classic: a-Arrhenius]. If one habitable planet exists every 10 LY, then the planet occupies about (10)-30 of the volume of space. A very large number of "spores" are needed to provide just one at each nearby planet. A magnetic driver has been proposed to send "pods" of such spores more efficiently to selected star systems [a-Zuckerman]. As seen, the driver would be very long and expensive unless speeds less than 100 km/s are chosen. Arguments for this program include: someone else may have done it to us; we can prepare nearby planets with programmed life for our colonies (over a long period to time); we ought to do it for the good of the galaxy; or the galaxy needs life, especially if we are the only life thus far.

While DNA can be constructed to contain a very large amount of information, it remains to be studied just how much is needed to take over an alien planet that has water, warmth, and nutrients. Plans to directly hit a detected planet seem virtually impossible, but panspermia could still occur simply by filling the vicinity of a star with spores. A solar sail program to do this has also been described [a-Meotner]. The 200 m 100 kg sail would move about 0.0001c and carry 10 kg of micro-organisms. The physical feasibility of directed panspermia seems high because it could be done at Voyager speeds and moderate expense. The biological success must necessarily never be very certain, but studies have been done on spore survival in space [a-Greenberg]. The first "seeding" would occur 100,000 or more years later. We might even acquire the ability to travel there first.

The feasibility of panspermia is important because if it is easy to do, we must research more carefully our origins [a-Crick]. Also, a new realm of ethics opens up: Are we justified in so dominating our part of the galaxy with our own DNA, perhaps competing with local life elsewhere and wiping it out. If we learned from extensive probes and other research that most habitable planets remain forever sterile, or that life rarely advances beyond rudimentary levels, or that some planets do not have time to evolve advanced life, these and other factors might affect a distant future plan for sending out our life forms. Some say that if life is very rare we have an obligation, and if it is very common, then our origin may not be independent. This is an interesting dilemma posed by life (ours) striving toward interstellar probing and traveling.

6.7 SELF-REPRODUCING PROBES AND DIFFUSION

One of the most fascinating ideas for interstellar exploration is the self-reproducing or von Neumann robotic probe. It is named for a design of a theoretical "machine" (on paper) which could reproduce itself from a stock of theoretical parts [b-von Neumann; a-Kemeny; a-Moore 1964]. The concept is far more general than interstellar probes. Very little hardware has been built, but much speculation has occurred. Limited chemical and mechanical systems (including blocks and toy trains) have been built which rely on a supply of building blocks at hand [a-Penrose], and a hypothetical general factory has been designed for Earth use [a-Moore 1956]. But thus far only nature has produced successful "machines" (life and some crystallization processes) which can assemble natural materials into working copies of themselves. The challenge of a sophisticated long-lived self-reproducing machine such as a probe is immense, but the payoff is also immense. Were one general von Neumann probe made as functioning hardware, all further work could be left to it. The only expense would be the very large program to develop and build the first probe. Were this problem even partly solved, probe and starship self-repair and self-improvement would be easier.

For von Neumann probes, the general problems are: design an adequate drive for which fuel and parts can be found near almost any star; package all the components of mining, manufacturing, assembly, storage, modification, quality control, computing, information, and more (in short, most of the apparatus of a high-tech culture) into a reasonably small vehicle; plan machine "intelligence" that can insure long-term survival, reliable operation, and adjustment to unforeseen events; produce one such machine to start, with superb long-term reliability; prepare the extensive computer codes (programs) to control this system; provide for the machine to copy its program as well as its hardware and get the two to start working in the new copy. If all this is done, then there are few limits to the number of generations of these probes, with a rapidly growing number in each generation. One wonders why the galaxy has not been mined-out long ago with trillions of these things running everywhere including coming here. Of course the purpose of these probes must be more than blind reproduction. They would also carry out all the previous functions of probes for exploration and making contact. They must look ahead from their base with large telescopes and choose their next destinations. Chapter 8 covers more technical limits on the idea of self-reproducing machines.

An elaborate paper design along these lines has been published [a-Freitas 1980]. Only some detailed description can convey the question of feasibility of this concept. The traveling probe is to consist of some thousand tonnes mass of very complex starter equipment. It is to travel by pulsed fusion à la Daedalus, at the same speed 0.1c and mass ratio 100:1, but also to stop, therefore needing another bigger stage and 10,000:1 mass ratio. Thus a standard 2-leg rocket mode is used. The initial vehicle, and its copy to be built later, would mass about 10 million tonnes, almost entirely D-He3 fuel. Thus shopping for fuel would be the major activity of the von Neumann probe. Other scenarios would be much more desirable on this score. Attempt was made to minimize functional probe mass for this reason, but 1000 tonnes seems reasonable to package much of a high-tech production system.

The first probe would be sent to a star with a gas giant planet with moons, and part of the probe would land on a probably airless moon. The moon must have a full range of natural minerals, and the planet provides the fusion fuel for the next jump. The landed probe would assemble a factory complex from stock parts and local material over 500 years, which would begin making as many new complete probes with fusion rockets as local material will allow, one at a time. The proposed cycle time is 500 years. A 400 Mw starter nuclear power plant is part of the initial equipment, and 5000 Mw of fusion power plants are to be built on site.

The proposal works out in detail how the full cycle might be done. Functions are assigned to about 10 specialties each with its robotic equipment, including mining, metallurgical, and chemical work, parts production and assembly, parts storage, mineral and parts transportation, fusion fuel acquisition, computer oversight, on-site supervision, correction of unforeseen problems, and quality testing for all parts and systems. Planning must be basic, starting with estimates of the amounts of every chemical element needed. Everything must be done in the proper order so that the parts for each sub-subsystem are ready and assembled before that system is needed. Clearly the whole system must be very reliable for a time of the order of 1000 years. After that, with sufficiently "intelligent" computers, it can be self-correcting and self-improving. It would be nice to scale this plan down to an affordable level that can be built and sent to the asteroids on existing propulsion, for a large boost in our slow space program. But ultimately the plan must be based on a moderate to fast drive, at least 0.01c. Effort to reduce the cycle time is unnecessary unless the travel time is similarly reduced.

Thought has been put into comparing the virtues of von Neumann and ordinary robotic probes for rapidity of galactic exploration [a-Valdes]. Most of the arguments center around SETI. The general conclusion has been that von Neumann probes would make more "contacts" than ordinary probes in long-term exploration involving very large numbers of stars. The probe speed has little effect on this result so long as it is over 0.01c. When cost effectiveness is considered and a von Neumann probe is assumed 1000 times more costly than an ordinary one (to start), then the above result still holds. If these results pertain to the engineering plans of other civilizations, then finding a probe from elsewhere permits deduction of the scale and distance of the civilization that originated it. Lack of evident von Neumann probes arriving here has also been taken (hastily) as evidence that other civilizations do not exist [a-Tipler]. An unusual suggested use of von Neumann probes is the reproduction of biospheres. The cybernetic capability of the probe would be turned to preparing an environment on a suitable planet so that organisms from Earth can survive and support the introduction of machine-nurtured humans [a-Fogg]. Or all organisms including us can be manufactured from local chemicals, using DNA and other codes. This "resurrection" machine is quite a different approach to colonizing the galaxy. Readers can ponder whether it could happen here.

Probes must proceed in jumps of 10 LY or so. Indeed, interesting stars would be skipped if jumps were larger. Probes would travel out in three dimensions from one point of origin (Earth) just as human interstellar colonization would expand.

Elaborate analyses of how reproducing probes (or colonies) diffuse through space have been published [e.g. a-Newman; a-Jones 1978], concentrating on how fast the wave of population moves. Some find probes or beings would require nearly the lifetime of the galaxy to disperse through it; others find that this could occur in a few million years, depending on assumptions [see Chapter 11]. Some assume civilization spreads almost as fast as travel at 0.1c, others that the recovery time before the next probe or colony mission is lengthy. But some elementary aspects of expansion spherically in three dimensions may have been overlooked.

A simple numerical case is given here (with pi approximated as 3). Suppose for simplicity that 4 stars of interest occur spread out approximately equally in 4 directions and are all at the first 10 LY shell from the origin [see Figure 6.3 & Appendix D]. The area of this sphere is 1200 square LY (SLY). At each of these stars the old probe makes more probes as needed with range 10 LY. Each probe can cover an area of 300 SLY sideways in seeking more destinations. The 4 existing probes can cover the whole region of space, and no more are needed at this shell. Suppose again for simplicity that the density of stars continues at 1000 cubic LY (CLY) per star and stars are all located on spherical shells at 10 LY intervals. The model can be made more realistic simply by varying star positions a few lightyears in and out so that their shells represent average positions. At 20 LY out, there are 8 times more stars or 32 stars, minus 4 already visited, leaving 28 new ones. The 4 probes look ahead and calculate that they must each produce 7 new probes (retiring the old one). Upon arrival each probe finds (in this simple example) that its range area is almost 1/16 of the area of this shell and therefore the 28 probe ranges overlap. Probes were overbuilt.

Nevertheless, the 28 probes look ahead to the 30 LY shell of volume 108,000 CLY where 108 likely stars exist, of which 76 are between 20 and 30 LY. This time some probes must produce 2 new ones and others 3 new ones to average about 2.7 new ones per old one. From the 30 LY shell 148 new stars are seen ahead at 40 LY in the 256,000 CLY volume. The 78 probes need to produce just less than double their number. At 50 LY each probe must produce 1.6 probes on the average. After the early burst of probe building, it should be clear that the needed rate of reproduction is decreasing at each jump of 10 LY and will approach 1:1 at very large distances, with only an occasional extra needed. The range sideways that each probe must cover also grows smaller. If at 20 LY probes had reproduced at 1:4, or even 1:2, instead of 1:7, no shortage of probes would develop.

According to this model reproducing probes is inefficient because the enormous investment in reproducing capability cannot pay back with rapid expansion of probes. There is no place for many probes to go. Better simply to have self-repairing probes, take occasional side trips, and skip a few stars. Or to have one or two large robotic probe factories near here at fuel sources and send probes to coast for hundreds or thousands of lightyears. For colonization, the waste factor is irrelevant because, unlike von Neumann probes, people and culture already reproduce easily. Later colonies will find little expansion room for them, just one or two opportunities farther out. New restraints occur when the spherical wave of expansion reaches the sides of our disk-shaped galaxy and can proceed only inward and outward in two dimensions.

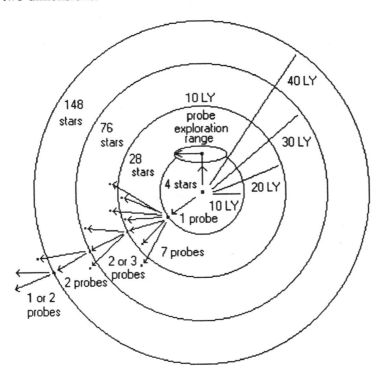

Figure 6.3 Spherical expansion or diffusion by probes or colonizing starships into galactic space. Cross-section of spherical volumes of space shown. About one star of interest occurs per 1000 cubic LY, and counts per volume are shown.

6.8 LARGE COLONY SHIPS

Robert Enzmann in 1964 may have been the earliest to propose an "ark" starship, indeed a small fleet of them powered by small fusion bombs [b-Ridpath]. The deuterium fuel would be a frozen ball of ice, presumably in a tank and located ahead of the habitat as shielding. Soon after Gerard O'Neill's rotating space colony concept was developed between 1969 and 1974 for local use [a&b-O'Neill; b-Brand], a proposal came to make one into an interstellar "ark" powered at lower speed by a variety of methods such as fusion pellets, ramjet, and solar sails [a-Matloff 1976]. From there the idea of "world ship" grew. This term is applied to starships with populations from a thousand to a million and implies the large closed ecology inside [a-Bond 1984]. While a million people is large indeed, it is hardly world-size, and the more modest term "colony ship" is used here for more modest proposals of the order of 1000 people. "Large" is appended where confusion with smaller human missions, likely to be colonizing, might occur. Some have said "nomad ships" [a-Jones], and "community ship" might be better. "Worldlet" is a tempting term, but a better term will surely be developed sooner than the giant starship itself.

Before Enzmann "arks" came Bernal spheres [b-Bernal] 10 miles in diameter with no gee, traveling "easily" at high speeds as colonies to the stars, made from local asteroid and moon material. The history of this idea is long and rich [a-Martin 1984]. Froman in 1962 proposed (in jest) using Earth itself as the worldship, burning 1/4 of the protons in its water to blast us out of orbit and keep us warm and lighted. More seriously, the biological and human aspects of large colony ships are too many to consider here. A few biological and social aspects are considered in Chapters 9 through 11. Developing large space colonies near Earth and around our Sun is important precursor experience for sending small and large colonies elsewhere [a-Singer].

A large colony ship solves the time, variety, resourcefulness, gee, and ecosystem problems by being large (but slow). There would be many people of many sorts, with plenty of room, an ecosystem sufficiently large to be stable and productive, and a habitat that can be spun for artificial gee. Indeed, each ship needs two counter-rotating colonies so that their angular momentum cancels to zero and does not cause various problems. The voyagers are not in a hurry, and are prepared to live comfortably through many generations before a destination is reached. Starting life on a hostile planet would need as many people as possible, backed with large equipment and many landers, the latter being difficult to make or carry. Conceivably, if new material and energy sources can be obtained along the way, the colony ship might prefer to cruise the galaxy for millions of years, perhaps stopping at rich or inhabited planetary systems just to visit, to build more colony ships, to replenish, to get directions, and then move on. This is largely fantasy because no sufficient energy sources are known for this lifestyle.

Although they travel slowly in style, large colony ships would require propulsion energy in proportion to their large mass [a-Matloff 1976]. They seem to be a much later phase of interstellar travel, after humans are living locally in similar colonies in space and have a society and economy large enough to work on preparing one for travel where there is no sunlight. Many shapes have been proposed besides simple cylinders, all rotating for gee, some filled with water [a-Bond]. A proposed "Mark 2A" design is about 15 km diameter, 88 km length, rotates once every 3 minutes, masses over 100 billion tonnes, optimistically needs just 5 times that mass in fusion fuel, travels at 0.005c, and operates at over $(10)15$ kw. Inside area is nearly 4000 square km, and could support half million people at very comfortable density.

Whatever the energy source (fusion is usually proposed, sometimes as full-scale bombs), light for agriculture would come from electric lights, and waste heat would surely be sufficient for warmth. Structural considerations show diameters could be at least 15 km before the slowly spinning worldlet is overstressed. As to be discussed, shielding from cosmic particles can be 1 to 2 meters of solid rock or slag supplemented by the metal shell and the fertile soil inside. The atmosphere is large enough for clouds and rain. If many other civilizations are doing this, we should see them flashing their way through space.

Studies have been done on the use of sails to pull large colony ships [a-Jones]. Some propose almost unthinkably large lasers for propulsion. Others seem a bit more practical in starting the voyage very near the Sun where many large sails would start pulling a cluster of sub-colony ships at 2 gee or more away from the Sun for an ultimate speed of 0.005c to 0.01c [a-Matloff 1981, 1985 twice, 1986]. Ion drag followed by lightsail drag is expected to decelerate the colony after a typical 1000 year travel time. The large problems with this mission plan have been covered earlier. Habitat power supply would seem to require developing fusion power. Supplemental drives to boost speed seem inadequate unless they are very massive and used as the starship drops around the Sun. But the sail-pulled large colony approach all in all seems as feasible as any.

Other big slow ways to travel include riding on comets (the icy nucleus without tail) [a-Jones 1983 twice]. Comets may be very common outside star systems and provide a chunk a few kilometers in size with 10 billion tonnes of valuable materials to work with. All would have the substances needed for life, especially water with its deuterium fusion fuel. Their rocky cores may have fissionable heavy elements. Their average separation is guessed to be about a billion to a trillion kilometers, making them difficult to find. A voyage would use several comets in a cluster, powered presumably by fusion rockets and collected from a large volume of space. Proposals to gather solar energy far beyond Pluto do not seem realistic however. Instead of accelerating comets, some might be found already traveling elsewhere, although very slowly. For faster travel modest starships might hop from comet to comet throughout interstellar space, using their resources for fuel at the expense of stopping frequently [a-Stephenson]. Comet colonies would better be societies in themselves than planned as missions to somewhere. Traveling comets would not last indefinitely, and voyagers must acquire fresh material and energy somewhere.

The need for colony ships might refer to a need to expand human civilization, or just to escape Earth, or to find a better life, or to be prepared to survive catastrophe on Earth, or to join with other civilizations (as an embassy perhaps). Catastrophe is not in the near future going to prevent colonization of space near Earth or around our Sun. The idea of colonization itself should be considered carefully. Interstellar travel will not be able to reduce population pressure because of the cost of travel. Other civilizations in some way must master their reproductive potential, else they would not last long enough to try stellar exploration. Human colony ships must control their population too, since their environments are much more limited than Earth's. Beyond colony ships is the possibility of human galactic empire, the result of an advancing spherical wave of colonization similar to the spread of probes discussed earlier [see also a-Hart & Chapter 11]. Empire is a subject beyond early interstellar travel except that other civilizations are expected to have occupied most of the galaxy and already found us [a-von Hoerner]. An intergalactic "ark" has also been described with galactic-size requirements for propulsion and a multi-million year journey [a-Fogg].

Chapter 7

Astrogation, Observation, and Communication

Guiding a starship requires knowledge of practical astronomy, both optical and radio, and much more. This chapter introduces some concepts and practices, partly to explore whether there are any reasons why we or others cannot astrogate over many lightyears. Distinctions between the operation of probes and human starships are usually not attempted here, but differences could be crucial. Some technical aspects of SETI are given here as part of radio methods. References to "Earth" usually imply the general location of Earth as part of a planetary system whose Sun is the only visible component detectable from afar. "Earth" may also refer specifically to a human-operated interstellar mission center, whether it is located on Earth or in local space.

7.1 LOCATION

Before interstellar travel there was only one "sky" for us, a dark apparently spherical background or "surface" on which a few thousand local stars appear as sparkling points. A starship will have its own sphere of space, with many more stars in different and very slowly changing positions. Despite the lack of perspective in deep space, stars will be more apparent as intense unvarying bright points distributed throughout an immense volume. But even at relativistic speed, the scene will change so slowly that voyagers may not acquire the feeling that they are moving among the stars.

No absolute reference or coordinate system exists for starship travel or anything else, with the partial exception of the uniform (to 1 part per million or better) background microwave radiation. A traveling starship shares our Sun's motion for most practical purposes, and a reference system could be located at the Sun. As discussed earlier, our Sun is moving about 22 km/s in a specific direction with respect to an average of the motion of local stars and gas. The stars around us are moving in various directions at various speeds and cannot be assumed fixed with respect to the Sun over times of hundreds of years. A typical star motion of 10 km/s relative to us results in about 0.003 LY shift in 100 years. This sort of motion 10 LY away results in a large shift of apparent position of the order of 6 minutes of arc. Star positions presently can be detected down to a fraction of 1"arc. The Sun along with local stars has a much faster motion of about 300 km/s around our galaxy, and our galaxy is moving about 600 km/s in another direction as discussed earlier [a-

Wilkinson; a-Dressler]. These motions do not matter for local starship travel and need not be measured or used in astrogation.

Common star maps list positions in terms of Earth-based coordinates (right ascension and declination, often of epoch 1950). A significant error comes from changing of the Earth's pole direction. This precession shows up as 50"arc per year change in the direction of North and a tilt of the equatorial plane. A several hundred year voyage would cover about 1% of the 26,000 years for Earth's axis to precess one turn. Converting known star positions to coordinates with Sun as center and Earth's orbit (ecliptic) as center plane requires a rotation of about 23.5° and seems the most useful procedure. This system is called celestial coordinates, with longitude measured from the direction of vernal equinox as before (where the Sun appears 21 March and ecliptic crosses equatorial plane), and latitude measured from the ecliptic plane. Galactic coordinates have also been defined with respect to the average plane of our galaxy and the direction from Sun to center. These are also angular directions as seen from our position and are not truly based on galactic center 30,000 LY away. Because astronomers have not planned on interstellar travel, no standard three-dimensional coordinate system in meters or LY has been established for listing star positions. But present spherical data can be converted.

Star positions as seen in the sky are measured now to about 5 digit accuracy, barely sufficient for astrogation. For a departing starship stellar distances, known only to 1 to 2 digit accuracy, are a major part of position information. Three-dimensional star positions must be measured much more accurately, at least to 0.0001 LY, before a probe or starship is sent [a-Richards]. The shifts in star positions as a probe travels are measured against background stars. Indeed the method becomes so accurate that background galaxies may be needed, but they are not sharp points of light. A probe which travels 1 LY from here permits measurement of nearby star distances to 5 digit accuracy (0.0001 LY) and distant 1000 LY stars to 3 digits without going to better optical technology. A bootstrap process is needed whereby probes traveling farther from our Sun provide the information to send the next probes more accurately. The first big starship should not be required to acquire accurate information along the way as it then may find too late that a major course correction is needed. Course corrections of the order of 1 part in 10,000 can be made as the starship proceeds, but larger corrections are costly at high speed.

7.2 ASTROGATION

A basic question is whether a starship can avoid getting lost. Early concern is not for voyages across the galaxy, so that our Sun is lost among 100 billion stars, but for local voyages and for probes. "Local" is assumed as before to refer to a radius of 10 to 16 LY around our Sun, but observers on the starship can see discrete stars across the galaxy, depending on telescope size. The brightest stars might be the best guideposts, and the farther the better. Of the 20 brightest stars we see, Deneb at about 1470 LY, Rigel at about 820 LY, and Betelgeuse at about 490 LY are the brightest far ones [b-Abell]. Distances to them are known poorly (to be improved upon during the mission) but their positions on the celestial sphere will change little during travel. A set of fixes on at least four stars over the whole celestial sphere is needed to track starship position, with those viewable to the sides being most useful.

Human voyagers can monitor starship progress and reaffirm the locations of Sun and destination. Probes depend only on computers and may have more difficulty avoiding ambiguous detection that results in loss of position of the Sun and perhaps loss of purpose (the human purpose originally impressed upon the probes). Fail-safe astrogation is especially critical for self-reproducing probes. Any probe which has an error in orientation due to malfunction of steering jets or gyro, or another failure, must acquire data on bright stars and sort out which ones are to be used for guidance before correcting its orientation. Probes must use an assortment of stars for fixes and might need to measure brightnesses and spectra and compare with prepared descriptions to identify them. Getting lost is unlikely for human missions closer than 100 LY but always a serious problem for probes. It also can happen to other beings traveling longer distances. However, no one has come by yet asking for help or a place to live because they had lost track of their home star.

Alpha Centauri is about as luminous as our Sun. Its apparent brightness is measured as a "visual magnitude" of about 0 as seen here from 4 LY away. A star is defined to be dimmer by 5 magnitudes for every 100 times less apparent brightness. Magnitude is measured on a logarithmic scale that depends on ratios. A star is about 2.5 magnitudes dimmer for 10 times less brightness, and 1 magnitude dimmer for 2.5 times less brightness. A star like Alpha, but 14 LY from our Sun and therefore about 3 times farther away than Alpha, would appear 10 times less bright than Alpha does to us, having magnitude about 2.5. The only near star seen from the northern hemisphere with luminosity similar to the Sun is Eta Cassiopeia (19 LY). It could also be used as an example here and appears at magnitude about 4. For our Sun to be difficult to find it must shrink to magnitude 15 or so, requiring it to be 100 times dimmer for each 5 steps in magnitude, or 1 million times dimmer than Alpha is to us. The distance must be greater by 1000 according to the inverse square law. Thus to reduce it to 15th magnitude, one must go about 4000 LY from our Sun, so that our Sun is not quite as obscure as sometimes stated. Less luminous suns can be lost much closer. [Appendix D shows how to calculate magnitude from luminosity; b-Baker.]

We can see stars unaided to about magnitude 6. Most sky surveys go to magnitude 17 or so, as does a 1 meter telescope. The Hale 5 meter telescope goes to magnitude 20 for eyes and 24 for film but is used mainly for extragalactic work. The Smithsonian catalog [b-Smithsonian] has about 1/4 million stars listed, without parallaxes and therefore without distances. A quarter million stars are expected within 400 LY of us, but catalogs go beyond this for bright stars, all the way to the "sides" of the galaxy about 3000 LY while missing many faint M and K stars closer. Many stars visible to a sensitive starship detection system are still uncataloged by Earth astronomers. The distances to most cataloged stars are still unknown. Many stars very faint to us would be brighter and important for astrogation near a destination star. Providing a starship with a nearly complete catalog of stars within 3000 LY (about half billion) seems very difficult, especially when distances are needed. A very large computer would be needed to track all these and present a selection of them on a map.

As the starship moves on a straight path (with occasional slight kinks as course is corrected), new apparent locations of stars must be accounted for [a-Moskowitz 1968]. Although angular positions of stars, such as celestial longitude and latitude seen in "spherical" space, seem most natural, astrogation calculations may need rectilinear (x,y,z) coordinates, where distance and the two angles are converted (with ordinary trigonometry) to 3 independent distances measured from three orthogonal axes centered on the Sun [somewhat as indicated in Figure 6.2]. Star coordinates as seen from the starship are then calculated from these with basic vector methods. Most of the work will be done by a computer, perhaps one dedicated to the task which can hold many visible star coordinates as a 3-D virtual star map. It can update its map with new observations, calculate new star positions according to their known motions, update those motions, calculate starship motion in real time (proper and Earth), and use relativistic mathematics for greatest accuracy.

Allowance for starlight aberration (apparent position change) must also be calculated from the speed, or if aberration is known, speed can be calculated. At 0.1c aberration causes an apparent shift forward of stars by about 6° for those located to the sides, and less for stars toward front and rear [see Appendix C for 3.3]. The apparent brightness is affected by high speed. At 0.1c intensity is increased about 20% in front and decreased by 20% behind [also Appendix C for 3.3]. These effects and others on star appearance have been calculated and depicted for selected fields of stars at various speeds from 0.1c up to ultra-relativistic [a-Stimets; a-Sheldon; a-Moskowitz 1967, 1968; see final section of this chapter for ultra-relativistic effects]. The view as the Daedalus mission would approach Barnard's Star has been depicted, and "snapshots" of the starfield including Orion during a long slow journey toward Eridani have been published [a-Moskowitz]. The latter includes "rear window" views showing how our Sun would appear in space at great distance, down to magnitude 7 after 93 LY, too faint for unaided eyes.

Since all stars in the background will move measurably and show parallax, no fixed background can be assumed. Highest position accuracy can be obtained only by allowing for their motion [a-Richards]. A recursive process is needed to know precisely where the starship is. The most accurate observations for astrogation must be made when the drive is off to avoid distortion of optical instruments by acceleration (particularly impulsive). Theoretical views as seen from starship (or anywhere else local) can be shown on a 2-D screen [recall Figure 6.2], but for some purposes 3-D projections in a viewing region may be of interest (perhaps requiring computation of holograms). Of course, the constellations of stars seen from Earth would be noticeably different from a starship after a few lightyears of travel.

Another way to measure the speed of the starship is to observe doppler shift in the spectral lines from any stars, not necessarily just ahead and behind [Appendix C for 3.2 & 3.3]. (Time dilation shift for stars to the side may be too small to detect.) Still another way is to measure the doppler shifts for known pulsars. These distant neutron star beacons send pulses of radio energy at a regular rate (6 digit accuracy or better) in the millisecond to second range. A pair of radio telescopes can fix their apparent position very accurately too. Unlike stars they have not been used yet for robotic astrogation and attitude control, but since each has a unique period, they could be. Another radio method is to receive a continual series of pulses from Earth. This provides a precise direction and, after allowing for dopplering by Earth's motion, a measure of starship speed. Several methods should be used for astrogation, with cross-checking.

Besides using observations of the outside, a starship will carry several inertial guidance devices, sophisticated 3-axes gyroscopes using ring lasers or other non-mechanical "rotation" methods. If accurate measures of acceleration are fed to a computer from the gyro in all three dimensions, it can calculate from prior information on position and velocity the present location without needing outside measurements (relativistic also). The key is direct measure of acceleration, since measurements of exhaust velocity and mass exhaust rate would have small unknown variations. Still, errors accumulate and this system needs frequent correction. Even the inertial guidance for Apollo missions was corrected often. Another astrogation device is an atomic clock which likely will have accuracy better than 1 part in a trillion. It will enable calibration of many other measurements, but data from Earth cannot be used to correct it because of relativistic uncertainties. More likely, comparing accurate clocks on Earth and starship would enable more accurate determination of lightspeed.

An assumption easily overlooked is that the propulsion force is applied in the desired direction of travel. If force direction differs from intended direction by a small amount, an increasing error in direction occurs. For example, misalignment by 1"arc results in 100 million kilometer error after 10 LY. Propulsion, such as fusion exhaust or photon reflection with large energies are involved, cannot be a perfectly-aligned smoothly operating system. Some fluctuation is likely, just as chemical rockets do not eject material with mass centered exactly on axis. The problem is not one of the starship arriving at a cant due to consistent steady off-axis propulsion, but of its course wandering due to random fluctuations in drive exhaust. There could also be undetected systematic bias in the propulsion direction. With telescopes locked to starship structure, changes in direction can be readily detected by observing a set of star positions. Unvarying bias in direction or structure may not be found, but is irrelevant. The average long-term error in direction can be corrected, but the short-term fluctuations should be only measured, not corrected.

7.3 MANEUVERS

Significant changes in course require substantial propulsion energy. A 45° turn would require a sideways acceleration to the same amount of speed the starship already has forward, resulting in 41% higher total speed if there is no braking [see Figure 7.1 & Appendix D]. A 90° turn requires stopping motion in one direction while acquiring it at a right angle. For small course changes the amount of additional speed needed is proportional to the angle (in radians). A 6° change requires about 10% additional speed (and about 20% more energy). No way is known to recover momentum from one direction and apply it to another. Aiming errors should be corrected as much as possible before speed builds up. Energy needed for mid-course changes can be kept to less than 0.1% of total propulsion.

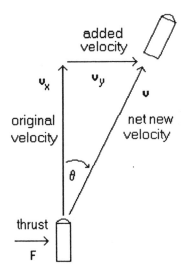

Figure 7.1   Course change requires additional fuel, thrust, acceleration, and velocity for a new direction.

Changing direction is not just a matter of rotating the starship by a small angle. It would continue in the same direction as before until the drive had operated a substantial amount. Going in a new direction requires acquiring significant speed in that direction. For all but the tiniest corrections the main drive should be used. One procedure, if shielding from dust and gas ahead is not of concern, is simply to rotate the starship 90° and apply thrust sideways until the desired new course is achieved, then aim the drive in the direction of travel again. Methods for tiny corrections include using an inertial wheel and using steering jets (probably chemical). Spinning a heavy wheel about the axis of the turn causes the starship to rotate in the contrary direction through bearing friction. When the new attitude is half obtained, the wheel is braked, and the starship should slowly come to a halt in the desired direction.

The possible use of interstellar magnetic fields to change course has been discussed in Chapter 4. The general result is that very much charge must be exposed to the very weak field to produce a weak force. The resulting Lorentz force is at right angles to the direction of travel, convenient for adding a circular arc to the course across the grain of the magnetic field, inconvenient for other maneuvers. The direction of magnetic field is not well known but generally is in the plane of the galaxy and azimuthal, good for maneuvers "north" or "south", inward or outward. For a radius of turn of 1 LY at 0.03c, slower than earlier examples yet still very fast, a 1000 tonne starship must use 1 million coulombs. To store this charge at manageable voltage, structure such as wires must spread over more than a million kilometers. Because of large wire mass and need for high speed, the whole idea does not seem as feasible as carrying more propulsion.

Another kind of semi-passive maneuver is braking, decelerating in the direction of motion. Chapter 4 has also covered various electric and magnetic methods of doing this. The important aspect in regard to astrogation is to locate regions of plasma (higher ion density) ahead for use in braking. It is not expected, but interstellar space may have enough "pockets" of plasma nearby to enable stopping a starship near a destination. Such ionized regions may also be useful as fuel in the event that a ramjet can be successfully built.

7.4 ACTIVE DETECTION

The best detection is usually done by observing information coming from distant objects (e.g. stars). Then there is no waiting time for a transmitted test signal to reach the object. But cold dark material emits negligible radiation. Even some gases do not emit radio or light waves despite ultraviolet light striking atoms or molecules. Radio or radar cannot make matter respond, but it can reflect or scatter. Energetic laser light, x-rays, electrons, and neutrons can cause response from distant material to help identify it and determine its composition, density, speed, and temperature. Obtaining sufficient intensity of x-rays, electrons, and neutrons is a much more difficult technical problem than for radio, radar, and light.

For detection of objects from large dust up to boulders, radio waves are best, with wavelength of the order of 1 mm. Two or more transmitting antennas could be widely separated to obtain fine resolution. If separated by 1 km, the diffraction limit permits a beam size about 1"arc, but the antennas are exposed to destructive dust from ahead. They are even more difficult to shield than a single 100 m reflecting "dish". Small easily-repaired or replaced dishes would be ineffective unless used in a wide-spaced array. A system with 1 km effective aperture can detect objects smaller than 1 mm and locate to an accuracy of 10m in 10,000 km. At 10,000 km/s, there would be 1 second warning to shift the direction of travel to miss an object. Two gees sideways would be required to shift by 10 m in this time, but this amount of drive is probably not available. Laser might provide a little more warning but not really solve the problem. Possibly the warning system might travel far ahead of the starship, pushed by an ion drive drawing power from the starship through a long cable. Demolishing an object ahead seems the only recourse for protection, either by releasing a cloud of dust ahead or zapping it with a high-power laser. There is time for the laser system to confirm the target before the main zap, if the starship motion is nonrelativistic. At relativistic speeds there may not be time for detection and response to objects ahead.

Objects farther than 10,000 km can be detected. Radar detection of objects is done by transmitting a narrow beam from a large dish. If the starship is rotating, a fixed beam can sweep sideways, advancing with the starship to cover all space. Otherwise scanning must be done with an electronic phased array system, mechanical scanning being out of the question. Such systems have been made up to 100 m in size, similar to a large dish, but are liable to damage from dust if facing ahead.

The detected object serves as re-transmitter, scattering about 10% of the radar power impinging on it, but in all directions including back toward wherever the starship receiver will be. Power is lost according to the inverse square of distance on the way out and on the way back. Radar detection must face inverse quartic diminution with distance. A practical limit is about 10 to 100 million km, as radar study of Venus from Earth has shown with a 300 m dish. Resolution can be 100 km at best at that distance, but much smaller objects can be detected without determining their size. Since smaller objects return less power, the distance of detection decreases as their cross-sectional area decreases. Long-distance radar (aimed ahead) can give 1000 second or so warning at 0.01c for large objects. More detailed estimates can be made with the methods shown later for communication. Detecting and avoiding a single oversize dust grain ahead seems an unusual task but is a necessity.

Laser detection of material ahead or to the side requires interpreting the scatter (partial reflection) of light from that material. Infrared laser can detect finer (still damaging) dust than millimeter radar can. For best scatter the wavelength should be as small as or smaller than the material size. Larger wavelengths diffract around small material with less back-scatter, as with Rayleigh scattering of light from molecules. Dust larger than a micrometer is of concern. This size is typical of dust grains and of the wavelength of infrared lasers. Detecting typical individual dust grains is not the goal but determining increases of general density far ahead is important. A large dust cloud could not be destroyed, only evaded. Lasers can determine more about distant material than radar can, but the working range is much smaller. The laser light must arrive at the material with sufficient intensity to affect the atomic states of many atoms to obtain a detectable return. Near-UV and visible laser light would excite atoms and cause them to emit specific wavelengths in all directions, enabling spectroscopic identification. The laser must be tunable over a wide range of wavelengths to allow for doppler shifts and to evoke particular spectral emission lines. Such lasers are not available in high intensity yet. An infrared laser can identify molecules by exciting them to emit series of spectral lines or cause Raman scattering. Polarized laser light may be helpful in identifying atoms and molecules.

7.5 PASSIVE OBSERVATION AND SETI

Besides locating and observing stars for astrogation, important observations would be done for scientific study as discussed in Chapter 6, for detecting dangerous zones (higher gas or dust density), for finding regions needed for maneuvers (magnetic field orientations, ionized regions), for detecting signs of other civilizations, and for the very important determination of planetary destinations. Unaided visual observation would not show the stellar background as colorful, but photography or its electronic equivalent can reveal splendid color in stars, galaxies, nebula, and so forth in the normal visual spectrum [a-Malin]. Carrying instruments sufficient to detect the wide range of possible evidence emitted by other civilizations would ensure the detection of a wide range of natural unanticipated phenomena too.

No portable optical telescope would be able to resolve planets near stars, but other instruments must work on this task as described in Chapter 6. If the mission carries people, good evidence for a planetary destination should supposedly already be at hand, and much observation would be devoted to refining knowledge long before arrival. If the evidence for planets is not good, then the whole mission is uncertain unless it is able to change destinations. Even so, it might fly by the first destination to acquire as much knowledge as possible before changing course.

Observation enroute will be difficult at high speeds because of interference from impinging interstellar hydrogen and dust. Most data must be collected from behind front shielding. Human observers (in space suits) cannot stick their heads out. If the starship is pulled by a solar sail, a blinding light would be seen forward during propulsion periods of many years or centuries, and behind during photon braking. Observations and astrogation would be affected by the contents of not-so-empty space. Charge (probably negative) may collect on the outside envelope of the starship, affecting many kinds of instruments. An electron accelerator may be needed to remove it by shooting it away. Interstellar magnetic field variations prevent straight-line observation of any charged particles and affect the polarization of light from stars and hot gas regions (HII). Local plasma would have other effects. Starlight is attenuated by dust and gas through absorption and scattering of light, with the redder light traveling farther. Dust may be found to vary in density so that the color and brightness of a star is different from different viewpoints. Lightspeed itself is slightly slowed by interstellar matter, and differently for different wavelengths. Relativistic effects on star position, brightness, color, and spectrum will be of great observational interest [see discussion in Chapter 3 and end of this chapter].

Deep space with plenty of time is a tempting place to Search for Extra-Terrestrial Intelligence (SETI) in the form of radio signals, as well as do regular radio astronomy. If a starship mission is intended to meet another civilization, detecting "their" radio signals would better rely on large space-based equipment near Earth, arrays of radio telescopes much larger than a starship can carry. A starship might be able to carry a radio telescope with 100 m parabolic reflecting dish. It can resolve or "see" about 600"arc at 30 cm wavelength (1 GHz) [see Figure 7.2] and serves to illustrate the main problems of SETI. There are about a million portions of sky 600"arc across in the whole celestial sphere. At 3 cm (10 GHz) there would be 100 times more. Checking them all would require a long time and even then a faint signal will be lost because the resolution is coarse. To equal present optical accuracy of about 1"arc, a radio telescope system for 30 cm requires at least two dishes about 300 km apart, not so practical in space on a traveling starship. (A fleet of two or more starships opens up interesting possibilities.)

Observing involves tradeoffs between resolution and signal strength, and between bandwidth and strength. Intelligent sources are expected to be point-like, and the resolution of the telescope should be as fine as possible. Besides a million or more places to point, there are about 10 billion of 1 Hz channels in the most favored radio spectrum between 1 GHz and 10 GHz where natural noise is lowest. A wider range up to 100 GHz might be in use. Bandwidth is the range of frequency used for one signal. Since we can achieve 1 Hz bandwidth or better, we expect that "they" might also. The more narrow the bandwidth, the more a signal can stand out from background noise and the farther it can be detected. Signals sent with high power in a very narrow band might be lost in general noise when broader band receivers are used. A broad-band receiver can only detect average noise power over its whole bandwidth and cannot readily determine a single spike of signal whose power is similar to the overall noise but is concentrated near one frequency. While we would not find it practical to use a 1 Hz channel, "they" might, and the search must be fine as well as coarse in frequency. Many channels should be watched simultaneously to accommodate the doppler shift of their moving source and/or our receiver. Source motion of 30 km/s causes shifts up to 0.0001, or 100 kHz on a 1 GHz signal. Whatever motion sender and receiver have in common is not registered as doppler shift.

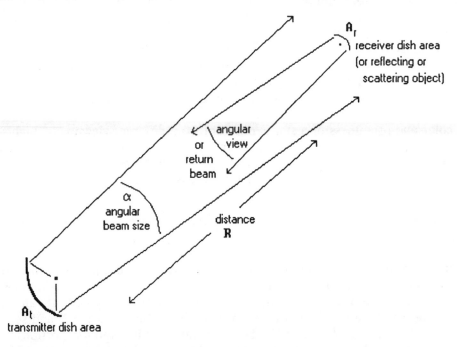

Figure 7.2   Directed radio (or radar or optical) transmission and reception in space. "Dish" antennas shown in cross-section give most compact beam or view.

Noise in deep space is due to synchrotron radiation from electrons all over the galaxy at low frequencies, to background radiation at intermediate radio frequencies, and to quantum noise at high frequencies. The latter comes from the inherent variability of energy at the small scale of Planck's constant [see Appendix D]. Anything warmer than 3°K emits more radiation than the background. The "window" of relative quiet is between about 1 and 100 GHz, ten times wider than available on Earth or on any planet with an atmosphere like ours. The low end of this range is best. This window would be known to all civilizations, although some may live in noisier areas. The lowest background noise is about $(10)-26$ watt per cycle of bandwidth passing through each square meter. (This amount is a defined unit of spectral flux called a "jansky" after the discoverer of the background radiation.) The radiated power of the background noise can be estimated from the temperature [see Appendix D] and also depends on the bandwidth. The smaller the range of frequencies listened to, the less noise is "heard". One photon at 1 GHz carries about $(10)-24$ joule. If received at the rate of 1 per second, the corresponding bandwidth is 1 Hz, and the minimum quantum power is about $(10)-24$ watt. If this one photon is received in a detector area of 100 m^2, its flux is comparable to the background noise. A signal cannot be weaker than one photon at a time, but photons can come less often than once per second to constitute a weaker signal. Photons arriving once per 100 seconds would constitute bandwidth of about 0.01 Hz.

'   Ubiquitous hydrogen atoms emit radiation at 1420 MHz (21 cm), but weakly. It was first thought in 1959 that civilizations would choose very near this frequency to broadcast since it is one of the most obvious, yet relatively quiet in the natural universe [a-Cocconi]. Extensive listening has been done at this frequency with no unnatural signal found. Detecting signal leakage from ordinary activities of distant civilizations is more difficult because the wavelength may be longer and the frequency in a lower noisier band. It has been estimated that if another civilization leaked TV carrier (1 Mw typical, 0.1 Hz bandwidth) and radar pulses like ours does, then our astronomers could detect their leakage at about 30 LY with our largest radio dish, 300 meters at Arecibo, Puerto Rico. [More detail is given on these topics in Chapter 11]. Starship radio dishes limited to 100 m diameter might be able to find similar radio leakage from civilizations just a few lightyears away, a marginal capability.

With possibly $(10)17$ to $(10)20$ combinations of position and frequency to study (and some other variables not discussed), 100 years of observation on a million channels at a time permits far less than one second to be spent on each. This is not enough time to find an intelligent signal which may be coming slower than one bit per second [a-Tarter]. The time factor also requires that any signals always be present or they will be missed. Encoding is also a problem in that efficiently-sent signals (the most information per bit) will resemble noise. One way to reduce the work is to aim only at visible nearby stars. With 20 million channel analyzers planned by NASA that can check 20 million adjacent frequencies simultaneously, all likely frequencies could be studied for a few stars in a reasonable time, either from a traveling starship or from Earth.

If a civilization is not transmitting in all directions (omni-directional) then it might aim at our Sun (and Earth) but would not know to aim at an isolated starship. A terawatt omni-directional transmitter could be detected across most of the galaxy, if "someone" wants to spend this kind of power. It is difficult to imagine a larger single transmitter, but a chorus of transmitters seems possible. If tightly beamed with a large dish, a kilowatt in very narrow bandwidth can reach across 10 LY. Our starship would need to pass through the beam to discover it, and those chances are very small unless such signaling is widely occurring. Needed signal power is discussed more in the next section.

Laser pulses are another way a civilization might send signals. These must be picked out of the light of their star, but study shows this can be done [a-Schwartz; a-Ross]. A few photons per square meter can be detected with a large collector (quality mirror not required) if they come in a very short pulse distinct from general background light. To cover 10 LY at least 10 kJ must be put into each pulse, a task within our technology now. Such laser signals will stand out more from distant stars than near ones. A pulsed laser beam is likely to be aimed at a star and not likely to pass through a starship position.

Chapter 6 covered some other kinds of unintentional radiation that might be observed from civilizations but not from natural processes. Starships from other civilizations can be detected if they reflect or emit unusual amounts of light, radio signals, radar pulses, or other radiation and particles, with unnatural doppler shifts and spectra [a-Viewing; a-Harris]. With the exception of reflected light from giant sails or strange trails in the interstellar medium, starships are unlikely to be detectable as far away as 1 LY. Chapter 11 covers more on detection of other starships.

7.6 COMMUNICATION

Communication between starship or probe and Earth vicinity is important to all missions. Earth staff probably cannot aid or control the mission beyond some early point, but maximum data on the mission is wanted at Earth for science and future planning. Humans on a starship are likely to want to be informed of news from Earth too. Transmitters must use reflector dishes to form narrow beams of radio waves to keep power usage reasonable. The larger the dish, the smaller the beam size, depending on wavelength. The same radio spectrum good for listening for other civilizations is good for our mission use. Discussion here continues with 30 cm wavelength (1 GHz) as an example (the longest likely) unless otherwise stated. NASA missions have used typically about 3 cm (X-band, about 10 GHz), allowing 10 times tighter beams because of less diffraction. The communication system must allow for the growing doppler shift between Earth and starship, shifting the frequency received by both parties downward. The shift itself is an additional method to measure speed.

A transmitter or receiver dish "looks" at a small usually circular piece of space in regard to where it sends or receives radiation [recall Figure 7.2 & Appendix D]. The size of the beam or view is best considered in terms of angular size. The minimum angular size is determined by the diffraction limit as before. A 100 m dish focuses on about 600"arc of space (at 30 cm). Ideally a transmitter dish can put all power into a diverging beam to that portion of space and has a large "gain" in terms of effective power, compared to a bare antenna. A 100 m dish increases the effective power of a microwave transmitter by about 1 million at 30 cm, 100 million at 3 cm. The gain depends on the area of the dish and inversely on the square of wavelength. As before, small wavelengths are better for radio communication equipment. A receiver with a dish does the same in regard to concentrating incoming power. It "sees" mostly the power coming from a small region of space, ignoring noise and signals from other directions.

The power a transmitter must put out to reach a certain distance depends on the square of the distance, the threshold of power that the receiver can detect, the bandwidth, and inversely on the gain of the transmitter dish used (therefore on the square of wavelength also) [b-Sagan & Appendix D]. The power threshold is limited by the 3°K background, but more practically by the ability to make low-noise amplifiers effective at about 20°K. The lowest detectable spectral power is now about 3(10)-26 watts per cycle of bandwidth. Improvements continue in operating "cooler" dishes and amplifiers, and putting receivers in space behind the Moon will help more, but an ultimate system cannot be as much as ten times better than present technology. More improvement can occur in extracting signals from deep in background noise, signals with 1/10 or 1/100 the noise power. The better the characteristics of the signal are known, including repetition, and the longer the duration of signal, the more noise can be removed from it. If all this must be done to find a signal from elsewhere, that is a discouraging state of affairs. Certainly signals to and from our interstellar missions can be made to arrive well above noise level.

A transmitter with 100 m dish must put out about 4000 w per cycle of bandwidth at 30 cm to be barely detected 10 LY away in a 1 m^2 receiver. Only about 1/2 w of transmission in a 1 Hz channel is needed for threshold detection in a receiver at the focus of a 100 m dish across 10 LY! But a reasonable data flow might require bandwidth about 1 kHz in a noisy channel, and perhaps 5 times less bandwidth when signal power is much greater than noise [see Appendix D]. The transmitter should be more than 500 w, again for threshold detection. This seems a very modest power compared to that estimated elsewhere, but must be augmented by 100 to 1000 for reception of practical signals not down at the level of noise. [See a-Jaffe for a different way to estimate transmitter needs.] As cited before, about 1 Mw in a 100 m transmitter dish was estimated to be the minimum power needed for detection over 10 LY when the power is spread over a bandwidth of about 7 kHz at 21 cm. The Daedalus project specified a 2.6 Mw transmitter to send nearly a megabit per second at 2 or 3 GHz from 6 LY with a 40 m dish [a-Bond].

A civilization which chooses to transmit radio omni-directionally must use about a million times more power, or about 1 Mw, to convey a barely detectable signal with 1 Hz bandwidth across 10 LY to a 100 m receiving dish. Here is where the expense of "travel" by radio comes in, whether "they" send to us or we to them. If we do not know where to send our signals, then we must use about a gigawatt to cover the minimum of 100 LY to the nearest hypothesized civilization that can receive it (with a small margin over the noise). For a more useful bandwidth, multiply by another 1000 or so. The power needed for an ambitious longer-range omni-directional galactic encyclopedia transmission station can equal that needed to accelerate a starship with a lightbeam! Another implication of unconcentrated transmitter power is the requirement that a probe or starship not lose track of where Earth is. The power needed to reach an Earth whose position is just 2 degrees uncertain is larger by 100 over that from the best tight diffraction-limited beam (at 30 cm).

A rudimentary theory of signaling ability considers the channel capacity for information flow, measured in bits per second. The power needed to receive a bit depends, as before, on the temperature of the background and receiver. At typical 20°K good operating conditions, about 2(10)-26 joules must be in one bit to distinguish it from noise. The receiver must be smart, because information coded this well is almost indistinguishable from noise. At best one bit per letter can be used to send text, but in practice about 8 bits per letter is used without considering other letters in a word. This amount of energy seems very small, but if many bits per second are to be received across 10 LY the amount of power needed at the transmitter is of the order of kilowatts to megawatts. Feasible bit rates require 1 to 10 kHz bandwidth, yet data collection, especially video, by high-speed probes can exceed this capacity by millions or billions during flyby of a new planetary system. Information must be stored and transmitted for years afterwards.

Information is imposed on a radio carrier frequency by some form of modulation. Possible modulations include on-off pulses of the carrier for binary data (with silence denoting 0, an unreliable method), binary switching between two nearby frequencies, binary switching between two polarizations of the beam, pulse-code modulation, and various unfeasible analog modulations which permit much more information to be crammed into a channel at risk of having it indistinguishable from noise. Modulation is indispensable but increases the bandwidth of the carrier to produce the meaningful signal, and it dilutes signal among background noise. Carrying 2000 bits per second, a very modest data flow, can require a bandwidth of 1 kHz. The narrowest-band modulations are best for interstellar communication. Humans can agree on particular data transmission systems which provide for automatic data checking and correction for bad bits, but we cannot easily guess what data checking system another civilization might use in its general messages. We are barely justified in assuming that others would use binary coding for interstellar communication.

As with radar, a 100 m radio dish is difficult to carry if it must be shielded from dust damage. The dish must be able to "look" in almost all directions to be useful, but it must be located behind a forward shield. The starship can be rotated to aim a side-mounted tiltable dish to cover most directions, just as Earth rotation lets the Arecibo dish with tilting receiver scan part of the sky. If there is no propulsion, or the drive is off, the dish can be moved behind to aim at Earth where the rocket exhaust aimed. The Daedalus plan employed the dead fusion chamber as the main dish. Otherwise to point at Earth the dish must protrude far from the side and be exposed to dust. Smaller dishes, more easily protected, would require more power. A gigawatt transmitter could be a strain on the largest starship power supply, and a very low data rate to Earth may need to be accepted instead. Small probes, even ones that consist essentially of large sail and dish combined, will have severe restraints on transmitter power and data rate.

Optical communication (via laser) has been studied [classic a-Schwartz, called "optical maser" then] and has not been chosen for any mission yet, not even Daedalus. Space optical telescopes are said to be needed for receivers, but light pulses could be detected on Earth with a simple imperfect reflector and photocell [a-Ross]. A UV laser would be desirable but the most efficient lasers now are infrared. One meter or larger aperture can be obtained by spreading the laser beam with a lens to illuminate a parabolic mirror which then forms a beam. Angular spread of a lightbeam from a 1 m mirror would be smaller (less than 1"arc) than that for radiowaves with 100 m dish, and laser power could be somewhat smaller. Bandwidth of the unmodulated beam can be very narrow, and data rate could be large. For radio and laser communication, the lifetime of critical parts is of concern. Transmitter tubes on present missions are proven for about 10 years and might last 50 years. Large lasers would surely require frequent replacement or rebuilding.

A nice but unlikely scenario is the need to maintain communication with another civilization during travel to visit them. Their transmissions would provide guidance for astrogation as well as extensive preparation for the momentous meeting of unrelated beings. More discussion on this is in Chapter 11.

7.7 MORE RELATIVISTIC CONSIDERATIONS

Further relativistic effects are discussed here to accompany the basic set introduced in Chapter 3 and found pertinent to astrogation above. Some of these are of interest only for ultra-relativistic travel, a prospect that seems unlikely. For communications, if the starship keeps accelerating to reach significant gamma, there is a point after which signals from Earth can no longer be received. For acceleration at 1 gee, this occurs when the starship has gone about 1 LY from Earth (by accident of numerical values) [a-Anderson]. At less acceleration, the limit is proportionately longer (e.g., at 0.1 gee, about 10 LY). The effect can be traced to time dilation. While a signal is traveling to reach the starship already 1 LY away, the starship is moving farther and faster and the signal is further red-shifted. At the limit the signal is dilated to nothing.

Considerable effort has been put into calculating how a starfield would appear aboard a relativistic starship going "north" at speeds up to 0.992c (gamma about 8) [a-Stimets] and toward Barnard's Star at 0.99c [a-Sheldon] and towards Orion and Eridani at speeds up to 0.9c [a-Moskowitz 1967, 1968]. The effects of doppler shift, aberration, brightness increase, and parallax as stars move in front of background stars, are included. Hot blue stars would be shifted into ultra-violet, dull red stars would become brilliant, many more stars would become visible ahead (almost 100 times more), and very few would be discernible behind. Dull red giant stars across the galaxy to the sides and nearly behind the starship would be seen as brilliant blue-white stars far forward. Blue-white would be about the only color visible.

Contrary to many published claims, stars would not become arrayed in a rainbow of colors or circular "starbow" from back to front [a-Oliver; a-McKinley]. A limit has been found to how bright or hot a star can be made to appear by speed, corresponding to a temperature of about 16,000°K. Beyond this brightness, increasing speed would make the star appear dimmer, a very nonintuitive result. At an ultra-ultra-relativistic speed even the microwave background would be dopplered up and compressed into a forward brilliant point about 1/10 as bright as the Sun is now to us, a sight no one may ever see.

Chapter 8

Technological Requirements and Hazards

Starship technology is one of several aspects where a starship project depends on details which can raise serious problems. A few critical technical areas are chosen for further discussion here to provide a feel for the present and expected capability of our technology. Technical and other problems of a rotating habitat are covered in Chapter 9. Other critical areas may not be realized yet. To broaden the technical picture some other less critical details are filled in here. Many technical limitations in propulsion have already been discussed. Technical requirements may differ markedly between probes and peopled starships, but most distinctions are not attempted here. Again, another civilization might encounter the same technological difficulties to the extent that engineering approaches, systems problems, and special materials are universal. The best standards for comparison with human technology thus far are the Apollo program and the scientific missions designed by Jet Propulsion Laboratory (JPL). In accord with earlier findings on propulsion, reliable operation for a thousand years is assumed to be needed for a starship unless discussed otherwise.

8.1 CLOSED SYSTEMS AND LEAKAGE

Many aspects of closed systems are descriptive of starship operation, but open systems are also pertinent. General theories of systems might provide guidance as to whether and how a technical system can function nearly perfectly for a thousand years. Our understanding of, and ability with, systems begins with material, energy, and information flows, and is limited only by informed imagination of scientists and engineers. Ecological and social systems are parts of the general starship system and also provide guidance for more understanding of systems. In the most general sense a system must include all materials and hardware, all power sources, all living inhabitants ("wetware"), all stored information ("data"), and all human-invented concepts that guide the system ("software").

No system is perfectly closed [see Figure 8.1]. Material and energy leak past its boundary or envelope. In the near-vacuum of space over hundreds of years, small leaks of gases and vapors amount to serious loss of material resources. Other leaks occur when old junk, broken tools, wornout shoes, and the like are heaved out the airlock or dropped into a rocket exhaust chamber. These material losses are to be avoided because after a thousand years the starship would be stripped clean. Another form of loss is the erosion or emission of material from the boundary of the system, whether by dust collisions, "loose" atoms from the outer skin or coating (avoid painting the starship), ions leaping off due to a charged ship, decay of radioactive nuclei in almost every material, or other means that must all be identified. Some measurement and processing work might be done "outside" with small amounts of material lost. Some heat radiators spray metal atoms from a jet to a collector and might lose a few. Matter is gained by means such as a scoop that collects gas and dust for analysis, scooping of material for fuel, attraction of ions to a charged structure, and collisions with interstellar atoms that result in adhesion.

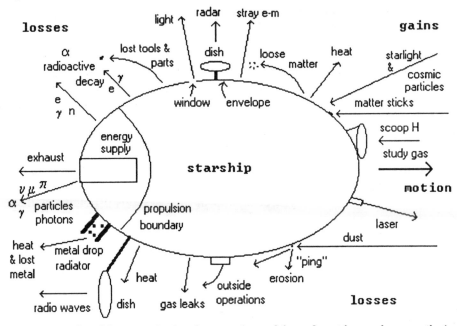

Figure 8.1   Starship as nearly closed system (propulsion exhaust is a major exception).

When energy flow is considered, a starship cannot be strictly closed. Because of mass-energy equivalence, any energy loss is a material loss. If propulsion is being used, stored energy and material is exhausted. To continue propulsion, a large store of energy and reaction mass is required, much larger in quantity than that for human habitation. To simplify discussion a boundary might be drawn between the fuel/drive system and all other parts of the starship. However, the rest of the starship is not strictly closed to energy flow because of leaks of matter, of radiation of waste heat and minor electro-magnetic leaks and emissions, and of electric signals sent between habitat and drive system. Every time someone switches on or off equipment, a small burst of electro-magnetic radiation disperses, partly to the outside to join general galactic noise.

Switching losses are negligible, but heat loss may not be. If insulation is so good that the envelope of the starship is at about 20°K, about 0.01 w/m² is radiates from a total surface of about 1 million m², for a total energy loss of about 10 kw. Over 1000 years (10)14 joules is lost, equivalent to about 1 kg of fusion fuel. Although total energy use might be only 10 times greater on a frugal starship, this loss is not worrisome because it is the final low-grade heat result of that usage, the leakage from high-tech power use that has also kept the environment warm. If envelope temperature were 10 times hotter, the energy loss would be intolerable. Energy is also lost as light through portholes, laser beams and searchlights, radio communication signals and radar pulses, and other still smaller effects. If extravagant power use resulted in average emission of 3 Mw for these assorted purposes for 1000 years, the loss would be about $(10)17$ J loss or just about 1 kg rest-mass loss, negligible compared to leaks but substantial for energy storage. By the same token, continual input of a collected lightbeam for operative power adds a negligible amount of mass.

A closed system can have homeostasis, that is, hold a stable state. This would be desirable in a starship in regard to maintaining a comfortable environment. Inherent stability would be nice, but without extensive human and computer supervision it is unlikely any major subsystems of the starship can be fail-safe and stable. Most subsystems have feedbacks, planned or unwanted. Unwanted feedbacks cause unwanted phenomena such as vibration, disease, or large variations in water or air quality. Most feedback loops are planned to keep subsystems in their linear range of operation rather than going to an extreme state ("saturating"). For example, carbon dioxide sensors signal to an absorption system if the atmosphere has too much and to the reserve release system if too little. Else some fluctuation might cause the level to run away to all or no carbon dioxide. The feedback signal cannot respond too quickly, or energy is wasted making micro-corrections, or so slowly that the system goes to an extreme. Feedback is important in food growing, course correction, propulsion thrust, and much else.

Problems begin when the many feedback loops in a complex system interact and cause some subsystems to be nonlinear or to enter chaotic unpredictable modes of operation. Indeterminism also applies. Theoretically, no matter how carefully the original state of the starship is specified and arranged, its evolution over time must move toward unpredicted states. Perhaps the time scale is far beyond the voyage time and there is little need to worry. But designers and voyagers must be ever aware that the system is changing with some amount of unpredictability despite increasing understanding and control of some kinds of chaos. It may be that only humans can perceive and reverse general deterioration in physical and social systems in the long term, if then.

A general feature of closed systems is that their entropy increases, which is to say that their disorder increases. This process is usually irreversible. Entropy is related to energy flow, and if there is input energy, either from a stored source or brought in by, say, lightbeam, then there are many ways to retard or reverse entropy change. Stored energy is very ordered, and its use results in loss of that order. Strictly, a closed system containing a large energy store still has entropy increase, but practically, it increases very slowly if the energy store is large. Some parts, such as the agricultural and human habitats and the repair shops, can have decreasing entropy while other parts, especially the fuel supply, have increasing entropy that more than compensates. Again, a boundary is helpful between energy store and operating starship. Inhabitants will see deterioration going on at the usual rate despite enormous tanks of fusion fuel nearby. In a robotic probe a similar balancing of energy supply against deterioration holds. Entropy can be calculated, as shown later.

8.2 DETERIORATION AND RELIABILITY

Everything except atoms themselves deteriorates over time. In this century, unlike some earlier times, most engineering is limited to equipment lifetimes of a few decades and dependent on continual maintenance. A starship might require near-perfect performance for a thousand years. The longer the mission the more chances for failures at all system levels. Most deterioration has a systemic aspect. Something fails a little, the failure provides a route or feedback for more failure, and so on until catastrophe occurs.

A chemical example is rust on steel, perhaps not as serious for a starship as other slower corrosions. Rust is a catalyst for the formation of rust, so that the process can run away, given other needed ingredients (water and oxygen). A mechanical example is a running motor. When its bearings have worn slightly, regardless of lubrication, the shaft wobbles, bounding around in the bearing and increasing wear. More wear leads to more wear in a runaway process. A fluid example is a dripping faucet. Once a small flow starts, ions in the water attack the metal parts of the valve, eating it away and increasing the flow until the washer can no longer seal the channel that has formed. Starships may not have metal faucets, but if plastic parts are used, they must be much better than most present ones. Still another example is the behavior of metals with high electric fields at their surfaces, such as needed in some propulsion methods. Beyond some reasonable level of field, atoms tend to migrate and pile up at places where there is already a little bump on the surface and a higher local field. Wherever the atoms pile up, the field becomes stronger (the "sharp point" effect) and more atoms join until there is whisker of metal sticking out [a-Bond 1986]. The field is ruined because electrons spray out and neutralize whatever makes the field. Or the whisker breaks loose and accelerates away to collide with something. If anything can fail, it will!

Corrosion and oxidation are of serious concern, partly because valuable materials are destroyed and dispersed. In closed space systems life-supporting oxygen is a dangerous gas (as NASA once learned tragically when using pure oxygen in Apollo capsules). Materials should not be used which react readily with oxygen and/or water vapor, but many common necessary materials do, from paper to gaskets to various metals. Unprotected materials must be planned for frequent replacement. Bins for growing food are most liable to corrosion and should be of durable plastic like teflon or of ceramic. A difficulty with oxidation is that the oxides must be refined with heat and chemical reactions to recover the valuable elements involved. Organic materials are not such a problem, since complete closed combustion with oxygen results in carbon dioxide, easily captured and converted back to organics by plants or chemistry. Rapid oxidation can become a runaway process called fire. It can be partially prevented by using lower oxygen-to-nitrogen ratio than on Earth, a requirement in conflict with using low nitrogen pressure. While few organic materials and fuels are planned aboard a starship, if a chemical or other accident starts a hot fire, light metal and plastic structures might start burning. Airtight doors, meant to save air in case of a leak, can also be used to stop oxygen flow to a fire. Fires may be more likely in large colony ships with forests [a-Smith], but fires and other oxygen damage can be prevented in robotic probes by not carrying oxygen atmosphere.

Deterioration occurs wherever there is interaction of matter, whether the rubbing of surfaces, lubricated or not, or the flow of fluids in piping, or the collision of particles with matter as in vacuum tubes (transmitter tubes, laser tubes, CRTs, x-ray devices, etc). Indeed, the performance of present bearing and lubrication technology is unknown beyond 50 years. Vacuum tubes also leak and are to be avoided, but some applications are likely to require them. They are the best known way to generate radio and microwaves at high power. Experience has been that most vacuum tubes last decades, but there is no reason to expect them to work for centuries. New alternative devices without vacuum exist for most applications but the lifetime of LCD screens, solid-state microwave power amplifiers, and low-power laser diodes, while apparently more than 10 years, is still uncertain beyond that. The solid-state transmitter on Voyager failed immediately, and its vacuum tube backup is working well. The backup receiver is also in use and has a partial fault. Micro-size vacuum tube devices are now being developed for durability and radiation resistance that solid-state devices cannot achieve.

The near vacuum of space causes some deterioration. Some atoms of materials tend to acquire KE to jump from surfaces exposed to vacuum and are forever lost in space. No vapor pressure impedes them. Many organic solid materials (polymers or plastics) cannot be used long times in vacuum, as their solvents "evaporate", leaving them brittle or without strength. (Same solvents would make an unhealthy atmosphere inside.) Most lubricants cannot be used, as most known fluids are not stable in vacuum. Materials not manufactured in, and exposed to, air do not acquire oxide layers which commonly prevent metals from sticking--really, welding--together.

As shown in Chapter 5, complexly engineered systems consist of layers or levels of self-correction in an attempt to be fail-safe. An example is an engine with moving parts. An engine that converts heat to mechanical power is a first level of engineering. It may well be used in a starship as a turbine to convert heat to rotation and then, with a generator, to electric power. Moving parts require lubrication, and the second level of engineering is a correction of a problem at the first level, lubricating the bearing surfaces by pumping oil to them through passages. The surfaces need more lubrication at higher pressure when the engine runs faster, and the engine itself can be used to drive its own lubrication, a third level and fail-safe. The lubricant pump itself needs lubrication and there are ways to make this inherent at a fourth level, by immersing all of it in its lubricant. This also makes it self-priming. In gravity the pump need not be left dry as other parts of the engine may become when not operating. This raises the matter of starting the engine with lubrication, possibly with an auxiliary lubrication system, an engineering addition to the hierarchy. If a separate pump is used continuously then it must be highly reliable and monitored, whereas an occasional failure to work during startup would not be fatal to the engine.

Then there is cooling the lubricant if the engine runs long, and distributing it accurately so that the last bearings in line do not get less than the first. These are fifth and sixth levels of correction. The properties of the lubricant become important, such as deterioration, cleanliness, erosion or corrosion of the engine, and loss of viscosity with temperature. Correcting these involves more levels of engineering. Using the engine to provide its own lubricant cooling is a fail-safe method. If the system is running, it is providing what it needs. Another set of levels of engineering is found when considering the operating temperature range of the engine, or the atmosphere in which it must work. What works under one condition does not work well under others unless corrections or compensations are made. This sort of engineering is the reason that up to 100,000 hours of nonstop operation can be obtained from semi-complex systems like car and jet engines and turbine-generators. This example is not far-fetched because rotating systems are likely to be put in starships. Chemical rocket engines, while less relevant here, have many levels of engineering correction, some of them fail-safe.

Temperature is a factor in deterioration. On Earth daily temperature changes have a major destructive effect on structures and systems. A starship should have nearly unvarying temperature, but there would be hot areas where propulsion, power sources, furnaces, and other such equipment are located, and cold areas for storage and where heat radiates to space. On the small scale, motors, solid-state power devices, and many other parts are used intermittently, therefore they warm and cool. The sizes of materials change at different rates with temperature, causing misfits and stress and long-term deterioration. Solar sails and propulsion units that are activated and de-activated go through temperature changes.

On the absolute effect of heat, the warmer a material is, the more agitated its molecules are and the more likely the weaker types of molecular bonds will randomly break. Beyond "room" temperature (300°K or about 20°C) the general failure rate doubles with each 10°C increase in temperature, a general engineering rule. Temperature drives unwanted diffusion of atoms within materials. Present electronics depends on various semiconductors sensitive to temperature. As solid state components operate at higher temperatures, the elements put in them for conduction begin migrating. Semiconductor electronics might work for 1000 years at room temperature and ten years at the boiling point of water. The lifetime of electronics beyond 50 years is uncertain, but elevated temperature is one way to simulate a faster rate of aging. However JPL was able to obtain meaningful tests only to 75°C. Beyond that, semiconductors develop hot spots and begin to fail from temperature rather than age.

Electronics near absolute zero is an untried technology, yet would be needed for some probe missions where no long-term heat source can be provided [a-Jaffe]. Work on computer components made from cold superconductors continues. Many valuable kinds of sensors already work at and depend on low temperatures, and the task is one of extending cold operation to all necessary electronic functions. Spacecraft structure is already proven at low temperatures, as many satellites and missions have shown. (Keep in mind that solar or onboard heat sources are used at present to hold electronics and some other components at "room" temperature, while the structure and parts of the envelope necessarily become cold.) It is very unlikely that computer and other electronics can be made which will function at any temperature between 3°K and 300°K or more. A new problem arises: keeping the electronics of a cold probe mission cold during its initial travel when much solar heating occurs. Perhaps a dual system is needed, with temporary onboard heat source or solar heating for the "warm" computer until the "cold" computer has fallen to its operating temperature and can start. This is further undesirable complexity.

Materials must be carefully chosen because some metals and most plastics become unreliably brittle when very cold (3°K). Titanium alloys, aluminum, stainless steels, copper, and nickel show little if any loss of strength and toughness at the coldest temperatures [a-Parfitt]. Fortunately these are abundant and therefore relatively cheap. While various "whiskers" in composite materials (e.g. for cables) would retain strength when cold, the binding materials presently used around them may lose elasticity. On the hot side, present experience with materials at high temperatures and future experience with fusion reactors will enable design of feasible starship propulsion systems that handle high-energy reactions. The metals just listed also can endure environments with moderate heating and cooling off, or being used hot at one end and cold at the other. [also a-Steinberg]

Reliability results from any or all of several approaches: multiple (redundant) backups, long-term testing during development to establish failure rate and mode, full understanding of the science and multiple-level engineering involved in a device, and identification of all operating conditions in which the device must operate. Reliability has become a large engineering field. An approach called probabilistic risk assessment handles large systems where big failures are expected to be too rare and/or too costly for substantial record of failure beforehand and thus relies on subjective judgment of risks. Another approach, critical states of dynamic systems, is covered later. Simple rules cited earlier for major successful space programs can be projected, with much caution, to starship reliability. Political interference was the major cause of misjudgment of space shuttle safety, causing engineering estimates of catastrophic failure at about 1 in 100 to be "reinterpreted" to a realm of 1 in 100,000 [a-Marshall 1986; a-Waldrop; a-Logsdon]. NASA has a list of 4000 critical items in the shuttle, the failure of any one likely to cause catastrophe. None are "fail-safe", almost none have been tested to failure, and yet shuttle flights continue [a-Marshall 1988]. NASA handling of reliability engineering is not to be extrapolated to all public space programs. JPL controls as far as possible its own clearly successful reliability engineering. A "single-point failure" approach was used for Voyager so that one failure could not jam more than a specified fraction of all operations [a-Jones]. This method is not likely to be sufficient for a starship.

Reliability for space and for nuclear plants is based partly on backup systems and partly on speculative failure analysis. One might think that two unlikely independent events occurring in sequence are too rare to consider. But in fact nuclear plant failures have occurred after four unlikely events. If, in a system like a starship with perhaps a million critical parts, each part has one chance in 1000 of failing in 1 year, then about three failures per day can be expected. A series of these can constitute catastrophe; for example, the water pump fails, then the power for the shop that fixes pumps fails, then a leak develops in the hull requiring the attention of the same people trained to repair the preceding. Other people depending on that water or that shop are halted in their daily work. (Aboard a probe without people, perhaps the robotic repair system can have less conflicts in what to repair first, but must have instant judgment, using a total view that humans are possibly better at.) Or suppose a hull leak occurs that produces a cold spot and loses valuable air. The cold spot might freeze and break a water pipe, resulting in loss of water in a flood. The water spray might cause several supposedly closed electronic packages to fail. And so on.

Thus several failures can occur in a short time which are independent initially but afterwards are interactive, making catastrophe out of smaller crises. A longer chain of failures could destroy the starship as a viable system. Furthermore, with a million parts, testing for each failure mode four deep (four independent failures at once) involves studying (10)24 combinations of failures, an impossible task even if computerized. More realistically, 100 parts might be examined for failures three deep, requiring a million tests or studies, easily justified for a starship. But unjustified assumptions are made that thousands of parts are too reliable to consider. The Apollo program attempted and achieved failure rates of one in 10,000 for all parts, for short (ten day) missions. Daedalus assumed 100,000 parts and required an overall failure rate of 1 in 10,000 over 50 years [a-Grant].

Incredible feats of reliability have been expected of other beings [e.g. a-Freitas 1983, 1980 p 95-]. Civilizations that send out probes are thought to design them to find planetary systems like ours and to remain operative for time periods from thousands to billions of years, hiding behind the Moon or in the asteroid belt, perhaps shielded by rock underground, collecting and storing solar power, sending off an occasional narrow beam message with a (large?) dish kept protected otherwise. Cosmic particles, solar wind particles, meteoroid, and asteroid collisions will do their work over that time span, with cosmic particles and local high-speed dust causing the most damage. Probes resting on or in moons will suffer other damages. Were we to find so much as a bolt or beer can left by others and date it, we would get a measure of how long technology can be made to function.

8.3 CONSERVATION, REPAIR, AND REPLACEMENT

Over long periods many materials emit vapors and supposedly non-volatile molecules to join the gases that can leak through the envelope that marks the boundary between inside and space and holds the recommended 1/3 to 1/2 of Earth atmospheric pressure. Leaks might best be detected by a network of leak detectors outside the pressure hull. These could be simple mass spectrometers analyzing the "atmosphere" that develops around the hull for content and density. As estimated in Chapter 5, total leakage from the starship must be kept below 0.1 milligram per second (about (10)19 atoms/s) if only 1% of the starship is permitted to "disappear" in 1000 years. With an outside envelope of about 1 million m^2, leakage must be kept below (10)13 atoms/m^2. Present vacuum systems are difficult to make this tight.

Another form of loss is the erosion of material from forward surfaces during high-speed passage through interstellar dust. Various kinds of shields have been proposed to stop this form of deterioration (as discussed later). Electric and magnetic shielding cannot stop uncharged dust grains as a solid shield can. To reverse the losses, material, mainly hydrogen, could be collected in space if ramjet-like technology can be made. Hydrogen is most likely to leak from the starship and most easily captured back from space, and a small "scooping" system might be necessary. Conversion of hydrogen to other elements is feasible only on a microscopic scale and not a good way to replace other losses.

Besides controlling leaks and not heaving old junk out, all materials must be conserved, recycled, and recovered. Thanks to entropy, materials tend to get mixed together and there is an energy penalty to separate and re-refine them. Human hair and nail clippings get mixed in with the metal shavings, discarded scratch paper, broken dishes and tools, clothing fibers, apple seeds, dead power transistors, paint flakes, and so on. All equipment with moving parts generates dust from wear. A list of sources of dust in a starship would be very long, but a necessary part of designing such a mission.

Inhabitants of Earth do not appreciate how difficult it would be to sort things at the level of dust particles, much less at the level of atoms, since we "throw away" most used things and sort a few old cans, bottles, and car parts. Suppose that after cleaning "house" one kilogram of dust is collected, consisting of 10 different elements in approximately equal numbers broken down to atoms. A kilogram of light elements consists of about $(10)26$ atoms. The entropy associated with them being all mixed is found by counting the number of ways they can be mixed [see Appendix D]. The entropy and energy cost are proportional to the number of components in the mixture, and fortunately to the logarithm (a slow function) of the number of possible mixtures. On the other hand, the log function is so insensitive that if the typical waste particle size is much larger than atoms, the amount of entropy would decrease by only a factor of 2 or so.

The amount of entropy involved is about 250,000 joules per degree for this 1 kg of waste. If this kilogram is sorted at room temperature, the minimum energy needed would be about 75 million joules. Actual refinement systems can be 10% to only 1% efficient. The typical energy needed to convert this material to a gas, perhaps as a first step toward refining, is in comparison only about 1 million joules. If over 1000 years, or $(10)10$ seconds, just 10 tonnes or about $(10)30$ atoms of the starship must be recovered from fully mixed trash, using 1% efficient processing, about $(10)14$ joules is needed, an average power load of about 10 kw, and tolerable. Recovery is not humanly efficient, and unless assorted efficient robotic micro-manipulators are available, several people might be occupied full time on this task. A mass spectrometer (operated in a section open to the vacuum of space) sorting the 10 tonnes of waste must operate at about 10 amps full time, a very high current for present technology.

The energy cost of sorting normal-sized simple hardware for smelting and remanufacturing is somewhat less than the energy cost of making new ones from raw materials. Farther down at the level of bolts, nails, and transistors the sorting cost is greater. A microcircuit has many special elements in it and might require a person a day with a microscope to separate some of them, aided by power-using tools. Atoms are almost a billion times smaller than these products, and the entropy of their mixing becomes much greater. A small explosion can disperse a chemistry store of elements into a hopeless mess. All the effort in the universe cannot completely restore the materials to their bottles. This humpty-dumpty problem could be very serious for small systems closed and isolated from planets and solar power for long times.

(A curious aside is that our Solar System and everything we are, have, and use, came from a random dust of elements in space condensed to form a sun and planets. The water, quartz, iron, gold, and many other useful compounds and elements sorted themselves out in and on the Earth, driven by generous internal and solar energy supplies. Some elements such as oxygen seem to have been released by life itself. Natural sorting to a rudimentary level seems easy, even automatic, on a gigantic scale with plenty of time, but not in a small starship with limited energy and tools, which themselves must be repaired and replaced.)

Repairs and replacements are the two main solutions to component failure. Space missions thus far have not been planned for repair in space, although some successful exceptions have arisen from emergencies. Present technology often depends on repair shops being much larger and more complex than the equipment needing repair. This aspect of technology is almost a basic rule, but one which must be broken if major space missions are to succeed. A starship must carry all its spare parts, or ways to make them. There is some doubt that enough people can have, maintain, and pass on specialized skills, making in small batches just the many simple items needed for daily life [a-Hodges]. Rebuilding almost all of the starship, even outside parts, may go on continuously during travel. Nothing can be overlooked for checking and repair. No subassembly can be taken for granted, lest a daring engineer out in the high-radiation blast from ahead be trying to repair a "whizmo" and discover that an unusual bolt head was used or that the adapter from one kind of cable to another is not part of starship stock. In other words, all components used must be dissected into basic parts and the need for all those justified and then used as a basis for tooling, for stockpile of material and parts, and for repair procedures.

Backup or redundant parts are a way to replace worn or failed parts. The conservative JPL designs used almost no backup or redundant equipment, although the computers were versatile and could replace one another or work around nonfunctioning components. A starship is likely to require multiple redundancy because few of its parts are expected to work for a thousand years. Banks of 10 or 100 identical units have been described for large systems such as propulsion. The ultimate, but more expensive, redundancy might be fleets of starships.

Repairability, also called maintainability, is an established engineering practice to a sophisticated degree. However, most of our present complex systems, which include mechanical parts and have complex maintenance schedules, have not been extended to completely reliable operation beyond a matter of hours or days, as users of cars and military vehicles can attest. (It is assumed here that a single part malfunction is critical rather than trivial and requires the system to be taken down for repair, for reasons of safety and to prevent a string of failures. This is not always the case. One or two seats in a jet could be repaired or replaced during flight.) Many individual parts could last 50 years and possibly 1000 in a controlled environment. Failure requiring repair is a problem at the subsystem level where thousands of parts have a small potential of failure in a long time period. Repair cannot be simply by replacement because even if every part is replaced just once, a doubling of starship mass would occur.

Repair includes identifying the subsystem responsible for the observed symptoms of failure, testing the sensing system that produced the report of failure, removing the subsystem or component from its position without disrupting or endangering starship operation, diagnosing the exact failure, understanding the failure, exchanging the damaged part(s), testing the subsystem, installing it, and making final tests. Alternatively, the subsystem must be repaired in situ, perhaps with the power on. Or the subsystem may be of a nature that requires total replacement with a new version. Part of the repair process is to repair the damaged parts and put them in store, or to send them down for smelting and to make more parts for storage. An easily-overlooked aspect of repair in a system that repairs itself with or without human agents is the matter of repairing the repair subsystems. Some of the tools, sensors, instruments, robots, and shop facilities will fail (or the technicians become injured or ill) during the repair of a subsystem. It seems, however, that self-repair can be a convergent rather than divergent process, since living organisms can do it. Some relevant cybernetic argument is given later.

Humans are efficient repair agents. When resting they consume (and dissipate) about 100 watts each while thinking up more ways to improve things. In action they use about twice this power to build and repair, as well as confer, study, help, and other progressive actions. A human can reverse a large amount of local entropy in a day. For example, in one day an experienced human can write a detailed repair or construction manual for a starship or robot subsystem which will work against entropy when copied and applied for possibly thousands of years ahead. It seems unlikely that a "universal" repair robot can be made which runs on 100 w and fixes everything necessary just to hold the status quo. (If ever not needed, the robot could be switched off and use zero power.) To be fair, the personal power used by humans results in almost negligible energy consumption, even over a thousand year line of many human descendants, as compared with total habitat energy needs. The few trillion joules used per person is a few grams of fusion fuel. The Daedalus project proposed robotic repair, with mobile "wardens", repair "shops", and stores constituting 30 tonnes, or about 7% of the mass of the payload [a-Grant]. With extrapolations from present aeronautical experience, the ability of the Daedalus mission to keep itself 99.99% reliable with one layer of backup and ten times better technology was found just barely possible.

Neither computer chips nor data storage media seem likely to last centuries or longer. Fresh supplies of each are needed, to be manufactured from worn out ones after refining the material. This seemingly simple requirement could be a very serious problem in implementation. The highest tech items needed on a starship initially were developed and manufactured by a large society on a large scale. Integrated circuit chips hitherto require a vast network of factories. An assembly line with a throughput of about one unit per day is needed, small, lowmass, fitting into a corner of one room. (Keep in mind that hundreds of other less high-tech items also need remanufacture in a limited space.) Effort is being made to condense and reduce the chip-making process on Earth, and a starship may be able to carry a limited chip-production facility. As few different kinds of generic chips should be used in equipment as possible.

In another approach, there could be return to small "discrete" parts (transistors, resistors, etc.) for some equipment. These have been very reliable but integrated circuits became better. Any faulty simple part can usually be identified and replaced in minutes. Repair robots would have much difficulty with present soldered-in chips with 40 or more pins, and a new fool-proof chip-socket method is needed. Larger scale use of integrated micro-circuits up to the scale of whole computers is also needed, so that chips perform a complete function with few inputs and outputs, permitting small simple connectors. Less sockets and connectors make more reliable electronics mechanically but make repair more difficult, a long-standing tradeoff problem. Digital equipment now requires 32 lines of signal and may jump to 64 or 128, worsening the socket and plug problem. Present practice of replacing (and throwing away) large complex circuit boards to fix one fault does not seem compatible with long-term efficient total recycling.

Before the assembly line there must be a tightly-closed smelting and refining material recovery process, with no leaks of toxic chemicals. Manufacturing should not involve complex and toxic chemistry, a problem which largely eludes Earth-based manufacture. Advanced chips and other electronics and optics involves very finely detailed "nanotechnology". It is difficult to envision processes that would make these items simply, although the new small table-top atomic force manipulation devices show promise of manufacturing things atom by atom, possibly automatically. If ultra-thin photovoltaic cells and metal films are needed in large quantities, manufacture may be outside a slowly-moving starship where metals can be sprayed on large areas.

Long-term conservation affects storage of energy. In electronics, in data storage, and in large energy storage systems, electric and magnetic fields are not viable long-term methods. Charges leak away no matter how well insulated. Vacuum is not sufficient. Stray electric and magnetic fields and charged particles eventually alter bits of magnetic field used in data storage. Only atoms held in stable rather pure, perhaps crystalline or covalently bonded materials promise long-term stability for microstructures and data storage. Energy storage as fuels often requires volatile organic fuels which evaporate. Hydrogen or deuterium seeps through very small pores. Radioactive elements with half-lives less than 1000 years are also not viable energy storage materials, and intense radioactive sources cause materials deterioration and other problems and dangers.

8.4 EQUIPMENT, STORES, AND MATERIALS

The equipment needed in a starship and at the end of its mission seems almost limitless (not a good sign for achieving low mass), and computers would keep inventory of all parts in use, number and location of spares, their ages, and who is responsible for them. Automatic warnings could be issued when expected service life is over and when stock has fallen too low. Despite needing to rebuild most of the starship, spare parts and materials must constitute a small fraction of its mass (less than a few percent). Obviously the starship cannot carry one spare of everything it has without doubling its mass. To reduce spare parts stores, minimizing the number of different parts is essential and contrary to present Earth practices. Many tools and furnishings can be made to last a thousand years, despite the present technology of early obsolescence. Equipment which must have moving parts must be planned for easy replacement of the bearing surfaces. Where possible moving parts should use frictionless, wearless magnetic fields. Electric motors can be brushless, with electronic switching of field windings. Machine tools, massive mechanical equipment with plenty of places for wear and failure, have been the staple of industry and high-tech society, but ways to avoid carrying them must be found. Cutting metal is too wasteful a process even with a universal computer-controlled machine.

Nearly every chemical element has been found useful in our technology, and a good many of their isotopes. While the list should be minimized, all elements on it should be stored for synthesis of existing or new compounds. The quantity carried would be in proportion to the initial composition of the starship and the expected dispersion during use. A list of 84 elements was generated for the self-reproducing probe proposal [a-Freitas 1980 p 251-], but surely some of these are not necessary. Some elements such as sodium are not safe to store but a stable compound is always possible. The number of possible compounds is so large that only the most common thousand or so can be stocked, and the rest must be made as needed.

An instructive example of the kinds of exotic materials chosen for present sophisticated space power equipment is found in the list of those used in the plutonium RTG generators for the Voyager missions [a-Heacock]. The plutonium is in an oxide ceramic form, covered with iridium shells, wrapped in graphite yarn. The thermal generator is machined from a huge slug of beryllium, coated with iron titanate. The silicon-germanium junction devices are attached with molybdenum shoes, coated with silicon nitride, insulated with 60 layers of molybdenum foil and quartz cloth. All of these special materials have purposes found from research in material realms not ordinarily encountered.

Similarly sophisticated specifications have been proposed for the components of the Daedalus pulsed fusion drive, and this trend should give the flavor of the sort of technology to be expected for a real starship. Since many materials are exotic and sometimes toxic, exotic repair, recovery, and remanufacturing methods are to be expected. As another example, titanium is an abundant strong low-density metal, very good for spaceplanes and such but long avoided by the aeronautical industry. To make the "stealth" bomber, enormous carefully controlled hot presses with high-pressure argon had to be devised to form this stubborn metal [a-Hayes]. Thus voyagers must be cautious about plans to duplicate or replace their spaceplanes or other large vehicles.

As discussed in chapters on propulsion, known materials seem to be required to perform near theoretical limits for most proposed very large technologies. Solar sails and their cables, launch track and its cables, some magnetic coils, and charged cables must operate as if made from the strongest known single crystals (whiskers), with stress levels about $(10)10$ N/m^2. Two large scales of equipment are proposed as necessary for starships, each 1000 times larger than the previous scale. The scale for sails, mirrors, grids, and some magnetic coils has been found to be about 100 km, 1000 times larger than a 100-person starship with compact propulsion (and 1000 times larger than chemical rockets and most terrestrial vehicles). The scale for charged cables or launch tracks is 100,000 km and up. Perfect superconductors are specified, not only for carrying huge currents at densities $(10)10$ A/m^2 near the largest present limits while withstanding their own high fields, but also operating warm, impervious to radiation while holding together under high magnetic pressure. When charges are needed, they are large, and materials are expected to support millions of volts. When any thought is given to maintaining and repairing such large sophisticated equipment during a voyage, no thought seems given to how to carry, or process, the enormous amounts of exotic material involved. Effects of planned high electric and magnetic fields on other equipment, on people, and on food plants are also often neglected.

Discovery of "cool" superconduction (at about 100°K) has raised hope of "warm" superconductors (about 300°K), a very large jump for nature. These could solve the problem of designing equipment such as computers which work here in warm labs and work later in cold space at very low power. Carrying two sets of equipment, one for warm and one for cold, is undesirable and less reliable. While electronics using warm superconductors seems remotely possible, obtaining high current density, high fields, and high strength from warm superconductors is unknown territory. There is much uncertainty about making and using "cool" superconductors, which still need extensive liquid nitrogen cooling, for large applications such as power cables, magnetic trains, and launchers on Earth. "Cold" and "cool" superconductors have reached only about 30 tesla steady-state, far less than needed for some large starship systems. Improvements in organic superconductors, now working up to 30 °K, cannot yet be expected to reach the range needed for high power, high field uses.

8.5 RADIATION DAMAGE, EROSION, AND SHIELDING

Radiation as discussed here is not to be confused with electro-magnetic waves. Radiation as a hazard commonly refers to high-speed particles, charged and uncharged, and energetic photons from four sources: material encountered by a high-speed starship, interstellar particles and photons, particles released from any particle or nuclear drive, and solar flares. Radiation from hitting hydrogen atoms ahead varies in energy and quantity, depending strongly on speed. At 0.3c, protons carry about 50 MeV and are very dangerous; at 0.1c, 5 MeV and still bad; at 0.01c,

50 keV and easily shielded. At 0.1c, interstellar protons are encountered at the rate of at least 3(10)12 particles/m²s, bringing energy at a flux (or intensity) of about (10)13 MeV/m²s or about 1 J/m²s. The intensity of radiation is approximately proportional to starship speed cubed.

About 0.1 mm or 0.3 kg of medium density material per unit area is needed to absorb 5 MeV, so that about 3 J/kg s is absorbed, or about 300 rads/s, or about 3 Grays/s [b-Rees; see Appendix A on units and D on radiation]. Particles of a given kind and energy have a certain "range" in material before 63% of them are likely to be absorbed. Radiation measured in rads depends on the thickness needed to absorbing this fraction of impinging particles, and the thickness needed depends on what material is absorbing it. A shield of a given thickness does not absorb all intended radiation, but as thickness increases, an exponentially diminishing amount gets through. Less than 1% of protons will pass through 5 times thicker material. Starship structure could take care of most leakage. Overheating of a thin low-mass shield must be considered. A typical metal heat capacity is 500 J/kg°K. In about 200 seconds a shield 0.1 mm thick increases temperature by one degree. Over a day enough energy would be absorbed to melt it, but it is thin and can radiate heat well too. For steady state the shield must radiate about 3 w/m², and will at less than 100°K.

Radiation in the form of cosmic particles comes from all directions and ranges from about 10 MeV (lower limit unknown) up to more than a joule (!) per particle. There are mostly protons but heavier nuclei all the way to bismuth have been observed and about 1% electrons [b-Friedlander; a-Linsley; a-Meyer]. (One might wonder how charge flow is conserved in the galaxy.) Observations near Earth are a poor guide because Earth's weak magnetic field protects us from cosmic particles below about 100 GeV and reduces the intensity by ten or more up to 10 GeV. Intensity peaks at about 1 GeV, is serious below 10 GeV, and decreases rapidly beyond that. The highest energy particles are so rare that a starship is unlikely to encounter one. The flux is about 10,000 particles/m²s with KE averaging about 5 GeV, equivalent to about 5(10)-6 J/m²s. The energy density (about 1 MeV/m³) of cosmic particles is comparable to the energy present in starlight and in magnetic field. As radiation absorbed in about 2 m of material, the energy content per unit area is about (10)-9 J/kg s or about (10)-7 rads/s. When rarer but heavier high-energy nuclei are considered, they bring up to 10 times more energy, so that the average cosmic particle KE is higher. They cause more damage in some cases [a-Birch]. Still, radiation from starship motion is much more serious than cosmic background by a factor of about a billion at 0.1c and still a million times worse at 0.01c.

Cosmic photons also come from all directions up into the GeV range, but cosmic gamma and x-rays are a small radiation hazard compared to protons and heavier nuclei. Whatever material helps shield from particles will protect from energetic photons. The neutrino flux is heavy but has completely negligible effect on a starship. The main products of a cosmic particle collision with matter at many GeV or higher are pions, gamma photons, and nuclear fragments, all of which cause more particles when they collide with more matter, producing photons, pions, neutrinos, muons, protons, and neutrons. The muons are difficult to shield against, but likewise do little harm since they interact poorly.

Radiation from propulsion is difficult to discuss effectively without knowing details of the particular method of propulsion. Many detailed considerations must await choice of a drive. A few preliminary observations are provided here. Radiation from the undirected output of an anti-matter drive is gamma at about 200 MeV and very nasty. There is also leakage of charged pions carrying about 200 MeV each which are easily caught in any nuclei around and cause reactions leading to more radioactivity. Upon decay pions produce muons carrying about 100 MeV, very penetrating but rarely causing nuclear reactions or other harm. A high flux would produce heat in material the muons pass through. An anti-matter drive annihilating 1 kg/s would produce a torrent of dangerous radiation in all directions. If, as is intended, all of the rest-mass is converted to radiation, about $(10)17$ J/kg is released, with about half directed away. If half of the remainder is captured in 10 tonnes of shielding, the level of radiation is over $(10)14$ rads/s.

Radiation from a fusion drive might be mainly 14 MeV neutrons, depending on the reactions used, and requiring up to a meter of light nuclei to stop. The normal exhaust contains He4 at about 4 MeV. A fusion drive exhausting 1 kg/s (approximately Daedalus level) might produce about 1/5 kg of neutrons per second, a dangerous amount even 100,000 km away. Another way to look at fusion is that it releases almost 1% of rest-mass as energy. If a major portion of this appears as uncontrolled radiation, the level is almost 1% of the anti-matter result, or about 1 trillion rads/s in 10 tonnes of shielding. Fission reactors also produce a range of dangerous radiation and their neutrons damage materials over time. Neutrons from reactors are likely to change reactor shield elements into other elements, weakening or ruining the shielding. Some metals such as nickel are very sensitive to neutrons and should not be used near nuclear drives. Electronics such as computer chips, storage media, and sensors are most susceptible since their operation depends on intact complex microstructure. Biological effects of radiation are discussed in the next chapter, and physical effects here, along with more detail on the shielding needed. During departure and arrival of a starship, solar flares or bursts of medium energy protons are a problem because the intensity is very high.

During travel, incoming protons striking forward parts of the starship are sufficiently energetic to cause nuclear reactions but most lose most of their KE before striking a nucleus [a-Martin 1978]. Thus they will not degrade a shield noticeably. The flux of about a trillion protons/m^2s would need millions of years to affect all the nuclei in 0.1 m (100 kg/m^2) of shield. Dust grains may be the main concern since they strike with much higher KE and can cause nuclear reactions. Electrons have negligible effects.

Present space missions use ordinary silicon electronics, somewhat radiation hardened and inherently resistant to mild radiation. This approach cannot continue because cold electronics may be needed and micro-circuit size continues to decrease. The smaller an electronic component, the more likely passage of a high-energy particle through it will permanently disrupt operation. In JPL's experience thus far less than 10 instances of radiation damage are known. Voyager spacecraft were designed to work in radiation at the level of 150 kilorads per year (about 1 million rems, depending on the particles), as encountered near the orbit of Io around Jupiter, and did not suffer significant electronic failures. Recent progress in using diamond films with or as semi-conductors may provide a new generation of more durable electronics.

Shielding from radiation can use massive materials (ice, rock, carbon, metal), "soft" methods like foil, foam, or dust clouds, magnetic and/or electric fields (for charged particles), and carefully selected materials. Whatever shield protects humans and biological systems (see Chapter 9) will protect most hardware. Some material shield is necessary because fields cannot deflect neutral radiation. Cosmic particles require typically 2 meters of dense material to stop them below 10 GeV [a-Birch; a-Matloff 1977; b-Johnson]. Earth's atmosphere is equivalent to about 10 meters of water and stops virtually all energetic particles at high altitude, but secondary particles from a high-energy proton hitting the nucleus of a gas atom often reach the ground. For large colony ships with atmosphere and agriculture, the rock and dirt floor at the rotating rim will be thick enough to shield from cosmic radiation and flares from below and the air nearly enough to shield inhabitants from above. The shielding that protects from radiation will not necessarily protect from dust damage unless it is a thick massive slab in front of the starship. Low-density shields made of clouds of dust or of foam blown from any suitable material can stop particles and dust and also cushion the shock of collision with a more massive object. When humans are not in the starship, the shield requirement can be relaxed considerably, with small thick shielding used for sensitive computers, data storage, and instruments.

The fifty tonnes of beryllium shield about 1 cm thick determined for Daedalus [a-Martin 1978] is needed mainly to protect from dust erosion. Upon collision the dust causes some nuclear reactions and radiation. Beryllium was chosen for thermal capacity and low density, not for lack of nuclear interactions. 10 MeV protons were expected to penetrate only about 1 mm. One tonne per year was expected to erode away. Such a fixed shield for the front prevents turning the probe or starship over for propulsive braking, or turning the exhaust to the side.

A lithium hydride shield is good for stopping neutrons within 1 meter. Despite being made of the lightest elements, it will be massive and should be used only where necessary around nuclear reactors. Tungsten 2 cm thick is good for stopping gamma photons from nuclear sources [a-Jaffe]. Shields can be consumed by conversion to other elements, and the thickness needed depends on the amount of radiation and the time it is irradiated. Cosmic radiation is much more penetrating than protons encountered at 0.1c, and a forward shield for dust and hydrogen will not protect the starship from cosmic particles. Instead the extra thickness (1 m or more) needed to stop GeV particles will stop all penetration of the much greater flux of 5 MeV particles from the front.

Besides hazards to people and systems aboard, some kinds of propulsion generate radiation hazard to those left behind. If an anti-matter (or inconceivably more powerful) drive were made and used, it would leave a trail of high-energy radiation far behind it. The starship could not be used near, or in the plane of, planetary systems. Fusion drives provide some hazard, if used on massive scale. The neutral radiation would travel indefinitely in all directions from the starship. The charged particles would be dispersed by galactic magnetic field surprisingly rapidly, about 1/2 degree in 1 second at 0.3c over a distance of about 100,000 km.

Electric, magnetic, or plasma shielding may be developed to deflect ionized interstellar gas aside, but it must be ionized first. Possibly a small ramscoop-like magnetic funnel could pull ions to pass by to the side of the starship (not through a tunnel down the starship). But a ramscoop system would immerse the ship in intense magnetic field. Various methods of producing electric and magnetic repulsion have been described earlier and in the references, particularly for large colony ships [a-Birch; a-Hannah; a-Matloff 1976]. Ways of building up and holding a space charge around the starship from interstellar matter should also be explored more [recall Chapter 4 on drag screens]. Possibly if the space charge region is large enough, it can ionize further incoming gas and dust and repel all interstellar matter, but the mass and scale of operation must be titanic and therefore unfeasible. If protons and other nuclei up to 10 GeV in cosmic rays (which inevitably are ions) are the primary radiation problem, then a potential of negative 10 billion volts (10 GV) would be needed to repel those protons. The few electrons can probably be left to a thin material shield, and the system would also repel protons from forward motion.

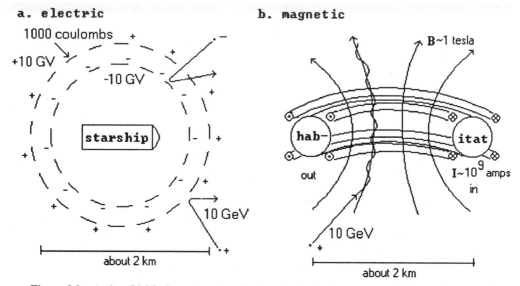

a. electric   **b. magnetic**

Figure 8.2   Active shields from cosmic and other ionized radiation, shown in cross-section.

One electric method encircles the vehicle with charged grids like balls of mesh about 1 km radius [similar to a-Birch; see Figure 8.2a]. Indeed, electric field from +10 GV is so high that a smaller radius of curvature is not feasible [see Appendix D]. At 1 km radius the field is about 10 million V/m, and the field can rip electrons from metals. The main grid must carry about 1000 coulombs. Because it attracts electrons like crazy, another grid with less charge and similar but negative potential is put inside it to repel electrons from hitting the starship. The combination appears nearly neutral from a distance and thus attracts little additional charge beyond what it encounters. Holding the grids together may work because the charge on them tends to stretch them. The grid wire must meet the usual high tensile strength requirement, and the mass may be about 1000 tonnes for 10 micron wire. Material connections from starship to grid are difficult because no material can be made to behave with 10 million V/m along it.

Imitating Earth to produce a large weak dipole magnetic field around the whole starship may be feasible. It would deflect some particles and collect others at the "poles", so a different geometry is needed. A simple dipole magnetic shield could protect a toroidal starship with a big hole in the center, therefore a large colony ship [see Figure 8.2b]. If current in superconducting coils is run around the major circumference of the toroid, a field is produced which wraps around the minor circumference of the toroid. Any ion aimed at the toroid crosses field lines and is forced into gyration that may keep it circulating around the toroid thereafter, somewhat as van Allen belts circulate around Earth. For a gyration radius of about 100 m and gamma of 10 for 10 GeV protons, the field needed is at least 1 tesla, very strong for such a large scale and over 10,000 times Earth's field. To produce it with a current loop 1 km in radius requires a current of a billion amps. At a current density of a billion A/m^2, near the upper limit of superconductors, the conductor has mass about 20,000 tonnes. The current needed is so large that unreasonably heavy structures may be needed to hold the coils. Other concerns are the strong field that might penetrate the habitat volume, and the intense electro-magnetic radiation from the gyrating protons until they lose their high KE.

An electric and magnetic method is the plasma core shield, again used with a toroidal colony starship [a-Hannah; b-Johnson]. The colony surface is charged to + 10 GV to repel positive ions and the whole vicinity is made to appear neutral to avoid attracting ions. A magnetic field is used to hold 1000 coulombs of electrons in a magnetic well (trap or bottle) in the center of the toroid. As with the starship in a charged spherical grid, getting into and out of the vehicle may require discharging it first, a long process requiring accelerating away the unwanted charge. A surface with 1000 coulombs on it involves a huge energy, about 10 trillion joules to charge it and apparently thrown away to discharge it.

To protect a 1 km toroid with a minor radius of 100 m with a layer of rock would require about 20 million tonnes. Clearly a magnetic approach using 20,000 tonnes has an advantage on this scale. But field methods are technically difficult and liable to failure. Material shielding (a passive method) is necessary at least in part but may require more mass than an active method using fields as the volume needing shielding grows beyond that of a 100-person starship.

Dust and neutral gas cannot be deflected by fields, but a foil "sail" ahead would ionize both as they pass through, so that fields could prevent erosion. Another kind of shield to prevent erosion from denser, larger dust (up to sand grains or rocks) was studied for Daedalus. Dust impacts, such as felt by Voyager near Neptune with dust density about 0.001 per m^3 [a-Stone], would increase in the vicinity of a star. A "shield cloud" of dust would be placed ahead to vaporize any substantial objects as the probe zooms through a planetary system at 0.1c [a-Bond 1978]. About 10 kg of fine dust in a cloud 100 m diameter some hundreds of kilometers ahead is expected to convert any object less than 1/2 tonne to plasma! An extra little propulsion system is needed to carry a dust packet forward. The cloud would need replenishing about once per "day", a rate that could not be continued long. Such a system has a steady mass loss unless it can be deployed only upon warning of debris ahead, and radar or laser may give such warning. Larger objects such as comets should be avoided rather than an attempt be made to destroy them.

8.6 COMPUTERS AND SELF-REPRODUCTION

Computers have been key components of space missions, robotic or not. Computers can calculate most numerical values needed for ongoing operations, and model most complex systems. For example, precise astrogation on the way out of the Solar System requires predicting the effects of many gravitating bodies (none spherically-symmetric) on the starship, along with the effects of dust, solar wind, and photon pressure, and fluctuations in the drive. Computers operating at one trillion operations per second (in parallel) are under development now [a-Corcoran], but even this speed is not sufficient for modeling complex distributed physical phenomena involving billions of positions in space and time and governed by several nonlinear physical laws. A pertinent example is the behavior of a fusion drive plasma. As parallel computing becomes the principal form of computing, both digital and in a return to analog forms, the available computing power will increase many more orders of magnitude. Lack of computing power itself does not seem limiting on either starship design or operation. Some limitations can occur if programs are too complex to validate, but it would seem that operation of a starship can use hierarchies of programs with clearly defined limits and purposes.

General purpose computers whose function is determined by software are more resistant to hardware failure but often slower than dedicated computer hardware that does one function. Either can include self-diagnosis and redundant parts so that a few hardware failures can be routed around. The question arises as to whether a fail-safe computer can be made. At present computers can, at best, reroute some functions around failed auxiliary chips, or perhaps switch from a failed processor to another one if it is fully wired in. Computers can diagnose some aspects of some of themselves. Whether they can diagnose their diagnostic system needs to be answered. The need for computers with arms with soldering irons, tweezers, and other tools to pull chips out of themselves and put others in from bins, while running, is not foreseen in Earth technology, but it probably can be done.

Some work on self-reproducing systems [b-von Neumann] is widely-quoted for providing confidence that computers can supervise their own reproduction and that of almost anything else, such as interstellar probes [a-Tipler]. Some cautions about this work should be kept in mind. In essence, a large mathematical theorem has been proved, showing with a theoretical model on paper that systems could be organized to reproduce [b-von Neumann; b-Dyson]. Four major components were identified, an automatic factory, a duplicator, a controller, and a program. A two-dimensional matrix of some 200,000 semi-autonomous interacting "cells" was found needed to organize a self-reproducing system, although no full list of cells and their functions was specified. The memory section adds up to about 83,000 cells. The system probably could be simpler in three-dimensions, as some work was expended on getting signals to cross over each other in two dimensions. This 1950s work was never finished nor has it been done experimentally (except in rudimentary form as described in Chapter 5). The published version includes extensive clarification, interpretation, and completion by a later scientific editor.

This abstract work is not purely speculative because biological systems using information coded in DNA follow a similar process. Indeed, the minimum number of coding elements (base pairs) needed by a virus that has its own reproduction machinery is about 200,000. But like a virus that works from already available biomolecules, human manufacturing systems work from previously purified materials and standard parts. A self-reproducing or self-repairing system landing (somehow) on a strange world must work from raw material, rock and dust laying about and little else [recall a-Freitas 1980 p 251-]. The study of cybernetic self-reproduction should be expanded because it is a key to long-term reliable performance of missions, self-reproducing or not. Von Neumann intended also to show that reliable operation can be obtained from unreliable components, a systemic effect enjoying some technical success even now.

Besides the real models discussed, a bit of the perplexing flavor of digital self-reproduction can be appreciated by considering a self-reproducing sentence [b-Hofstadter]. Here a human who can read and write english is needed to complete the cycle. No matter how long and precise, a sentence is unable to command a typewriter or pencil to make a copy of itself. A tangible (hardware) result is wanted, not just software that consists of virtual instructions that do nothing by themselves. Consider a sentence that says: WRITE THE FOLLOWING SENTENCE IN CAPS PUTTING QUOTED SENTENCE AFTER COLON: "Write the following sentence in caps putting quoted sentence after colon:". The part in caps is the rule to be carried out. It is software specifying action in words to be distinguished from the software in quotes which consists of words not to be meddled with, just copied. (Attempts to distinguish hardware and software here would be more confusing than productive.) Note the need for rigid literal observance of the rule and for careful attention to the meaning of symbols, whether they are instructions or material to be worked on. The period must fall at the true end of the full sentence, and other sentence ends are "understood" (left ambiguous, rarely a good idea). The dual level use of the word "sentence" is asking for paradox, and there is confusion in how to refer to a sentence, but even computers can distinguish what is in quotes from what is not. When the word "sentence" is copied, its contextual meaning jumps. The agent here is also expected to provide a colon symbol as directed, thinking of it as a sentence end, and to drop quotes after using what was in them rather than create an infinite series of quotes (but to keep quotes to denote a quoted sentence).

Many subtleties occur in this simple example, requiring many more rules than fit into the sample sentence. A fully self-contained and rigorously working system might well require the large number of unit cells found by von Neumann. The reader is urged to create more elaborate and/or precise versions, discovering that greater precision is likely to require more elaboration. A more realistic example would provide the material in quotes as an assortment of words, or even as bins of letters from which the rule would pull what was needed for self-reproduction. If a word processor could read and follow instructions, this sentence would be reproduced until memory was full and the system crashed. If all details are precisely specified in available computer terms, then some common programming languages will reproduce this sentence. All self-reproducing systems must distinguish software and hardware (also known as rule and seed, subject and object, gene and cell, meta-level and working level) as shown by von Neumann, biology, and computer technology. One does not work on the software while copying or using it, just as no ordinary text written with a word processor can reach and affect the operation of the word processor. Fortunately a third level of meaning (meta-meta-) does not seem needed for practical work yet, but mathematicians have found higher levels necessary and it could happen. Two levels seem difficult for a mind to deal with although it can imagine an indefinite hierarchy.

If a computer (really, a more general cybernetic machine) could duplicate itself as von Neumann seems to have proven, it certainly could fix itself. There would need to be in effect two computers working together, either of which can shut down the other and fix it. Working on a live computer may be impractical, and the computer must be able to control its power switches, too. When the power is off, the computer in effect does not exist. Paradoxes lurk around the corner when an entity (even a human) can control its own existence or diagnose itself. Consider us, or a computer, tracing the flow of information during the processing of that information, or tracing the tracing. We do it in a limited way and call it consciousness [b-Hofstadter]. Paradoxes are more than cute; when embodied unknowingly in blind hardware and programs, they can jam or "crash" a system and prevent recovery or further work. Starship computers probably need unjamming routines, if such can be designed to recover from any kind of programming failure. Proving a probe program to be fail-safe could be a very difficult problem.

Besides computers monitoring each other, further redundancy is needed. Several more computers may be needed to ensure sufficient redundancy in this most vital subsystem. Then the question is how to establish priority. Present methods include simple voting. The result of each computation is whatever 2 out of 3, or 3 out of 4 computers gets. These matters are being worked out now as computing enters three new realms: parallel processing, net processing with some degree of indeterminancy, and distributed networks of many autonomous computers. These kinds of questions are raised because starship proposers may have been hasty to suppose that we can design computers to handle any complex real unknown situation for 1000 years without significant failure. Multiple and parallel processing can solve most of the long-term hardware failure problem provided the number of computers is not exhausted, but if humans are present it would be better to repair any failed computers.

Beyond blind automatic self-repair or self-reproduction is the idea of artificial intelligence (AI). Controversy begins with defining AI, but the concept implies versatile computers which recognize new situations and attempt new solutions to them.

Probes without human guidance must be adaptable and make good decisions from vague information, even evolve to improve themselves. This capability is widely proposed as necessary for long-term missions that must do new things at new planets, even "talk" with new civilizations encountered, decide what to tell "them", and determine whether they are any danger to us. Many arguments about, but few impressive results of, AI have been presented [e.g. b-Hofstadter; a-Churchland; b-Dennett]. The paradoxes and problems of AI are beyond the scope of this book yet bear directly on the success of most kinds of interstellar missions and on the question of whether other beings are able to probe the universe looking for civilizations like ours. Bit by bit some AI capability is being developed--recognizing objects in visual scenes, recognizing the near infinite variety of handwriting, inducing rules from raw data--but it is too soon to say definitely whether any system besides a human brain with ten billion neurons operating according to well-known physical and biological laws can do all this and also be creative in new ways. Readers wanting to explore more of how mathematical limits such as Godel's Theorem might affect long-operating starships could start with the reference [b-Hofstadter]. There are limits to what we can do with logic systems (real computers or software) but no limits to the variety of situations in the Universe that those systems must cope with.

8.7 DATA STORAGE

Vast information storage is required, especially for a one-way human colony mission. The equivalent of at least a major university library is needed (10 million volumes including journals, or 10 terabytes, or (10)14 bits), plus a wide sample of Earth arts and other material. A very large effort would be required to reduce this to the information most needed, as would the effort to record digitally the large amount now available only printed on paper. Videotape is presently our densest storage medium but not a long term one and easily erased or damaged. The optical compact disk is a read-only memory which can be pressed from a master by a factory on Earth, for long-term library storage. But a quarter kilogram package stores only 1 hour of music, far less of video signal, and about 100,000 pages of words. Redundant storage is vital, preferably on different media. All of the permanent library should be in triplicate on very reliable media. A working subset of the library could be on magnetic tape or disk. Many people will prefer books and journals as being more convenient, but besides the low density of data storage they will find the best paper products to fall apart in a century or so.

Optical digital recording is very promising [a-Kryder], with 1 gigabyte on a 13 cm "compact" disk. It provides high-density storage, is not subject to electro-magnetic erasure (in most forms), and is durable and lasting. But plastics are used, and long-term stability needs study. Thousands of larger disks would be needed, and many that are erasable. Each one hour of low-definition mediocre color video would require several hundred disks or much data compression. Eastman Kodak Company is one of several companies that sell optical-disk library systems. The Kodak system is closet-size (about 4 m^3, 1 tonne), stores 1 terabyte on 150 disks, and accesses any disk in 6 seconds [a-Kodak]. A disk is 36 cm diameter, masses about 1 kg, and can hold 160 color high-resolution pictures. Laser write-and-read-only disks consist of layers of glass, dye-polymer, and gold on aluminum, and are guaranteed for 30 years, required to be clean, and limited to a narrow temperature range for use. Twenty of these systems, at 20 tonnes, would be needed for a minimum starship library in duplicate, very limited in pictorial and video storage.

This state of the art system for Earth use must be pushed much further to support an interstellar mission. It must be smaller with higher capacity for visual information, be able to rewrite data, and be more durable. Write-and-erase laser disk systems are also commercially available. These use the magneto-optic effect, which makes them sensitive to high external magnetic fields. They also use motors, never a long-term reliable component. Work has begun on an electronically-guided laser beam read-write holographic optical data storage using crystals in three dimensions [a-Pollack]. Each erasable "frame" could hold a book or picture, and each small crystal could store 100,000 frames or nearly a terabyte. (To change one bit, the whole book or frame would be corrected in memory, then recorded again as a whole.) Perhaps much more storage capacity is needed than 10 terabytes. Even now a few space satellites bring in data by the terabyte. Such large amounts are useless unless processed on line, but in some cases, such as a one-day flyby, temporary storage might be needed. Then processing can reduce it to an amount that can be sent to Earth or elsewhere at the slow rate permitted by the narrow transmission channels possible.

Beyond optical storage limited by wavelength (micrometers), a limit on information storage is determined by the size of atoms, or more practically, by the precision with which atoms can be manipulated (nanometers). Atoms are now being manipulated individually by laser and direct electric forces on a small scale. If a cubic millimeter (about 10 million atoms thick) is used for information storage with each layer of atoms laid down as one kind or another to encode information, it can store 10 million bits. A cubic meter (mass about 3 tonnes) could store on this basis $(10)16$ bits or 100 times the above requirement. If information could be stored and retrieved atom by atom in the cubic millimeter, it could hold about $(10)21$ binary bits, enough to satisfy thousands of high-tech civilizations. These are the nearly ultimate limits of information storage.

8.8 BREAKDOWNS

Breakdown can be discussed at the theoretical and the practical levels. The ideas of systems theory and the methods of establishing reliability can be extended to an analysis of the starship system as a dynamic system--a simple or complex non-linear system with many kinds of interactions or feedback loops. This kind of study is now being done on complex systems on Earth from earthquakes to economics, with a view toward predicting or preventing catastrophic failures. It appears that dynamic systems can be triggered to breakdown by some minor event because they are self-organizing for such an occurrence [a-Bak]. A simple and metaphorical example (and experimentally tested) is the addition of sand grains to a pile. The pile builds itself to a critical state. Then one more grain causes a large region to cave in and avalanche. Normal operation resumes in the form of steady addition of grains. At some unpredictable moment another different avalanche occurs. The behavior of critical systems resembles a weak form of chaos that increases slowly. Full chaos is unpredictable behavior by a supposedly well-defined system. Thus large fluctuations in the state of "well-being" of the starship may be unavoidable, along with large variations in physical performance of those subsystems that can be classified as dynamic systems and organize themselves to critical states. The concern for starships is whether recovery from catastrophe can be effected from the inside, without support from outside agents.

Major breakdowns can occur. Being "stranded" in space (at high speed) is a chilling prospect. The principal design steps for preventing breakdowns have been long-term prior testing of all subsystems, use of multiple backups (where each backup is an equally viable system), use of many identical units in parallel, and, to prevent a catastrophe from a breakdown, provision for multiple sources of power, tools, and parts for all conceivable failures. (Although if a failure was conceivable, a solution to prevent it should have been developed.) Technical limits are known for all these steps, and therefore breakdowns remain possible.

Main drive or not, there should be three or more large power sources available for emergencies at all times. An unused carefully stored and designed fission reactor might be started up in the event of drive failure, but it may require hours to prepare and start. Perhaps fusion reactors will become available in suitable size and capacity. For emergency outages hydrogen and oxygen could be "burned" in fuel cells for electric power. Or a radioisotope thermal generator (RTG) could be carried, but its hot isotope cannot be controlled or switched off and may only last decades (else it would not make much power). Or oxygen and some organic fuel could be burned in an ordinary boiler-turbine-generator system vented to space. RTGs produce only about 200 w/kg. Fuel cells have the highest efficiency of conversion of chemical fuel to electric power but are no better on power per unit mass.

There should be several sources of daily and emergency electric power, distributed throughout the starship so that no more than a few percent of capacity can be lost in one accident. Power supplies must be multiple, with separate wiring so that there is no way that the starship can lose all operating power. Diodes are a low-loss method to connect one device to multiple d.c. power lines if voltages are moderately high (100 to 400 volts). The line with the highest voltage present supplies the power. Long-term reliable low-mass storage batteries, a technology that still eludes us, are also needed to reduce the chances of loss of power to computers to essentially zero. Batteries and fuel cells are unlikely to work much colder than "room" temperature, a problem for cold probes.

Some far-fetched methods have been suggested for getting emergency power. Large coils in the weak interstellar field have been considered, but magnetic induction in a 1 km coil produces only about 10 volts per turn at 0.1c, and proportionately less at lower speeds [recall Chapters 4 & 5 & Appendix D]. Tapping energy from the starship's motion is possible via its passage through the interstellar partial plasma medium, but there will be more drag. If electrons tend to adhere to the starship, the protons they attract can amount to only about 0.01 amp at high speed if a collection grid is 1 km in area. Even at 100 kV, only 1 kw is obtained. Spinning storage carried from launch time or spun up by lightbeam on a "windmill" has been proposed [a-Matloff 1985] but is not fail-safe. If it breaks, much damage might be done. If the energy is lost, it cannot be replaced.

8.9 TRANSPORT

Transportation between starships in a fleet and between starship and planet, moon, or other destination poses some serious problems. At high speed there is radiation hazard and a "shuttle" that jets between starships spaced, say, 100 km apart must be well shielded too. Assorted transport vehicles are additional large pieces of equipment to be carried in multiples, requiring storage, repair, and fueling. Strapping about six of these to a starship adds to its size and mass, and they too must ride in a shielded area. Little thought has been given to whether other beings will face difficulties similar to what we face in developing easy, safe, cheap transport between planets and space orbits. Many unusual methods have been proposed, and each requires a much larger study in feasibility than can be provided here.

Space transport has not been well-solved for operations near Earth. Thus far chemical rockets are the best solution but enormous masses of fuel are needed to place small payloads in low orbit. Landing involves skipping a small heat-resistant capsule along the upper atmosphere to dissipate the large orbital speed, or gingerly gliding a heat-resistant unloaded shuttle into the atmosphere. Solutions for transport between starship and an Earth-size planet elsewhere can only be speculative. The starship may need to carry several versatile shuttles, more like the hypothetical space-plane (NASP) that would save most fuel mass by ramscooping oxygen from air to burn hydrogen [a-Morrison; a-Heppenheimer]. A space vehicle departing as a plane in air requires a runway and uses aerodynamic lift to reach high altitude before it must resort to rocket power. Such a space-plane must be designed for flying in atmospheres with a wide range of pressures and compositions and must endure intense heating to gain sub-orbital speed in air. For Earth atmosphere US NASA has begun work on NASP, France on Hermes, a smaller one, and Britain on HOTOL, a small robotic one [b-McAleer]. Very sophisticated materials will be needed [a-Steinberg]. The technically very difficult space-plane will not be proven until working versions are demonstrated, but a small space-plane without extra tanks would seem to be a better design than the present large shuttle.

Chemical fuel is marginal for Earth-size planetary use and out of the question for larger planets. Much must be used to depart the planet again. Reaction mass heated by anti-matter is a remote possibility for local transport; it develops high thrust and almost any speed from a low mass ratio and gram-size amounts of anti-hydrogen. There must be fuel supply and refueling method in space and on planet, probably by finding and breaking ice/water into hydrogen and oxygen. Landing a fully fueled shuttle or rocket is an unlikely procedure. The present shuttle needs three external containers of fuel to lift off and could not land with that cumbersome load. An unprotected fuel tank would surely rupture and explode as it heated up. The destination is unlikely to have a landing pad or even a hard surface, and probably not a long salt bed for a runway. The number of landings would be large, and shuttles or space-planes must make tens or hundreds of round-trips to space without mishap. As a baseline estimate, if just 1000 tonnes are to be unloaded from a starship, 40 trips with our present large shuttle would be needed. While present shuttle experience has shown that heat-resistant tiles can survive ten or so hot landings, a simpler heat protection covering should be sought. Most equipment could not be dropped with heat shields and parachutes, and if it could, it may never be returned to the starship.

Landing on a Mars-like planet may require a different vehicle than landing on an essentially airless moon. Only rockets can land on airless bodies, and would require much fuel for one larger than Mars. Apparently a planet as large as Earth without air is unlikely, and transport for this situation need not be planned for. Some experience with small 1 tonne retro-rocket landing has been gained with human landings on our Moon and robotic landings on Mars. Unlike methods shown in sf films, it is not known what sort of vehicle can handle travel deep into a gravity well with thick or thin atmosphere without extensive ground-support facilities. Before any landing extensive study of ground, weather, fueling possibilities and much more must be done. Some planets will prove unsuitable for colonies and the voyagers must be able to launch themselves and their equipment back into the sky.

Laser-heated rockets which need a huge ground-based laser with power supply are another unlikely possibility. Many still less orthodox methods have been proposed which use materials to their limits and/or involve immense projects. High-strength cables to lower and raise people and supplies between ground and synchronous orbit have been discussed for Earth's gravity well and found marginally possible using the highest strength materials [a-Pearson; b-Forward; classic a-Isaacs; see also Arthur Clarke's novel]. At least four cables up to 150,000 km long must be made at the destination or carried from here, each massing perhaps 25 million tonnes. Because the cables extend from ground to far enough above synchronous orbit to be self-balanced and under tension, they are rigid and have also been called "orbital towers". Should a mission find a planet with low gravity yet long-term stable habitable atmosphere, the method becomes more feasible but manufacturing the cable would be a major space operation requiring the whole crew, asteroidal material, and other landing vehicles to assist. Shorter and more daring is a rotating cable that swoops down from low orbit every few hours to deposit or pickup passengers [b-Forward].

Balloons, conventional towers, mountains, launch tracks, and jet airplanes can give a space vehicle altitude and/or speed before rockets take over, if these facilities are available. An electro-magnetic launch track on the ground has limited capability, needing 100 km of track to reach less than 3 km/s at 3 gees. An evacuated launch tube to orbit, suspended by balloons and containing an electro-magnetic track, is an interesting combination [b-O'Neill]. An orbiting ring of electro-magnetically driven material (iron) can provide a set of stationary platforms in low orbit, much easier to drop cables from [a-Birch]. Then there is the fountain of magnetic objects shot skyward with magnetic drivers to hold up a platform in space [b-Forward]. Many of these methods could be very efficient, recovering gravitational potential energy for the next launch and approaching theoretical minimum launch costs.

Chapter 9

Biological Requirements

Some requirements for living in a starship have been established earlier, and new considerations and limitations are provided here. While there are some organizing principles, discussion cannot be definitive because of incomplete knowledge in biology. Considering biological factors often requires additional study of physical factors, so that further physical aspects of starship design enter here. More speculation occurs on biological aspects because less is known. A deep understanding of biochemical and ecological systems on global and micro-scales is needed to prepare a starship mission if the starship habitat is to carry about 100 people and operate 1000 years. Again, in the discussion here one should consider what factors might make it difficult for humans or other biological beings to survive long periods in starships. Generally it is expected that if the physical problems of interstellar travel can be solved, determined humans are likely to solve the biological and social ones. Dealing with human life in a system, especially a closed one, raises social and ethical questions which are taken up more in the next chapter.

9.1 PHYSICAL REQUIREMENTS OF LIFE IN CLOSED SYSTEMS

The starship is nearly a closed system with physical exceptions noted earlier. For life it is essentially a closed system, although there must be some energy input for life to survive, either from outside or from stored energy. Energy flow through the system, exhausted in a lower grade less ordered form, is required for life to be able to maintain and increase its order and reverse entropy. The system is called an ecosystem, small ecology, or habitat and requires a definite boundary or envelope between itself and empty space to retain its material intact. The parts of an ecosystem are very interdependent, and some have gone so far as to say that the whole system is a living one. Its state needs to be stable but not at equilibrium; that is, its ongoing processes have many kinds of order during continual use of energy and change in the material parts. In a starship the organization of the ecosystem is imposed by the physical resources of the starship and by its human and cybernetic controllers.

Heating the ecosystem should not be difficult. A human idles at about 100 watts and uses typically 200 w when moderately active (one cannot be very active in a crowded starship). Use of equipment by people could be limited to a predictable average value of, say, 800 w per person including lighting (non-agricultural), cooking, computers, shop tools, lab instruments, and media, for about 1 kw total. Most habitat heat production would come from regularly operating equipment for human support such as computers, instruments, water pressure pumps, ventilation, and environmental recycling, and from lights for food growing. The total has been estimated at 1 Mw for 100 people, similar to US society but on the high side. With a total habitat envelope area of about 5000 m^2, at steady-state, power produced (from energy stores) must equal power radiated outside to space, giving an intensity of 200 w/m^2 at the envelope. The envelope would warm to about 200°K to radiate this power. Since insulation can perform much better than this, there is clearly plenty of heat to keep the habitat warm, and the excess can be sent to storage areas to make them more tolerable for occasional use.

Most light is used for food plants, and humans need moderately low levels except for fine work. Food plants need at least 1/10 of Earth's solar intensity, or more than 100 w/m^2. The spectral quality and efficiency of lighting are important, and no present light sources seem suitable either for plant growing or human living. Sources with ultra-violet or strong blue components should be avoided to avoid damage to skin, eyes, and materials, and most present light sources have such damaging wavelengths. Green plants need different light than people, intense red and blue and no other colors. People need weaker broad spectrum light with an emphasis on the red end ("warm") and no strong spectral lines. The literature has overstated the need for light both in Earth buildings and in space colonies. It would be very difficult to light a large colony ship with ecosystem as bright as the Sun does because of power required. Light is unfortunately not a substance that can be retained in a large enclosure. It all gets absorbed as heat and more must be provided continually.

Humans need at least 1/40 of solar intensity for an hour each morning to reset their diurnal "clocks" to avoid "winter" depression and several other syndromes [a-Wurtman]. The typical human clock free-runs at about 25 standard Earth hours per day. Humans have a choice of day length in a starship or colony ship and may choose 24, 25, or 26 hour, or longer "days" (few would want shorter). Human voyagers are likely to use several different wake-sleep cycles equally phased around the "day". Humans would sleep in, say, 8.33 hour shifts, and share facilities and duties similarly when awake. Couples and families would likely want to be on the same cycle. Three shifts are needed to monitor starship function continuously and to make the habitat seem less crowded. Times can be found when all voyagers can assemble briefly. People can choose to change to a different cycle, although they should change gradually an hour per day instead of abruptly as on Earth.

Humans need oxygen at a pressure of at least 13,000 pascals, somewhat more than half of standard Earth atmospheric oxygen at 23,000 pascals [b-Johnson; 1 pascal is about 1/100,000 of a standard atmosphere]. Nitrogen, relatively inert for humans but vital for some organisms, is needed at 27,000 pascals or more. The total pressure would be 40,000 pascals or about 4/10 atm and might best be raised to 50,000 pascals to allow oxygen near Earth normal. CO_2 pressure must be kept less than 400 pascals to prevent triggering human CO_2 asphyxiation responses. A low humidity about 10% to 20% (relative as used in weather, not as percentage of the air) indicates about 300 pascals of water vapor and lets people keep cool without sweating much. Other gases in the air from biological activity are mostly detrimental to quality and are covered in a later section. Air pollution from inorganic sources must be prevented also. If gee is zero, extra effort is needed to make air circulate and to pull all air through the sensing and filtering systems. Separating gases from condensed water vapor also requires special equipment.

Habitat volume for 100 people has been estimated at 30,000 m^3, giving volume per person 300 m^3. For a cylindrical habitat shape, total floor area (an unnecessary concept in zero gee) is about 10,000 m^2 in 14 levels or about 100 m^2 per person. In Rome there is 40 m^2 per person and New York has 98 m^2. NASA suggests about 50 m^2 and a volume of about 800 m^3 per person [b-Johnson]. (NASA was studying space colonies for this estimate which incorporate high overhead space.) The mass of air at 1/2 atm (density about 1/2000 water or 1/2 kg/m^3) in this volume is about 15 tonnes. Additional but isolated habitat area is needed for growing food as discussed later, up to 15% more volume (1 to 2 levels) to be added to these preliminary estimates.

For variety in starship environment, the temperature, humidity, light level, air motion, air composition, and perhaps other variables should be varied regularly or randomly, depending on preferences. A temperature generally somewhat greater than 20°C would permit light clothing, saving on cloth requirements. A clean safe environment would permit bare feet, reducing the need to make shoes (from what materials?) and making use of the natural low-wear properties of footsoles.

As with electric supply, the water supply serves biological and other purposes. Purified water must come from waste treatment discussed later and can be of several grades: drinking and cooking; washing bodies, clothes, habitat, and equipment; watering plants; medical and laboratory; and industrial processes. Micro-organisms and minerals must be prevented from cycling through most of these uses. Deionized water would prevent corrosion of any metal parts in the plumbing system. Taste for drinking can be restored by adding small amounts of minerals necessary for humans to the water supply for consumption. Pressure should be low to minimize leaks and failures. Waste water should be collected in noncorrodible pipes too. Up to 100 tonnes of water is estimated as the total need (up to 1 tonne per person), with most water devoted to the plant watering cycle. If industrial work or power generation involves large amounts of water, including conversion of water to hydrogen and oxygen and back, this water must be added.

For completeness, a sample of ordinary daily life habits (not all done every "day") in a small closed system might convey in another way what is involved in sending people to other stars. People must wake, toilet, eat, wash body, clean special parts (nose, ears, hair, nails, teeth), clean living area, snack, work on familiar and new tasks, toilet, converse, make love, wash, cook and eat, rest, exercise, work, eat, meet, converse, repair clothes and other personal equipment, snack, wash hands, have fun individually or with others, toilet, and try to relax and sleep after another domestic day in a small starship hurtling through space. In zero gee all these activities would be different, with some harder, some easier. There would be no sinks, showers, tubs, or flush toilets in zero gee. Washing anything in water rather than with a sponge would require using a glovebox or entering a tank of water in toto (an unlikely luxury and a problem for breathing). Sleep and sex might be fun since one can be comfortable in any position. No part of the habitat would be a "floor" and all surfaces could be used for living and working.

9.2 ARTIFICIAL GEE

One result of earlier discussion is the difficulty of producing artificial gee. In a medium-small starship with 1000 m^2 frontal cross-section area, there is not room for a rotating habitat that provides at least 1/2 gee. It would need to rotate so fast as to cause a noticeable coriolis effect. An example of the coriolis effect is that an object "falling" toward the outside rim (as if gee were acting on it) would accelerate toward the side as it gained speed. The human equilibrium system in the ears notices this effect and becomes confused. Rotation rate may need to be once per minute or slower to avoid motion sickness and illusions [b-Johnson; a-Murphy]. Only a colony ship about 0.5 km in radius can use a slow one-minute rotation and generate 1/2 gee. Higher gee would require a proportionately larger radius, and the starship becomes very massive. A design that uses less mass for shielding might be a rotating toroidal habitat with mass all around it [a-Matloff 1977].

A rotating habitat or wheel has many technical problems to be solved besides shielding and strength. The axle and bearings would need to be massive and very reliable. As the largest mechanical system in the starship, the wheel would not be easily stopped and repaired. A method of spinning it up, adjusting speed, and stopping it would be needed in any event. The bearings should be "frictionless" using superconductors for magnetic support, but may still transmit some rotation to the starship. (The massive starship should not be rotating if propulsion, astrogation, and observation are to be kept simple.) Entering and exiting the habitat require special hardware at the axle. Vibrations may not be completely suppressed. Any change in starship direction would evoke the gyroscopic effect of the wheel, complicating astrogation. Massive rotating systems can play other subtle physical "tricks".

If a rotating habitat is not feasible, the biological consequence is that humans may need to live in zero gee. This may change them permanently and make them unfit for planetary work. Humans may evolve to adapt to zero gee and then would build zero gee space colonies at the destination. Effects of zero gee on present humans include: resting posture becoming a crouch, blood redistribution with blood puffing up the face, water loss, space sickness with motion and orientation problems, muscle loss, bone loss, cardiovascular deconditioning, and irregular heartbeat (arrhythmia) [a-Murphy; b-Lorr]. Digestion and blood circulation are designed to work regardless of gee in the short run, but long-term changes may be serious. Most weakened muscles can be exercised strenuously without weight. But it may be difficult to keep the calf muscles' ability to pressurize blood. It is uncertain whether strenuous exercise with special equipment can keep bones and blood vessels and other parts fit indefinitely without gravity. Working all limbs against springs for two hours every eight on a machine would require much time, food, dedication, and many machines. Elastic suits that pull the shoulders toward the feet to load the skeleton and suits that apply negative pressure to the legs would be very awkward to wear most of the time. Such methods have been tried but not yet shown effective [b-Lorr].

If linear acceleration (in one line) could be present continuously, it would best be near 1 gee. Even 1/2 gee may require a supplemental program to maintain body structural health. No known feasible drive can provide continuous acceleration at 1 gee, nor possibly at 0.1 gee, the lower limit with significant biological benefits. If a 1 gee drive were found, it probably could go higher to 3 or 4 gee and achieve significant time dilation for long periods. But a new set of problems would arise. Humans would have health problems at more than 1.5 gee continuous. Immersion in liquid tanks might help tolerance of high gee for part of the time. Hibernation might be compatible with liquid tank travel. Other beings may be accustomed to higher gee and, if they had the propulsion, could cover thousands of LY in a few years of proper time (as measured in Earth years). Some beings may adjust easily to zero gee also.

9.3 FOOD AND ECOSYSTEM

Humans live to eat and eat to live. A much larger part of human activity on Earth is driven by the daily need to eat a reasonably balanced diet than members of high-tech societies realize. On a starship producing and preparing food must be very efficient and reliable so that most voyagers are free to spend most of their effort on other activities. (If people become bored they might prefer to spend more time on food growing and making.) Biological metabolism is surprisingly efficient in animals such as humans. If a minimal 8 MJ per person (2000 kilocalories) is taken in daily (as measured by total combustion in laboratory equipment), the average power input to the human is about 100 w. By conservation of energy, average input power must be the same as the average power output of 100 w expended by a human in near idle wakefulness. Long periods of strenuous activity require proportionately more food energy input.

Most humans are accustomed to eating meat from animals, but this custom seems unfeasible in a small starship. It has been demonstrated many times that people can survive well on grain, vegetable, and fruit diets, even without dairy components. [See a-Richards 1976 for a complete vegetable diet, but the growing areas listed seem wrong by 100.] Human societies have survived on diets with little variety but it seems possible and safer to provide a wide variety of foods. Proteins would come mainly from grains. A few grains and legumes already provide all amino acids, and corn and wheat are being altered to provide the full range and to make their own nitrates. New plants could be invented before a starship mission occurs which provide the equivalent of meat with more of a protein taste. Soybeans alone could provide most of the range of human needs but little effort has gone into making more palatable products. Chinese have made "milk" and "cheese" equivalents from soybeans for about 3000 years.

Daily food input per person requires less than 1 kg dry mass (and equals dried waste output after slight allowance for growth of hair and growth of children). Wet mass plus peels, husks, seeds, etc would amount to at least 5 kg of plant-produced food per person [b-Johnson]. Daily food grown and processed for 100 people must be about 0.5 tonne wet. The active biomass of stalks or branches, leaves, and roots is some 2 to 20 times greater and lasts until plants are finished. Some such as corn produce food once, some such as tomatoes produce continuously for a long period. Food preparation is a major operation and indicates why agricultural tribes are seen gathering, cultivating, processing and pounding food all day long. If food were all grown as corn in a conventional field, not one of the most efficient food producers, about 80 million J/m^2 is embodied as corn itself. (This is like harvesting 2 kg of hydrocarbon fuel per m^2, but the corn mass is about 6 kg/m^2.) At this rate 10 m^2 of crops must be harvested for each day of food. Since crop cycles might require about 100 days, about 1000 m^2 must be devoted to food growing.

Soil mass a meter deep would be about 1000 tonnes, a large mass although low-density materials would be used and partly saturated with water. A growing method for small plants that uses more technology with less mass instead of wet soil is described later. The biomass carried depends on the mix of plants (and trees) used and would be up to 100 kg/m^2 (similar to a rain forest) or about 100 tonnes total, including the water bound in it. In addition up to 100 tonnes of circulating water would be needed, much more than for human uses, although plants would use as little as 1/10 of this per day. NASA estimates about 20 m^2 of growing area per person, and this larger amount would provide a safety factor. The minimal 10 m^2 is used in further discussion here because much effort can be put into more compact, more efficient plants. If the diameter of the growing area is about 30 m, then food growing requires more than one full level or "floor" added to the habitat volume, not enough to negate earlier estimates of starship size and mass. Some vegetables could be grown in several levels of one floor.

The food estimate includes use of solar energy at full strength (1000 w/m^2) for half-time. Some plants can use light continuously, others need a dark cycle, but most food plants need intense light. Power required could be about 1 Mw. By conservation of energy it is not surprising that 1 Mw of light at typical 1% plant efficiency is needed to grow food for 100 people needing 100 w each of food power. Providing light at solar intensity is a separate technical problem. Plants need light in two narrow spectral bands in red and blue for photosynthesis by chlorophyll, and proper light sources (to be developed) could use less than 1/10 of the power needed for broad spectrum solar light, or about 100 kw total. Plants also need a source that provides far red light at the right times to trigger various parts of growing cycles (detected by the molecule phytochrome). There should be at least four separate isolated growing areas, and their light cycles could be out of phase so that full light power is never drawn at one time. Plants need a variety of temperatures to produce food, mostly on the warm side. Plants should be developed for cool food production to minimize water emission (transpiration) by leaves. Narrow spectrum light with no stray heat would also minimize leaf heating and transpiration.

More efficiency of volume, mass, and energy can be obtained in at least three ways: redesigning plants to grow more food per m^2 with maximum nutrient and energy content (perhaps using less water and light), growing in water solution (hydroponics) to eliminate soil mass, and manufacturing "food" directly in complex chemical machines. Natural photosynthesis is so inefficient that improvement seems possible by as much as a factor of 10. Fungi, molds, bacteria, and other simple food-producing organisms are more efficient but must start from organic materials. Plant-like organisms such as algae, plankton, and seaweed need not. Oceanic food producers have not been studied as much as their potential indicates. Genetic engineering is already being used to alter some of these to produce nutrients for humans, and these organisms might be used in nutrient factories, especially if space gardening proves unfeasible or breaks down. Genetic engineering promises the biotechnology for setting up simplified food ecosystems, and further genetic manipulation is likely to be done during their 1000 years of operation. Methods besides ordinary plant growing are likely to result in reduced variety for human tastes, whereas maximum variety of food kind and taste is desirable for making starship life enjoyable.

At present no significant food is manufactured from, say, petroleum, although some flavors, colors, and non-nutritive "additives" are. Manufactured food from some unspecified hydrocarbon feedstock is an unknown area. Starches, oils, and sugars are relatively simple, cellulose or artificial fiber more complex, and proteins or their 20 amino acids much more so. Fourteen (or more) amino acids are what human bodies need (with phosphorus and nitrogen included) but stomachs are designed to digest them from proteins. Trace compounds (21 established "minerals" or elements, 14 or more vitamins, others) can be added, but these are many in number with present requirements not fully understood. Providing flavors that convince the palate that the "food" is fully nutritious is another large unknown area. Dairy-like products could be made from grain protein and/or micro-organisms in a semblance of the processes a cow uses, but the efficiency must be less than the grain itself.

NASA has begun serious funding of experiments on small closed ecosystems in anticipation of a space station and Mars voyage. These are now known as Controlled Ecological Life Support Systems (CELSS) [a-Moore]. At the other extreme, large colonies have been proposed with towns, fields, lakes, and forests [b-Johnson; b-O'Neill] with some detailed but theoretical studies of ecosystems diagramed in this literature [also a-Richards 1981]. Figure 9.1 shows a simplified form of ecosystem flow chart without numerical values. Two or three different ways of controlling small ecosystems are envisioned, human, computer, and natural, in increasing order of complexity. Probably some combination would control a starship ecosystem, with sensors and nutrient controllers spread throughout for human and computer monitoring. Since humans are part of the ecosystem they would control, the possibility of paradox exists, but does not seem a serious problem in this case. The study of optimum closed ecologies can help Earth as well as interstellar travel, showing what must be done to feed people with minimum environmental damage and just how many people can be fed long-term without system breakdown.

CELSS are envisioned without animals, although some meat animals could be considered since poultry are nearly 25% efficient, as are dairy cows [a-Janick]. Fish are among the most likely vertebrates for food despite the mass and hazards of water tanks. Poultry for eggs could be carried if a low-cholesterol egg is developed.

Milk may be easier to manufacture than to obtain from a herd of animals necessarily smaller than cows. Another study [a-Richards 1981] has shown that animals provide a small net gain in productivity in a food ecosystem, and other arguments can be made for animal systems. Without animals the ecosystem has no ruminants to use plant waste and no carnivores to recycle animal parts that humans will not eat. An all-plant system with humans has poorly closed loops. On the other hand, with animals, bones become a major waste material requiring special treatment to recover their minerals. Animals carry another set of diseases and micro-organisms requiring control. Ethics may prevent carrying animals for slaughter. A few pet animals may be accepted in the ecosystem. Carnivores like cats would provide more pleasure than benefit to the ecosystem since there should be no food for them to catch. A variety of animals could be raised from stored frozen embryos if this is so decided.

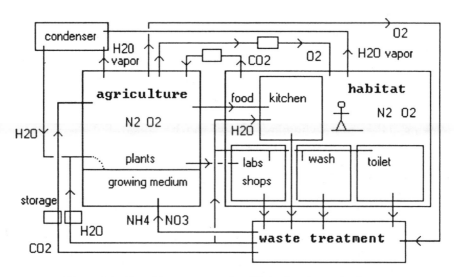

Figure 9.1   Starship ecosystem simplified, showing major loops.

"Gnotobiotic" ecology would grow plants without the micro-organisms needed to give them nutrition from soil and would require precisely-controlled water solutions. Agriculture can be as sterile as hydroponic tanks, or as natural as a garden plot of drained soil a meter deep. It is impossible to sterilize people because we depend on a wide variety of symbiotic micro-organisms to live. In view of the massive buffering capacity of soil and the difficulty of keeping human-carried and other micro-organisms away from food plants, carefully-controlled soil ecology seems more likely than sterile hydroponics. Earthworms may be the highest animal included for plant growing. Soil need not be dense, and non-sterile hydroponics can be used for small plants. Present incomplete knowledge of soil systems should be much more complete before a soil-based food system can be sent in a starship with confidence.

Overall, a closed ecology centers around biochemical cycles of carbon and nitrogen, with water and oxygen playing major roles. Carbon and nitrogen would cycle endlessly between oxidizing (low-energy state) and reducing (high-energy state). CO2 in air is the main feedstock for plants, and they take up CO2 as waste gas from humans (about 1 kg/person-day). This loop is not in balance because humans supply by respiration only about 1/3 of the CO2 that their food plants need. Much of plant production is inedible stems, leaves, peels, husks, etc which must be digested and oxidized separately, along with human waste, to produce the missing CO2 for another cycle. Nitrate NO3 and ammonia NH4, the other feedstocks for plants, come from human digestion, urine, and other anaerobic decomposition of waste. Some is inevitably converted to nitrogen gas and must be recovered from the air. Large energy consumption occurs if carbon is made to cycle from its reduced hydrocarbon fuel state CxHx (x denotes various numbers) to fully oxidized CO2. Plants also release more oxygen than people need (about 1 kg/person-day), and the extra is used to oxidize waste. Water inputs per person-day are about 2 kg drunk or eaten and about 1/2 kg as carbohydrates (literally carbon with water, made by plants). Water outputs per person-day are vapor from breathing (about 1 kg), urine (about 1 kg), and sweat and excretion (about 1/2 kg). Plants take up and emit far more water and dominate the cycle. [b-Johnson; a-Richards 1981]

Many starship habitat cycles interconnect with the agricultural ecosystems. Total isolation by airlock would be difficult, but partial isolation is needed. Plant growing affects water, oxygen, nitrogen, carbon, acidic, and many mineral cycles. Some water moves as vapor, and plants generate high humidity. Semi-tight airlocks to the growing areas might be preferable to using power to remove excess water from the general atmosphere. Water moves as liquid put into the soil and returns as food, cellulose, oils, and other plant products. Moderate water usage is for washing, cooking, cleaning, and laboratory and industrial work. Carbon comes from the growing area in food, plant products, and waste leaves and roots. Plants and/or people also emit CO, NOx, SOx, HxCx, H2S, and HCl, and plants emit control gases such as ethylene that affect reproduction and fruiting. All the details of the cycles need not be traced here to appreciate that complex and careful accounting is required. The return sides of the cycles involve waste treatment, covered in the next section.

Seeds may be planted everyday to keep regular harvests coming. Genetic integrity of foodplant seeds is another difficult problem, hitherto relying on vast farms for hybrid seeds. Seeds should be selected from the most successful of each generation of plants and used for later generations. Unvarying genetic monocultures should be avoided, and genetic variation should be permitted as long as food does not become unusable. Seed planting and food harvesting will always be difficult to keep in step, and shortages and surpluses will occur. Ecosystems are complex nonlinear systems whose outputs can oscillate regularly or chaotically. There should be hundreds of plant varieties so that voyagers always have many food alternatives. For security, the amount of food being grown may need to be up to twice as great as can actually be consumed, and in times of plenty food grown with great difficulty may be partly dumped as waste. Alternatively, freeze-dried food may be stored until several years of food have accumulated, then production scaled back to the minimum necessary.

NASA and others have proposed a no-gee "salad bar", "vegematic", or compact greenhouse system with lettuce, carrots, onions, cucumbers, even tomatoes, growing "above" and "below" chambers that nourish roots [a-Briefings 1990; a-Leggett]. Roots can be in wicks that draw water from circulating solution, or a mist of water and nutrients can be blown through a root maze. Roots are kept dark to promote root growth into nutrient chambers, and sheets of efficient lighting between layers of plants cause vegetative growth toward light. Gaskets at stems prevent nutrient from leaking out. Small food plants can be stacked several layers per floor. Plants per "floor" area may not be greater than soil methods, but volume density is. Surrounding light sources with plants is efficient in energy as well as volume. For growing tall plants such as grain, corn, dwarf fruit trees, the illuminating ceiling must be raised and stacking is not practical. Growing food may require artificial gravity, perhaps as low as 0.1 gee. A small "wheel" for growing food is to be considered, but it cannot be so small that its mass does not bring serious structural problems. Its angular momentum would also affect starship orientation. Possibly a few separate starships in a fleet might be spun and used only for growing food, with more compatible air and lighting too.

Plants and humans depend on many elements down to microgram quantities. All necessary nutrients and their functions must be identified for long-term successful ecosystems. The content of humans includes 21 or more elements ranging from 1.5 kg of calcium down to 1 milligram of cobalt, with daily intake requirement about 1/1000 of these amounts, give or take a factor of 10 [a-Scrimshaw]. Many trace elements are vital parts of enzymes needed to carry out many digestive and other physiological functions. Plants need a few elements that humans do not (e.g. boron). Entropy makes these difficult to recover from waste, although cycling human waste through a growing system might recycle some of these adequately without an intentional separation. It seems that some amount will be lost from the system, concentrated in soil or dispersed as general dust (from vegetable peels, seeds, hair clippings, etc). Continual measurement of trace elements in food and humans must be done to determine how far concentrations are deviating from norms. If of the 1 gram to 1 microgram of an element needed daily, only a nanogram of an element is lost from accounting in the system per person per day, then less than a gram of that element (often a nontoxic compound) must be carried to replenish it over 1000 years. There is no difficulty replenishing much larger amounts lost as long as extra supplies carried are measured in kilograms rather than tonnes.

A reserve of food should be stored by several different preservation methods including freeze-drying (not to be done by exposure to space else water is lost), by ordinary freezing, by irradiation, and other means. While carrying a variety of meats and unusual vegetables and fruits from Earth in low-mass freeze-dried form is desirable, their mass and volume for millions of servings would still be prohibitive. Storage is better used for a large variety of seeds, not only for food growing in the starship but also for planting at a colony. Seeds must be carefully stored, but freezing them is uncertain. A wide variety of pollen, sperm, and ova (gametes), zygotes, and embryos, enough for a zoo, should also be carried and stored much colder than food. Many useful bacteria, fungal spores, insects, and invertebrates could also be carried this way. No electrically-cooled freezer should be needed to store biological material and food because heat radiation to space can cool a special chamber to a few degrees K. However, it must be shielded heavily to prevent cosmic radiation from mutating vital seeds and animal embryos. The possibility of long-term storage of seeds, embryos, and germ cells by whatever means is quite uncertain but should be solvable because DNA itself can be kept stable indefinitely. The full diversity of life on Earth, involving millions of species, would likely not be carried because of limits on storage. Bio-technology would likely progress to develop equipment which can generate simple lifeforms from their DNA codes.

Another factor is pollination. Some food plants need no pollination, some are self-pollinating, and some require pollen to be transferred between plant genders or between different plants. Flying insects are not foreseen yet for space ecosystems, and some efficient regular simple mechanical pollen transfer method is needed. Other techniques can simplify closed ecologies. For example, silicon compounds are ubiquitous in agriculture yet unneeded (they make leaves rough). When eliminated from the system, waste processing is easier. And the ecosystem might produce (and recycle) other products needed in the starship. For example, cellulose can be made into paper (using safe processes) for hygiene and medical treatment, and a moderate amount of waste paper treated like human waste. Other fibers can be grown for making cloth. Chemicals for analytic and industrial work such as alcohols, oils, and acids can be made or refined from ecosystem material.

9.4 WASTE RECYCLING

More difficult than growing food is recycling all waste products from growing and eating food. Recycling human waste would probably be done with separate technology rather than directly by micro-organisms in soil. By conservation of matter, about 0.5 tonne of sewage and food residues must be processed each day for 100 people. At least 10 tonnes of water must be recycled, but most of this comes from plants via air or runoff and can be returned to plants with adjustment of nutrient content. Some of the 1 tonne or so of other waste water is already separated into water from food processing and urine from human metabolism. Urine is normally sterile and almost the right mix for fertilizer as is. Major concerns are biological toxins from people and the nearly indigestible cellulose and lignin from plants. Collecting waste is another problem area, especially if zero gee prevails. NASA has put much effort into human waste collection systems using forced air (space toilets). Unused material from plants is simple to collect, but washing hundreds of dishes may not be simple in zero gee.

Four kinds of waste recovery systems are being considered for spaceships: the fast physical ones of incineration and wet oxidation, and the slow biological ones of activated sludge and composting (very slow) [a-Moore]. In the first, organic material is burned with oxygen to produce CO_2 and other oxide gases, the exact results being sensitive to conditions. Or it is combined wet under heat and pressure (100 atm) with oxygen to achieve less complete combustion and done in quick batches in high-strength tanks. Or, in a biological process, organic material is combined with much water, oxygen, light, and certain bacteria or algae as in present sewage treatment. Or it is mixed moist with appropriate fungi for oxidation. In all cases the process is aerobic, breaking down proteins and fibers and fully or partially oxidizing carbon and other elements. Some of the output from semi-natural biological processing is less ready for feeding plants since it is oxidized less fully. Less energy is used, but carbon compounds must be oxidized to CO_2 by another way such as wet oxidation (which is mainly self heating). Incineration should be done on dry material, since removing water as steam is very costly. Methane-producing anaerobic systems do not seem useful here. But in imitation of ruminants, bacteria are already being developed which convert indigestible plant cellulose into edible carbohydrates, bypassing some need for waste processing and increasing energy efficiency by at least double.

Some set of these processes must be selected with several waste processing channels for different chemistries. The same process that makes CO_2 also makes NO_2 which is bad for everything. Carbon and nitrogen cycles must be separated. Waste processing tends to form many compounds which plants cannot use, binding up the scarce phosphorus. A system to control sulfur has been described [a-Macklin]. If silicon is present and heated, it binds up nitrogen and phosphorus. The many salts are difficult to separate but sodium chloride must be removed and not let to concentrate. Most gases to be removed from the air dissolve well in water, which aids their separation. Hot waste treatment corrodes many metals to contaminate the system, and stable but replaceable plastic, glass, stainless steel, or other nearly inert plumbing must be used. The waste treatment systems must be completely repairable like all other systems. Thus there must be at least three of each kind of subsystem to avoid having all of one kind of processing down at one time. There must be tanks to hold waste in the rare event that no processor is available. Another technology to be investigated is the membrane that can separate various gases such as oxygen from nitrogen. An enzyme that converts nitrogen gas to ammonia exists in nature and is now being studied [a-Moffat]. Typical of enzymes for living systems, it uses an atom of iron and an atom of molybdenum in each huge protein molecule. (Nature has found metallic elements essential for moving electrons around to make organic reactions occur.) The problems of handling trace metals have been indicated earlier. Combustion would make unusable oxides of them and should be avoided.

Bacteria and viruses in human waste are a major cause of health problems when sewage is not separate from water supply and agriculture. Over time, or by design, everybody in a starship may acquire the same bacteria after living closely and sharing communicable diseases, eliminating problems of bacterial contamination in the ecological cycle. But heat and chemical sterilization should remain part of the waste treatment cycle. Humans also rebuild/replace most of their bodies about once per decade. Hair and skin loss from the outside is a small part of the number of tissues being steadily broken down, excreted, and replaced. However, human waste treatment can be minimized by developing plant foods more closely matched to human metabolism and by avoiding meats.

Since plants may not need the CO_2 from processed waste immediately, the waste must be stored until CO_2 is needed from it. CO_2 cannot be stored in the air because humans cannot live in high concentrations of it. Substantial CO_2 storage would be difficult and require a major extraction process from the air. The time cycle may be up to 100 days. The total carbon mass stored may need to be similar to the biomass in use (about 100 tonnes or about 15 times human mass). Counting soil, water, treatment machinery, storage, and material reserves, but not structure, the entire ecosystem mass seems to be about 1500 tonnes for 100 people. Ecosystem mass per person has also been estimated at a low 4 tonnes per person [a-Richards 1981].

Power for recycling is difficult to estimate without knowing the principle cycles selected. The assumption here is that most power is expended on the bulk of the material handled and that no minor cycle requires an unusual amount of power. Pumping 10 tonnes of liquid per day at several stages requires a few kilowatts. Pressurizing a tank of daily waste requires less than a kilowatt (average). Since a 1 kw blower can move about 50 kg of air per minute, this is sufficient to circulate the 15 tonnes of air at several stations. Boiling and condensing all water flow each day must be avoided because hundreds of kilowatts would be needed. Radiation to space should be used for condensing vapor. With many more small pumping and heating cycles as indicated above, at least 10 kw (average) seems required for all waste treatment but surely less than 100 kw (1 kw per person). Some processes would be done briefly each day at high power.

9.5 GENERATIONS AND LIFETIMES

Present estimates of travel time converge on a journey of about 1000 years to the nearest stars. This exacts a severe toll on humans making the journey and on those back home who began the project. There are several biological implications. Some 30 generations of people would be needed to complete the mission, due to the long time for maturing and training each generation and the short time thereafter for best reproduction. Some proportion of people would be incapacitated near the end of their lives at age 70 to 100 and be a burden on others, raising social and ethical problems. Some who recognize that their mental or physical abilities are not contributing may opt for voluntary euthanasia, and the remainder may be so few as not to be a burden, especially if medical technology can prevent or cure circulatory and brain diseases.

Population can be controlled by selecting a couple to have a child after the accidental death of someone and by authorizing couples to conceive at a rate anticipating the average death rate. A prioritized waiting list of couples to produce offspring is likely. All sexually-active voyagers would practice contraception carefully, and in rare cases of failure the couple or larger group could support the extra child. If the gender ratio goes out of balance, gender selection could be used during fertilization. After several generations inbreeding is an apparent problem, but deleterious effects have been shown infrequent [a-Bittles]. Testing can select voyagers without various rare genes which are aggravated by inbreeding. A variety of sperm and ova from other Earth people would insure genetic variability. People and genetic banks must be protected from mutations from radiation or other agents to maintain the quality of the limited genetic stock. A wider variety of genetic (and associated ethnic) backgrounds can be represented in this way than in the original group. Voyagers would be selected for lack of serious genetic defects, and cures for many may be available by the time of the mission. If the starship is part of a fleet carrying 1000 people, then genetic problems would be minimal. Quality of fetuses would be insured by safe tests early in pregnancy. No facilities and no surplus caretakers are likely to be available for defective children. Social aspects of mating and reproduction are covered in the next chapter. Generally, a rational and dispassionate attitude toward population and the quality of health of that population seems required for mission success.

There is no evidence yet that active human life can be prolonged beyond about 80 years for all persons, although semi-active life may reach 120 years for a few [a-Olshansky]. Human cells seem programmed to divide about 50 times and stop, but if this is the case, it should be possible to identify the mechanism and alter it [a-Hart]. Aging of cells is likely to be separate from the degenerative diseases that usually lead to death by old age [a-Gibbons]. Another general result from Earth biology is that there is a definite relation between lifetime and body mass and between lifetime and metabolism. One might wonder if beings from other planets and ecologies could have much longer lives. A long-lived animal on Earth is big and slow, but in lower gravity a creature might be mentally and physically lively yet have slow metabolism. Humans in space might select themselves for slower metabolism or slow it with biotechnology.

If aging as a fundamental biological problem is solved, interest in interstellar travel would change. Popular predictions of human immortality are quite premature, but lifetime might be extended to hundreds or thousands of years so that voyagers could live much or all of the journey. It has been said often that the longer people live, the more concern they will have for their safety, with immortality dissuading most from considering the hazards of interstellar travel [e.g. a-Drake]. Be that as it may, if ancient people cannot be healthy and active but require caretaking, then long lives would be a liability on a starship and would prevent replacement with new generations. There would be a tradeoff between scarce medical and ecological resources and the mental contributions and wisdom of the very old people. A committee may need to decide for the few who cannot decide for themselves to cease burdening the eco- and social systems.

Motivation to continue the journey would be reinforced if 1000-year people can learn to cope with long lives in cramped quarters. Some humans on Earth would also choose long lives, including many scientists and others who begin the starship program. They would live to see the journey completed, having been very busy with other projects in the meantime, such as sending out more in a regular series of starships. Alternatively, safe hibernation might be developed, relieving people of witnessing the entire journey and much reducing the need to produce food and power enroute. Present partial successes with cryogenic preservation of cells and with understanding and duplicating animal hibernation are not sufficient yet to project that humans can be supercooled and revived to normal brain function. Frogs are the highest animal yet to survive being frozen solid [a-Storey].

The sad subject of human death raises several biological problems. If 15 generations or 1500 people massing about 70 kg each die during the voyage, that is 100 tonnes of organic material to account for. The ecosystem can perhaps not tolerate the loss of this material by dumping overboard. People might opt for several solutions. Some would want to be incinerated and recycled in the ecosystem. A few might prefer their bodies be jettisoned into space, and this could be done. Some might want to be frozen and carried for burial, and this could be accommodated on a small scale. The voyagers would decide policy if individual preferences are not compatible with mission survival.

One of the most bizarre proposals on human biology has been the sending of human embryos by probe to distant planets. Proponents have supposed sophisticated robotic machinery which can bring fetuses to infancy, then nurture and teach them to be autonomous humans to start a society on a habitable planet. The idea is to reduce the size of the starship and to eliminate the tedium and complexity of sending adult human colonies. But the complexity of the robotic equipment and the amount of supporting equipment and landers would seem to comprise a substantial starship load anyway, assuming that automatic equipment can function so long. And there is plenty of evidence that children developing without committed human parents come out lacking some essential qualities. A human colony that we would recognize as such might not be the result.

9.6 RADIATION AND OTHER HAZARDS TO LIFE

As before, radiation here refers to energetic particles and photons that affect materials and living organisms adversely. At 0.01c the forward radiation has energy about 50 keV per particle and a flux or intensity much greater than cosmic. At 0.1c forward radiation at 5 MeV per particle is very serious. The radiation in rads absorbed is more severe in biological materials [see Appendix D]. For protons dose in rads is to be multiplied by a Relative Biological Efficiency (RBE) of 10 to obtain the biologically effective dose in rems. RBE is 20 for light nuclei as particles and possibly 1000 for heavy nuclei. Light particles like electrons and photons have RBE as low as 1 [b-Johnson; b-Rees]. This is another way of indicating that the effect in living material is more severe than just the energy dumped in tissues or the ionization caused there. Fragile and vital microstructures such as DNA are disrupted. The average cosmic radiation in local space is about 10 rem/yr if it reaches living material, equivalent to about 100 simple medical x-ray pictures.

The permitted dose for people is about 0.5 rem/year, or 0.05 rads/year for protons. (In SI units this biological dose limit is 0.005 sievert per year.) Time averaging works up to a limit, so that a higher dose might not cause problems if followed by a suitably long period without exposure. Some recent work has shown that even lower limits should be set, 5 rem for life, or about 0.1 rem per year, 5 times lower [a-Briefings 1991]. The Earth background exposure is about 0.1 rem/yr and medical exposure is similar. On a starship there would be many other ways to receive low or high doses of x-rays and other radiation. Any nuclear reactors (fission or fusion) irradiate those who must work on them and those near the shielding. Planets with magnetic fields have belts of trapped energetic particles that must be traversed to leave or reach the planet. Near Earth the dose is about 1 rem per day [b-Lorr]. Some stellar objects, unlikely to be visited soon, emit intense energetic radiation. Solar (or star) flares of protons, an occasional and severe hazard on the way out of and into planetary systems, can give doses of hundreds to thousands of rem over a few hours at the distance of Earth [b-Lorr]. Such doses are fatal and millions of times greater than the permitted dose. Death is likely after 500 rems in any short time, whereas 500 rems spread over a lifetime is not likely to cause problems although not clearly safe. NASA has suggested astronauts might tolerate about 200 rems if received over several years.

While cosmic particle radiation provides less total dose, the higher energy can cause many kinds of secondary particles flying through the starship after the primary particle is stopped by thick shielding. Cosmic particles are dangerous, come from all sides, and require at least 2 meters of solid shielding all around living organisms. Earth's atmosphere provides the equivalent of about 10 meters of shielding. The shielding requirement is determined by protection from cosmic radiation. A layer of cosmic particle shielding will absorb all forward protons and capture dust grains too. Another effect of cosmic particles is permanent damage to retina and neural cells which do not replace themselves. At the rate astronauts saw flashes in their eyes in unshielded spacecraft, they could lose most vision and brain function in a few years.

Forward radiation was calculated in Chapter 8 at about 1 rad/s at 0.01c and about 10,000 rad/s at 0.1c as deposited in the top fraction of a millimeter of material. This sounds very worrisome but is not. The former will not penetrate the skin and the latter would not penetrate a spacesuit, plastic, or any substance between person and interstellar space. Hence this radiation cannot have significant biological effect despite the intensity, and it would be misleading to estimate the biologically effective (RBE) dose. By comparison, solar flares can deliver GeV protons in the same energy range as most cosmic particles but at much higher intensities. Increase of energy accounts for most of the increased radiation danger because GeV protons or their products will penetrate several meters of material.

Shielding the starship is a problem because material 2 meters thick amounts to about 5 tonnes/m^2 all around the habitat. Yet this mass is barely sufficient because the shield itself enhances the amount of secondary radiation. With a minimal volume of habitat and stored materials used around it as part of the mass, the mass required is still huge, over 10,000 tonnes, or as much mass as was planned for the entire starship. (If the starship carries a massive drive with fuel then this shielding is a negligible part of the total, but as payload it requires higher mass ratio.) No electric or magnetic shielding seems suitable unless a way is found to ionize the interstellar gas before the starship crashes into it. (More details of shielding were covered in Chapters 5 and 8.)

The danger of high electric and magnetic fields to human life is well established, but the danger of very low fields is unknown. Some present claims find harm at magnetic intensities less than Earth's field. The need for very high electric or magnetic fields for starship propulsion has been discussed. If these are present, shielding is difficult because small leakage still constitutes a biologically high field.

Present technology is fairly good at electric shielding but not so good at magnetic. Lasers at all powers and high-power radar and radio transmitters are also well known hazards needed on a starship. If a solarsail or collector is used, the intensity of any propulsive lightbeam is ruinous to eyes and may also burn people and materials if exposed.

Other hazards to health resemble hazards on Earth, such as noxious chemicals, injuries from equipment, electrical shock, electric and magnetic fields, fires, and other disasters. All equipment should be designed to be compatible with people: no sharp edges and corners, no swinging doors, no hot spots, no slippery handles, tethers for use in zero gee or in space, no release of loose particles or droplets in zero gee, no toxic emissions, no electrostatic charging, and many other factors. Because of the sealed environment, extreme care must be taken not to bring materials into the habitat portion which emit noxious vapors. Air pollution control must be more stringent than on Earth. Plastics, glues, fabrics, and other manufactured materials retain low levels of solvents and release them over time. Active filters might be needed for general removal of hydrocarbon, mercury, and other vapors. There must be detection and removal of simple trace gases such as CO, CH_4, NH_4, NO_x, SO_2, H_2S from the ecosystem by means of filtration systems specific to each gas. Furnaces and other heating equipment for manufacturing and scientific work may need to be located in a separate air system so that unusual organic, acidic, and metallic vapors are not released into the general system. Open fires must be strictly avoided because of the variety of noxious vapors released.

9.7 HEALTH

All important aspects of medical science cannot be surveyed for a starship study at this time, but the future of medical science is crucial to long missions. Besides problems of reproduction, genetics, radiation treatment, and accidental injury, there are up to a million human diseases to keep a medical staff busy. More will continue to be discovered, and new ones might appear by mutation. It seems impossible to screen all voyagers to prevent introduction of diseases [a-Crosby]. Bacteria, viruses, fungi, and other carriers of disease from Earth would enter the starship habitat in many ways even if all animals and insects are excluded (unlikely). Humans carry a host of tiny parasites which are benign but could evolve to threatening forms, especially in the semi-sterile starship with many open niches. Micro-organisms which entered the starship in personal and general equipment might find niches to evolve, with enhanced radiation hastening their evolution. At a new planet with a strange biosphere new medical and disease problems could be negligible or overwhelming.

Many voyagers should be trained in a variety of medical practices while two or three are medical specialists (and trained in other supportive abilities). Passing on this knowledge and practice will be difficult, since a major medical school, hospital, and government laboratory cannot be carried. It is to be hoped that large computers dedicated to detailed models of human physiology will be developed. Many more noninvasive and compact medical instruments should also be available in the next century. Carrying or making every medicine and vaccine presently in use seems unfeasible for a starship with limited laboratory space and essentially no test subjects, human or animal. A different approach to medicine seems necessary, perhaps one based on a small set of medicines, full understanding of human genome (our 3 billion unit DNA code) and immune system, and a computer model of human physiology that can specify individual drugs to be synthesized. A universal biochemical synthesizer may be possible, an extension of present research methods in biomedicine. Genetic engineering is likely to assist rapid production of antibodies on a lab-scale when an immunity boost is required. It may even become possible to locate, interpret, and correct mutations in reproductive cells. While the reproductive cells for "bringing forth" lab test animals would be carried, it is to be hoped that better test and culture methods will be available than cages of white rats.

The combination of radiation exposure and zero gee (and some other physiological and psychological stresses) is likely to retard immune system operation and make voyagers more susceptible to external infections, to "opportunistic" infections from normally benign micro-organisms, and to self-produced malfunctions [a-Murphy]. More radiation causes more mutation of bacteria and viruses to produce more unexpected diseases. Continual testing of environment is needed to detect new forms early. A preventive measure is filtering of air to catch spread of micro-organisms by air circulation, with testing of filtration products. Any epidemic would be a starship disaster.

General prevention is the simplest route to minimize health problems, and voyagers should take care on diet, hygiene, exercise, and safety. Poor sleep is another cause of stresses and poor health. Astronauts have had difficulty sleeping in space because of zero gee, spacecraft noises and vibrations, crowded schedules, and crowded quarters. A regular "day" on a starship that includes proper bright light exposure and private quiet sleeping rooms may permit normal sleep cycles. The more massive the starship the more people will sense, in quiet moments, that their world is large and therefore secure, even if huge propulsion is throbbing in the distance.

The most dreaded effect of radiation is carcinogenic, the final insult to the nuclei of cells after several kinds of damage to cell DNA have accumulated. As discussed, quite low radiation limits may need to be set to reduce incidence. The mutation rate is about $(10)-6$ per rem per year. Single high-energy particles can damage a cell nucleus to start problems. Radiation seems to have cumulative effects, hence the need for keeping average dose low. After too much damage to DNA, carcinogenesis starts and is presently fatal in about half the cases. A low average dose of radiation can be as fatal, after several decades, as a single high dose. Total early cure or prevention of carcinogenesis is a major requirement for starship missions. Many other factors in high-tech lifestyle may be causes and therefore of concern in a starship. One way to reduce shielding requirements by 10 or more is to be prepared to correct the occasional medical problems that result (unpredictably) from the higher radiation levels. Very high-radiation doses near the 500 rem limit will disrupt too many vital functions to be survivable, and large medical facilities would be required to care for such cases after a radiation accident.

If the starship environment has zero gee, the medical facilities must cope with counteracting its effects. Monitoring of many body parameters is required. Bone calcium is lost steadily over time, muscles weaken, cardiovascular systems weaken, and many hormone systems are changed by zero gee. Readjusting to planetary gravity after a lifetime of zero gee could be a larger problem than adjusting to zero gee, due to permanent changes. While present space medicine recommends 0.9 gee continually, it seems possible humans could adjust to 0.5 gee and readjust back, given full understanding and treatment. However, half gee does not relieve the problems of providing gee in a small starship.

9.8 LARGE COLONY SHIPS

A starship mission, small with ecosystem as described above, is likely to be a colony, first in travel, then in final purpose. Large colony "worldlet" ships are starships for thousands of people with extensive growing areas maintained not just for food but for enjoyment of nature during long journeys requiring tens or hundreds of generations. Their ecosystems are still tiny in comparison to Earth ecosystems, which can be stable for millions of years. They may prove more resilient than those in a smaller starship, or they may be more problematic. At the end of a voyage, the likely plan is to colonize a planet, or possibly to build more colonies in space from available material. Not only must the ecosystem enroute be self-sustaining, in the end it must be capable of being scaled up to planet size. Its human occupants must be able to increase rapidly in numbers for some time after arrival. This does not mean massive extra supplies must be carried, but rather more knowledge, technology, and stores of seeds and reproductive cells of more species.

Many descriptions of large and traveling space colonies have been published [b-Johnson; b-O'Neill; b-Brand; a-Matloff 1976; a-Bond 1984]. One major purpose of these has been to transport stable complex massive ecosystems, which require light (therefore heat), gee (by rotation), area, air, and water, at a minimum. Rotating colonies must be paired to keep angular momentum zero. Another life-related requirement is shielding from radiation, either 2 meters of rock and soil on the inner surface, or complex and formidable electric or magnetic fields. Colonies with ecosystems are massive, and large ones reach the limit of known materials to hold them together against air pressure and rotation (at about 15 km radius).

The minimum living area needed by people (say, 100 m^2 per person) can be used to estimate the capacity of a colony ship. A rotating cylindrical colony of 1 km radius and 10 km length has an inner surface area of about 60 million m^2 (plus end caps), could support up to 600,000 people, and masses about 500 million tonnes shielded [b-Johnson, fuel not counted]. Most of this area could be open space, fields, meadows, forests, perhaps lakes, and a more reasonable population set about 100,000. At its destination this colony ship would use local sunlight for energy, so there must be two compatible ways to illuminate it, interior as if by a giant tube lamp, and from outside by means of mirrors and large windows. Other designs than a cylinder might solve this problem better with less loss of land area to windows. However, if such a colony is successfully built and operates stably, moving it to a nearby star where an Earth-like planet is unlikely means that it must dwell permanently in space there, with more like it built from local material. (Just landing all those people on any planet would be a major problem.) Besides cost, a major difference in the near future between staying here to live in space and going elsewhere to do the same may be political, a declaration of independence from Earth control.

Besides propelling such a massive vehicle, lighting its interior seems to be a large problem. The electric power needed to provide 1/10 sunlight on 60 million m^2 with broad spectrum 30% efficient lamps is about 20 Gw. If the lights are slowly dimmed to darkness and slowly brightened again after 8 hours of "night", energy use is improved at best a factor of 2. The 20 Gw is an underestimate anyway, since this light intensity is somewhat weak for food growing. It would not help to suspend the lamps close to the plants and trees, except for small vegetable gardens. Carrying the equivalent of 20 nuclear plants seems excessive for lighting. Moreover, the light degrades to heat as it is absorbed by plants and other surfaces. This heat cannot be collected and stored but must be dissipated at the same flux of 300 w/m^2 to avoid roasting the colony. The outside must reach about 300°K to dissipate this power into space. The 100,000 inhabitants themselves need "only" about 100 Mw to lead high-tech lives. Again plant growing dominates the energy picture, but an improvement seems possible. To save colonists' eyes and reduce light power, light sources can be developed which concentrate power into the narrow ranges used by plants. The light would be pinkish unlike any stellar source but would improve efficiency by at least a factor of 10. The only way this reduced 2 Gw of power can be obtained (since nuclear plants cannot be carried and run 1000 years) is as a small diversion from whatever propels the colony ship, a propulsion still uncertain. Until such propulsion is established, the transportation of large forests and fields away from a sun seems unfeasible.

Once colony radius exceeds a few kilometers, real weather becomes possible inside, although it is uncertain that clouds and rain will keep the interior well watered. Dry patches due to weather or irrigation failure or water shortage would be liable to fire, a possible catastrophe in a closed system. The point of using a large colony is to avoid having to build a titanic recycling system, so that a means to clean the air of unplanned pollution would not exist. There should be provision for inhabitants to use indoor, people-only air and water system during short crises. Less oxygen pressure than Earth would reduce fire danger. CO_2 might be difficult to balance without a large moderating system, since a large colony still might not be sufficiently large to buffer gas flows. Used or dead biomass would need to be oxidized at precisely the same rate as plants are using CO_2, or humans might suffer. A colony-wide conventional sewage treatment system may also be needed. Doubts have been expressed that other cycles can remain stable without special elaborate controls and that large colonies can be large enough for successful ecosystems [b-Brand].

Chapter 10

Personal, Social, and Political Considerations

Many human traits combine in the desire to visit the stars, including curiosity, adventure, and escape. Humans have often reached much farther than their grasp and achieved new goals. Many expect that interstellar travel will be just another extension, a very long one, of our reach, and a conquering of a final frontier. This chapter examines some psychological, social, and philosophical aspects of such an endeavor, including the political and economic. A starship provides a special chance to consider human social structure in a microcosm. In view of earlier technical findings, it is assumed here that the mission must function socially for as long as 1000 years. The discussion here also attempts to convey an impression of what it would be like to participate in such a voyage. Again, one should wonder if humans have some special abilities needed for interstellar travel, or major disabilities for such voyages, or both. For missions without people, the first few sections of this chapter are largely irrelevant, but later sections are pertinent to questions of how society is to launch and benefit from robotic probes.

10.1 PEOPLE FOR AN INTERSTELLAR MISSION

If estimating technical requirements for interstellar travel to better than a factor of 10 has been difficult, the more so for estimating personal and social requirements. The number of people needed in a mission is a major factor in estimating the size of a starship and therefore its drive. The duration of the journey also matters. One kind of people would choose voyages that last only a few years in proper time, another kind would tolerate voyages that take most of a lifetime, and still another kind would elect to begin a mission that requires many centuries. The technology for ultra-relativistic and ultra-reliable travel is not foreseen now, and the likely mission would be very long and risky. Regardless, if the call went out for a hundred or a thousand super-multi-skilled brilliant specialists and generalists, plenty of volunteers would appear for this ultimate trip.

Pioneer couples have survived on Earth, but in space there would be too many advanced technologies for two people to master. Ten people may be the lower limit for a fast spartan round-trip in one lifetime with an ultra-reliable starship and nearly automatic food production. Two of the people might need to be expert medical doctors, and everyone would need to know several major technical and practical fields and many more minor ones [see also a-Hodges]. It should go without saying that both genders would participate in about equal numbers although most litera-

231

ture still refers to "manned" interstellar missions. Space explorers will resemble less and less the groups of men (only) who undertook historic sea and land journeys. Those who decide to go would consider how many other people are going, the gender ratio, the range of ages and interests, the safety and comfort, and many more factors.

The initial composition of the population is open to many variations. The people might be mostly young, 25 to 30, and mostly paired as male-female couples. Or the ages might range from 25 to 50, a mixture of singles and couples with a few older couples leaving their grown children on Earth. Some couples might bring their children. Or children might not be permitted until the mission has had 5 years of success. What married couples would volunteer knowing they may never be permitted to have children? What couples committed to the security of their children would volunteer to bring them? Since an interstellar mission will occur in the context of extensive local space exploration and colonization, there may be families to choose from which are already experienced at living and working in space.

When the journey requires many generations, the population must be much larger than 10, not only to carry the child-raising workload but also to insure that the next generation is large enough that every member commits to learning some of the needed skills, with overlaps. The typical minimum population in other sources and here is about 100. The 100 is assumed to consist of adults, fully trained and responsible, but there must also be children. The distribution of ages is complex to predict or control. The number of generations required is also difficult to count. Here the active lifetime is assumed 70 years with some living as long as 100, if there are no breakthroughs on youthful longevity. The next generation must begin at least 20 years before the previous one starts dying, to receive maximum training. But people cannot wait until 50 to reproduce. Reproduction should start no later than age 35 if parents are to feel young and active for raising children. The unfortunate demographic result is that the starship must support a larger fluctuating population than the 100 planned upon, so that all facilities must be expanded.

A simplified example illustrates the population problem. Suppose that 100 voyagers at age 30 begin the interstellar journey (called Generation 1; see Figure 10.1). At age 35 let 50 couples each attempt to have one child. Within a few years there are 50 small children to raise; these are Generation 2. Now the population is 150, but Generation 1 has reached age 70 and is diminishing in number. If each new couple in Generation 2 follows the previous policy and has 1 child to start Generation 3, the population would soon fall to 75, too low. Therefore when Generation 2 reaches age 35 each of the 25 couples has 2 children. Now the population is 200 but will fall rapidly. When Generation 3 reaches age 35, the population has fallen to 100 and each of the 25 couples has 2 children each, raising the population to 150. Thereafter with each new generation the population fluctuates between 150 and 100. Since each generation comes 35 years after the previous one, a total of 29 generations is needed for a 1000 year journey. After the initial reduction in child-bearing, the system settles down to each couple having 2 children to replace themselves.

More realistically the initial voyagers would have a wider range of ages (25 to 50) and child-bearing would be spread over a decade or so from age 25 to 35. Four generations would be present at any time, and population must still surge to 150 temporarily. Up to fifty children would be present with a variety of ages in the first century of travel but after that children and young trainees (ages 0 to 20) would be present in approximate proportion to the average lifespan, or about 1/4 of the population. In any given year there would be 1 or 2 people with any given yearly age, and part of a new generation would start in every year. The number of generations becomes difficult (and useless) to determine but must average near 35. 25 to 40 children is a large number for about 100 adults to look after, and child-raising and training will clearly be a major task in the starship. This is a consequence of the succession of generations to keep a nearly constant human population. After the first century total population variation would smooth out to less than 10% so that problems with living space and agricultural production should be minimal. The human biomass may become nearly constant.

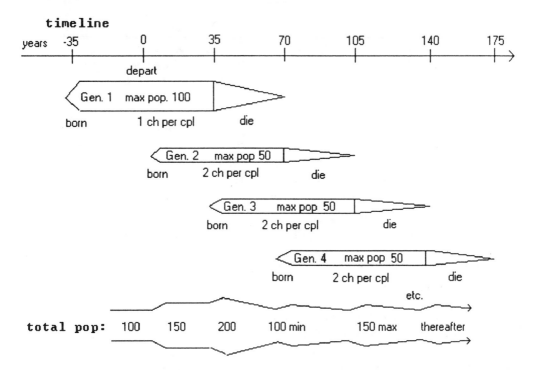

Figure 10.1 **Fluctuations in population unavoidable in multi-generation starship, shown for a nominal 100 person mission departing at time 0.**

A major consideration for population is genetic variability. One hundred people is an uncomfortably small gene pool. If young people are at all stringent about partner selection, the number of eligible partners for mating will be very limited in both genetics and personality. For every pair of people which does not successfully mate, two other couples must have two additional children to maintain the population. This would further limit genetic variation unless each female mates with a different male for each child. Another genetic effect is regression: if the original voyagers were selected for high intelligence and other superb abilities, repeated reproduction would shift the general level of ability back toward (but not reaching) the general human average. Preparing for a new planet can occur 20 years early if there is room to produce extra children and train them for colonizing tasks, and if the planet is known to be habitable. On the planet the small population must expand as quickly as possible. The genetics can be improved by carrying and using a wide range of human reproductive cells. Colony ships with many thousands of people pose different and easier population and genetics situations. But colony ships are unlikely in the early age of interstellar travel unless technology for voyages shorter than 1000 years is not developed.

The background of interstellar explorers will surely include experience living in, perhaps building, space colonies near Earth. Such people will be more comfortable with space and with living in smaller, closed social and eco-systems. Very few from Earth would likely apply or be accepted. The decision to make such a journey would include the perception of personal benefits. (Financial benefit, except for support of family members left behind, would be very unlikely.) It is difficult to say how anyone would personally benefit after the glory of starting the great voyage wears off in years or decades and isolation sets in. Good voyagers might be people who like a simple, small world to relax in, enjoying themselves and solving small problems well and raising children well to do the same. They must also react resourcefully and courageously to major problems that may arise. Generations later, those who are alive to see the approaching star and planets and make plans for landing and colonizing may feel very fortunate if things are going well and a hard but successful life appears to be ahead. If contact with another civilization is the final goal, voyagers well-educated in the history of the mission would surely feel an incredible mixture of joy and anxiety at the prospect.

The skills that voyagers need come from the sciences and technologies involved, and from most other human fields. Almost all Earth fields are represented, and some new ones may be developed for starships. The need for a minimum of about 100 people is clear when many relevant fields and subfields are listed. There are about 20 engineering fields (including design and systems, mechanical, electrical, electronic, optical, materials and metals, fluids, astro-, rocket-, and aerodynamics, nuclear, communications, computers, robotics, manufacturing), more than 10 subfields in physics and chemistry (mechanics, thermal, electro-magnetism, atomic, acoustic, optical, solid state, molecular, plasma, nuclear, elementary, relativity, cosmology, astronomy), about 10 relevant ones in biology and planetary studies (including ecology, exobiology, molecular biology, microbiology, physiology, neurology, and genetics), more than 10 other professional and practical fields (including mathematics, statistics, education, medicine, counselling, exercising, agriculture, environmental, instruments, media, plumbing, welding, construction), and about 10 other softer fields (including psychology, sociology, philosophy, history, several arts). The last set may be considered extra expertise carried by many people in addition to their primary fields.

Clearly, each person may need to know more than one field or subfield, and many should be generalists too, that is, people who can combine and use the principles of many fields. In a bind, they can readily study and master new specialties. All skills should overlap; that is, two or more people should have experience with each specific skill, and each person should know several different skills. It is difficult for one person to master more than two significant subfields. In addition, many people must be skilled at fabrication and repair of the starship subsystems and at making small items.

It is premature to produce a list of traits that voyagers must have. A short list would include all our finest, including intelligence, ingenuity, imagination, determination, patience, honesty, cooperativeness, stability, robust health. Selecting for and against traits will be a difficult problem because on the first hand the full range of human potential might be needed for interstellar travel and subsequent colonization, but on the second hand some human traits seem incompatible with long confined travel (as shown in the next section), and on the third hand we do not now know or understand the full range of human traits. There seems to be no scenario in which evolution (testing and survival of the fittest in adverse conditions in large populations) can aid the acquisition of traits most suited to interstellar travel. Starships cannot tolerate revolution and killing as some early sf novels proposed.

Since the voyage will require many human lifetimes, the humans must reproduce at a replacement rate somewhat greater than the expected death rate. Hopefully, illnesses and accidents will be a very rare cause of death. Couples or others who do not replace themselves (or choose not to) would leave room for other couples to have more children. The first generation or two might more or less grow old together, but after a century deaths will spread uniformly over time. In a population of 100 (plus children) with average life expectancy 80 years, at least one person will die per year. Since deaths cannot be predicted in advance, reproduction rate must be set at an average level expected to balance the death rate. Problems of caring for aged who cannot contribute mentally or physically to the mission must be considered. The solutions should be in the hands of the voyagers who must judge how many people can devote effort to taking care of how many old (as well as young) in view of the specific resources, work load, and crisis level of the starship.

Children must be raised and trained for specific tasks despite genetic predispositions. Technically-skilled people are needed far out of proportion to the present incidence in the population. Voyagers must have time to educate children carefully, but education can never be adequate because students cannot have first-hand experience with many skills needed. Some skills for using and landing space-planes and building colonies must be transmitted along many generations until starship arrival at a planet. With a total population near 100 there is serious question as to whether a sufficiently wide range of interests and aptitudes will be available in the next generation, and whether sufficient education can occur [a-Wachter; a-Hodges]. Then there must be room for children to safely explore and play. Time may be plentiful and child-raising a blessing that occupies voyagers in a very human and rewarding activity (unless the mission has many crises). The mid to late 20th Century phenomena of teenagers may be irrelevant on a starship. There is no place for adolescent restlessness, rebellion, irresponsibility, or alleged trauma. As in earlier times, children may begin learning adult jobs at age 6 to 8 and begin semi-responsible work soon after puberty when they acquire young adult status. Training may last until age 20. Selection of partners for companions or families would start before then and would not be simple in a population where there are only a few young people at one time of similar ages. The idea of one-to-one pairing may not survive.

The people who leave Earth will not be the ones who arrive at a different star, barring some very major improvements in human longevity, body preservation, or starship propulsion. Each new generation must take on the full responsibilities and dedication the mission requires. Changing missions as the population changes (and changes its collective mind) is unlikely because of constraints of energy, propulsion, and supplies. The grandchildren of the generation that decides to give up and turn back may prefer to go forward again. Multi-generation commitment beyond anything yet seen in human society seems required. Unless much better technology permits rapid and return trips, voyagers must leave understanding they will never see Earth again. Some kinds of propulsion would permit turn around and return after a technical or social failure in the first few years out.

10.2 PSYCHOLOGICAL CONSIDERATIONS

People would be able to choose freely to begin an interstellar voyage; that is, candidates who apply or are invited would have the nature of the voyage explained at length for an informed choice. Coercing anyone to go would surely lead to disaster later. Balanced against a substantial list of expected hardships, including the need for and difficulty of raising a small family in a starship, would be the eagerness of some people to undertake such a momentous voyage. People would choose to take the long journey for positive reasons such as scientific interest, need for new adventure and challenge, or deep curiosity (not all would have a scientific frame of mind). If signals or an invitation were received from a reachable other civilization, the range and intensity of interest would be greater, yet those who depart would not be able personally to meet the other beings. People may leave Earth or its vicinity very reluctantly or they may be glad to go. Those who take the voyage for negative reasons may not leave those reasons behind. For example, if no social system on or near Earth suited them, the system in the starship may not either.

People would choose to build a starship if it were at all feasible because some people (mostly men thus far) like to build large difficult, dangerous, systems which move fast and amplify and project personal power. A starship can also be viewed as an ecological womb nurturing people on the way to the stars and giving birth to new human civilization, but attempts to genderize technology do not seem fruitful here. While women have not in the past instituted big technical projects, they are likely to in the future. Even now women are beginning to show almost as much interest in engineering and piloting as men do. Women have gladly gone into space and are demanding an end to being kept from some military jobs such as jetpilot.

At present it appears that an interstellar mission would be very restricted in mass and volume and would travel relatively slowly. Many generations would be born, be trained to operate the starship, and die before arrival. Not being able to witness an entire journey is rather new to human experience. While most pioneers arriving at the American continent crossed only a small fraction of it, many could envision later generations living and traveling across it. The prospect of not seeing the end seems somewhat sad, hence the interest in space colonies which remain here. One hundred young people might think nothing of moving their favorite possessions into their small quarters and departing from the vicinity of Earth forever, not really thinking about how they and their children must learn to enjoy the same small volume seemingly forever. The young tolerate most kinds of stress better than older people. When they reach age 60 and are guiding their grandchildren while their friends grow old, they may wonder why they embarked on such a crazy idea and may miss the security of their original home. Earth will be a dim and yet haunting memory. Not being able to share their lives with outside family, friends, and all humanity might take more of the fun out. Two-way communication with Earth at best would require years of waiting. This situation would be rather different if ultra-relativistic travel were possible. Then there would be plenty of energy, many stops could be made, and many stars would go by quickly in a lifetime. A likely permanent home would be found, but forever cut off from an Earth left thousands of years in the past. The option would remain of returning to Earth's future, when they could hope to adapt to the new culture they would find.

Interstellar travel is likely to become lonely unless thousands of people are on the journey, but this would be a technically more difficult undertaking. Prospective voyagers would get some idea of the hardship from accounts of early pioneers who travel to Mars and back, who explore the asteroids in perilous adventures, and who spend a decade traveling to and among Jupiter's moons. A human mission might be sent to Neptune, Pluto, and beyond just to accumulate experience in coping with such travel. Voyagers are likely to be drawn from space colonies rather than Earth, since a large community of space colonies seems prerequisite to preparing for a mission to a star. Experience will accumulate of humans coping with the desolation of hazardous local space voyages, and the social experience of people in space colonies will also contribute. But there can be no preparation for dealing with the first 6 decades of life in a small complex ship, dependent on just 100 other people until one dies. The next generation may have it easier, not having known any other life. But the library of images from Earth may haunt them too.

Living in a starship requires accounting for all human needs at all times in an environment where natural and genetic tendencies cannot necessarily be fully expressed or supported. Some people want to do some of the following some of the time: run in the open air, ride horses, drive hot-rods, make love, raise children, fight, get drunk, abuse medicines, abuse trust, hoard, collect, be moody, argue, be irresponsible, be alone, and much much more. Planners must admit that these sorts of propensities (and worse) are in many people, else modern society would operate more smoothly. Extreme fluctuations in psychology and sociology occur over time in otherwise stable people, in ever-changing social combinations of people, and sometimes in carefully planned genetically new people. Manic irrational behavior, addictions, stealing, serious cheating, sabotage, and killing could occur. No starship population can be selected to eliminate nearly all of the most destabilizing behaviors. Possibly alcohol and other abused drugs can be omitted from stores, but any range of cultural material in the library will contain references to these and other aberrations of behavior and intrigue some people, especially those with unrecognized propensity for the same. A basic question is whether a relatively "pure", but not puritanical, microcosm of society can be assembled and kept that way.

Human vanities are certain to be carried to the stars, along with diverse cultural interests. Low-mass environmentally-compatible pursuits must be encouraged. Personal possession must be minimal. In a virtually closed system, the supply of nonessential goods is nearly independent of demand. Some voyagers will want impractical clothes, special ways of grooming, musical instruments, hobbies requiring unusual or natural materials (wood, stone), pets, old Earth books, and so on. Our culture has instilled in at least a few people a range of aesthetic needs, ranging from beauty and harmony to communication and intimacy. A starship would not be a return to pioneer days with simple survival in harsh nature. Its high-tech environment would reinforce some present habits and needs and some new ones we cannot predict yet. It definitely would provide an orderly environment, which bothers some people. People have need of a diverse natural environment as it occurs on Earth. The vital clues about a natural environment that our bodies seek (e.g. variable weather, variable sun, steady gee, leg room) would need substitute stimuli. Except in a colony ship, visual and spatial variation as in a large space colony would not be possible [b-Johnson]. Few people want to be confined to the equivalent of a small tight building in a vacuum and restricted to a definite routine all their lives. As indicated in earlier chapters, some "natural" variation in environment, still safely controlled by starship systems, would contribute to people's well being. Another vital aspect of nature is growth, which can be witnessed in a starship in regard to food plants, children, and everyone's mental and emotional development. There might be some rebuilding of the interior too.

Can people be sensitive, careful technologists 99% of the time and not create major problems the rest? No amount of training and good intentions can overcome the strain of lifelong dependence on a complex fragile dangerous starship. There is plenty to worry about, such as: social breakdown, ecosystem (food) failures, air leaks, propulsion failure, hitting a pebble, excessive erosion, zero gee medical effects, a region of high radiation, communication failure. Few people can be dedicated to one small set of purposes all their lives, and such a demand usually results in psychological stress that leads to medical and social problems. Some biological stresses have been discussed (zero gee being of greatest concern), and these can lead to and interact with psycho-social effects [a-Murphy]. Other space stresses are to be expected, such as the "feel" of the limited, isolated environment sensed visually, acoustically, by smell, by orientation, by vibration, and in other ways. Staring into deep space might disturb some. An impulsive drive "engine" might be a bothersome background.

Life in a starship need not be a microcosm of modern high-tech society or of any particular society; instead the social structure and practice could be new or everchanging. But some recent developments in our society are lessons for thought. For example, the high divorce rate here indicates that couples in a starship may not be stable for life. Perhaps organization into couples is not needed, or will be changed to a more communal arrangement enroute. Irrational jealousy, especially over partners, is an old behavior that may spring up anew. Pairing of people into couples is both a source of stability and contentment and a source of strife that can never be fully eliminated. Almost all people need intimate companions for long-term happiness. A few seem to need intimate partners of the same gender, a development which modern society is accepting reluctantly but which must be totally acceptable in a starship if a certain percentage of voyagers is born with that orientation. Some people are comfortable with changing partners often. Without group pressure and support, "divorced" couples will accumulate and constitute a set of people annoyed with each other or feuding. Partnering may need to be casual so that rearrangements are casual rather than traumatic. The very limited choice of fresh partners also indicates that tight bonding might be paradoxically destabilizing.

Another observation from our society is the continued repression of sexuality, media distortion of it, and propagation of myths. Along with the arts, cooking, and growing plants, sex may be an important pastime in a starship, since harmful stimulating experiences such as drugs and violence must be eliminated [a-Tanner]. Major personal and cultural attitude shifts are needed if voyagers are to be comfortable with this basic and deep human gift which can provide psychological and social renewal if people avoid possession and jealousy.

Just about everyone in the starship must be comfortable with just about everyone else because everyone depends on the skills and dedication of most others for survival. Yet harmonious small tribes on Earth have been very rare. A few people can be acknowledged recluses as long as they do their tasks. A few or more can be very gregarious, but most will be at the center of a series of circles consisting of a few intimate friends, a few children they guide, a few close friends, a few occasional friends, many congenial acquaintances, and a few they loathe. The trend in some human societies toward individualism and competition does not seem compatible with a starship mission. Nor is blind following of charismatic or authoritarian leaders (more in the next section).

If humans achieve long, nearly immortal lives, or hibernation, psychological and social effects on travel would be large but as yet uncertain. The need to raise and train many short-lived generations in a starship would be eliminated. It is often said that near-immortal people would fear for their safety at all times and therefore would not undertake hazardous interstellar travel. Who would want to die in a preventable freak accident after living 10,000 years? Hibernation may be risky too, but long-lived or hibernating people may accept some risk in return for seeing the end of such a journey as no ordinary mortal can. Another concern might be boredom, but it is difficult to say how a near-immortal would view 1000 years of memory aboard a starship. Perhaps the brain could not remember the whole voyage. Perhaps there would always be fresh experience to savor. Perhaps a long-lived person would become tired of it all no matter how fascinating the setting. Much depends on the details of the long-lived body. If people can look, feel, and be young instead of caught in a stable aged state, if they can have intimate relations in starship society and look forward to reproduction at the end of the journey (after being without children for 1000 years), then near-immortality would be heartily accepted. Meanwhile near-immortals back at Earth would await patiently the results from the mission they personally began centuries before. [see also a-Hart]

10.3 COMMAND AND COMMUNITY

With humans aboard, command and control is a complex matter. Modern society now has an assortment of command structures but which one is appropriate for the starship micro-society is not clear. Most people do not like others to have much control over their lives, but some crave it. The social structure must keep the people at peace and doing their tasks regardless of technical or social crises while taking the very long-term view that the mission will continue as planned for the next decade, the next century, and for 1000 years. The social structure must also be compatible with the technical requirements of the mission. A military structure with commander, lieutenants, and soldiers is often thought the best way to operate a technical system [e.g. a-Bloomfield] but may not endure socially. Life must be left enjoyable, with a minimum of rules, or there will be rebellion. A leader in a democratic structure should be replaced regularly to avoid accumulating stress on the leader, loss of interest, or conflicts with others. Women should never have been disdained as leaders because women seem more easily to provide an interactive style which shares power and full information with others for group decisions. One point is certain: mission controllers on or near Earth will not be able to control what happens in the starship. Once the time delay in communication exceeds a few days, voyagers will not wait for orders from Earth (unless there is a problem which only Earth can solve). Eventually no voyager will carry out tasks that are not justified by the voyagers themselves.

Social organization has connection to lifestyle [b-Johnson; a-Lacey]. A homogeneous small society of people (genetically unlikely and undesirable on a starship) may prefer order imposed on them and may then dislike those few among them who are different. Historically most human societies have had hierarchical structures. If they do not have a chief, they might have strict majority rule democracy. An individualistic society would have every person for him/herself, incompatible with cooperation for survival. The likely mix would be autonomous pioneers who cooperate as needed! Such a heterogeneous society would need to be egalitarian and recognize people as each having some differences and some commonalities (although ethnic groups on Earth have fought more than cooperated). Life can be more lively as people try to accommodate their differences while cooperating to survive. Rules could be minimal and all related to common survival. Lifestyles could vary from the solipsistic to the communal, with plenty of time to work out relationships. Formal governance could be minimal, but much time and effort might be spent by committees on problems (risking invocation of the rule of thumb that the smaller the problem, the greater the effort). Ad hoc changes in governance would be likely, keeping life interesting. Many sorts of quasi-democratic systems have been developed with elected or voluntary councils and committees.

When an emergency develops, however, an immediate shift might be needed to a prearranged militaristic command because there would be no time for committee deliberations.

Should responsibility be traceable? Responsibility should be individual, with no one passing blame to someone else or to a group or system. In one sense there is a chain of responsibility, with a leader or a council which initiates action and verifies that it was done. In another sense, if someone goofs, should all know who did it so that a reputation is tarnished for a long time? One rule seems certain: no one should have the sole responsibility for any given area or task. Either two people should share and trade the same duties, or the person on duty should be able to call a replacement at any time. In some cases a duty must continue at all times, requiring three committed people (plus backups) who carry it in shifts. This is very costly in people power and creation of such duties is to be avoided wherever possible. Monitoring general starship operation is one such responsibility requiring three people (in shifts). The load may be sufficiently heavy to require three assistants in shifts. Some scientific observations may require continuous attention for many days but surely not longer. Some repair work may require continuous support by two or three teams of 5 or 10 people each. Some duties are "on-call" and the experts can do other things while being available at all times for an emergency such as medical attention, communications from Earth, or technical breakdowns detected by computers and passed to the monitor on watch.

Science fiction has preceded sociological study of small isolated high-tech societies and has tended to predict inevitable breakdown, especially in "generation" starships [a-Lacey]. The social shift from modern society to starship society is like the difference between living in a large city and living in one of its apartment buildings (with a big greenhouse) completely isolated from everything else. A quasi-experiment has begun in Arizona with 8 people (both genders) living in isolation. In a 3 acre complex of buildings and greenhouses, they are partly dependent on their own food growing, recycling, and company. A broad society of tenuous dependencies, to which we have culturally evolved, would be replaced by a small very interdependent group somewhat like an isolated tribe, although the analogy should not be stretched as far as anthropologists have attempted [a-Tanner]. Social problems predicted for space colonies, such as ethnic conflict, would apply to a starship with diverse population, and problems of rigid lifestyle to one with homogeneous people [Hardin in b-Brand].

Several general kinds of social failure seem possible. Some arise through conflicts between people or between factions (groups of people united on one or two issues). Intellectual ability would be a major genetic component of all voyagers, leading to many ego conflicts as the many "experts" disagree on a subject. Placid people who are experts at operating a starship is a contradiction in description. Some failures can arise through systemic instability out of the control of any one person or perhaps any group. Situations can occur where everybody's best intentions leads to inaction or even breakdown through unexpectedly destructive interactions. (Earth examples where most people are said to sincerely do their best while the system fails to achieve its goals are education and legislatures). Each new generation may want to change the rules on population, on behavior, or even the destination. Paradoxical or meta-failure can occur when the issue becomes one of changing the rules, or changing the rules for changing the rules.

All in all the psychological and sociological problems in a starship of any size are probably solvable but sometimes much effort would be needed. Traditional authoritarian organization is unlikely to succeed except in an extreme emergency. Democratic procedures are uncertain because a disgruntled minority that has been outvoted by a majority can be very disruptive on a starship even if the minority tries to be cooperative. Consensus decisions can take a very long time if ever, and some problems will not wait. Appeal to rationality, evidence, and scientific procedures can work but sometimes the data needed will be lacking. Appeal to the major goals of the mission to settle disputes is unlikely always to work because the goals are remote. The next generations will realize that they did not develop the goals and may disdain them.

A number of new methods for solving personal and social problems have been developed in the 20th Century, and more methods seem likely to be found. We can hope that human nature will be better understood before the first interstellar mission. Thus far the new tools available include rational analysis of social situations (such as the use of game theory), multi-dimensional behavioral models, conflict resolution methods, and group psychology. In game theory, people in dispute analyze which course of action leads to maximum gain or minimum loss for both. In multi-dimensional approaches, the worth of each human participant is absolutely established by identifying each person's unique contributions. In group therapeutic methods, people exchange ideas and emotions without coercion to appreciate and empathize with the points of view of others. The most complex human problems also require a philosophical base, an ethics, for solution. Voyagers will have a wide choice of old and new ethical systems from rigid rules to situational ethics, and a library of their applications. Ethical decisions for which science provides no clear answer, such as how many incapacitated people should be kept alive, are the most difficult and are better made by a conferring group. Also difficult is the decision to execute someone who sabotages or kills.

10.4 SOCIAL AND PHILOSOPHICAL CONTEXT

The voyagers would at least partly be chosen by people other than the voyagers, by planners not making the journey, although once some are selected their approval of their future companions might be crucial. It has been suggested that the population be diverse [a-Lacey; a-Tanner] and others that it be homogeneous [a-Bloomfield]. Genetics, skills, traits, and motivation have been mentioned as major determinants in selection, with equal numbers of each gender and representation of much Earth ethnic diversity. By that time the few groups which remain distinct should still be educated in science and technology and probably living in assorted space colonies. Besides criteria such as interest, safety, long-term purpose, and prestige, a social factor that will affect who chooses the voyage is the stability and/or openness of starship microsociety. People might not want to commit to a voyage until it can be shown that social mechanisms will prevent breakdown, rigid imposition of lifestyle, or incompetent starship management. Along with engineering a fail-safe starship might go engineering a fail-safe pleasant society for it, and possibly within the context of a deteriorating, unpleasant, rigid, or disorganized Earth society.

Voyagers would commit themselves knowing that the mission is one-way and therefore must be a colony mission. They would never see Earth again. This situation would change if much faster (yet affordable) travel is developed [a-Hodges]. Another social effect of the possible development of faster travel is that no one would want to take an early slow journey if a faster one will occur later and arrive sooner. The first arrivals share in the glory and possible wealth, as well as the worst hardship. By this reasoning no starship might ever depart.

Instead of choosing the voyage positively, some people might choose it to escape problems on or near Earth. Earth may become unbearable environmentally, politically, or demographically to live on, but this would not be a reason for interstellar travel, just a reason to live in a colony or asteroid (if one can afford the trip up). But some people still would not find a compatible place to live, or would have difficulty surviving economically, or would feel oppressed by population or irresponsible governments. They could not expect ultimate relief themselves on a 1000 year voyage, and the starship probably should not carry people from the fringe. It would be better to choose among the many who feel ambivalent about living in space colonies or Earth nations than to take the few who feel driven to leave.

Several other more unusual aspects of travel to the stars should be considered. The mission is unlikely to be a suicide one where the starship is boosted to high speed without the means to slow it or without a habitable destination. Almost no one would sign up for such a mission and much social trouble would occur along the way. Similarly, it is far-fetched that mission planners would claim to provide a mission with a safe destination, then program the starship to explode after it has accomplished whatever secret purpose was planned.

Social breakdown on Earth as a result of environmental or other catastrophe seems sufficiently likely that an interstellar mission should not rely upon Earth itself for any part of the construction, departure, or support programs. For long-term success the mission would require social stability and commitment from a group of space colonies (really, space communities), therefore political stability including safeguards against intercolony war [gen refs: b-Brand; b-O'Neill; b-Johnson; b-Heppenheimer; a-Chernow; a-O'Neill two; a-Jones]. At the least, plans beyond human lifetimes require pervasive unselfish attitudes. It has even been proposed that a cosmic religion would be needed to motivate and sustain an interstellar travel program [a-Bainbridge].

One motivation that would not apply to interstellar travel is population pressure. Much more so than in Earth history, the concept of colonization is to be disconnected from population pressure (and from exploitation). If humanity were to spread among the stars, nearly all those people would be born out there, to a small group of hardy voyagers. Carrying them from here would be impossibly expensive.

Even getting them off the Earth by presently known means would be very expensive and destructive to Earth's environment. Population that grows too large in space colonies can build more space colonies almost without limit. More important is whether humans can control their population on Earth, in colonies, and in starships. We have sufficient contraceptive technology now, but some large groups seem unable to use it well (not to mention the impoverished who often do not obtain it).

Sending an interstellar mission raises many ethical problems and can be an ethical problem in itself. Besides the questions of who gets to go and who is made to pay for it (taken up shortly), there is the matter of whether voyagers' descendants should accept the mission they find themselves in some sense forced to be on [a-Regis]. The answer seems to be along the lines of considering whether people can object morally to any setting in which they find they have been born. Certainly the first parents in a starship will have given much thought to whether they are providing a quality environment for their children. Occasional isolated groups on Earth have survived well and comfortably, and space voyagers are likely to have grown up in space colonies of various sizes.

Another question is whether we should be colonizing distant habitable planets. This and plans to take over the galaxy are at one end of a spectrum of interaction with our galaxy. At the other end, few could argue over the value or ethics of sending robotic observation probes. A case could be made against panspermia, the spreading of Earth micro-organisms on distant planets. A serious case can be made against self-reproducing probes which could come to dominate the galaxy if successful. Landing people on a habitable planet is most worthy of debate, since such planet is likely already to have indigenous life. Even if the presence of intelligent beings or of advanced life which might soon evolve intelligence is quite unlikely, humans would interfere with local low-level life (if not wiped out by strange diseases first). No one can say beforehand whether humans with their terrestrial plants and animals can coexist with entirely different lifeforms, or whether competition would be inescapably detrimental to the native life (e.g. changing the atmosphere or another large part of the biosphere). These and other questions will be debated as if they were crucial when an interstellar mission is funded and begun, but it is unlikely that all kinds of missions will be canceled for ethical reasons alone. Human groups have rarely made the most ethical choice when considering a new technical venture.

10.5 POLITICAL AND ECONOMIC CONTEXT

The connection of human philosophy and politics is deep and crucial. Starting a nation is perhaps an analogy to sending out a colonizing starship. Participants attempt to take a very long view and establish a successful program, knowing that results may be several centuries away. The later result of interstellar travel might be to start a new series of civilizations. If one colony can be established, then the technology is proven to send more both from here and from the new colony. Some have said that interstellar travel must have goals beyond scientific exploration and starting one new civilization. Keeping such far-flung goals in mind before the first starship program seems premature and probably will carry little political weight for starting the program. Only a tiny part of present human society can identify with taking over the galaxy. There may come a time when a starship program is given top priority, for example, an impending disaster to the Solar System such as a predicted nearby supernova or a passing star that disrupts orbits, but these are sf for the time being [b-Asimov]. Evacuating the Solar System appears hopeless, and the only goal would be for a microcosm of human society to survive.

The size of human civilization needed to begin and support an interstellar mission depends more on the propulsion method than on the type or size of mission. As shown earlier, it is likely to be easier to build a large mission than to make it travel faster. Only probes in the tonne range, if feasible, would be much easier to send out than human missions. It may be that probes are all that is feasible in the foreseeable future. Although the most political glory comes from human missions, there is ample glory (at much lower cost) in a probe program if glory is to be had back home from exploring the stars at all. This discussion of political and ethical aspects pertains to many aspects of probe as well as human missions.

The politics of interstellar travel depends partly on its economics. Politics and economics are almost inseparable, but general politico-economic considerations are given here and detailed cost estimates are considered in the next section. In all economic discussion the unit of value is taken (at considerable simplification) as the present US dollar. The dollar symbol "$" is used as a physical unit like any other, usually placed after the numerical quantity. With large values, any other scheme produces symbolic confusion. The usual prefixes apply; for example, one trillion dollars is shown as "1 T$".

Paying for a starship program can be accomplished if civilization is rich enough. For simplicity, the concept of GNP (Gross National Product) is used here, and its expansion to GSP or Gross (Solar) System Product is also considered. For comparison, the US GNP is about 5 T$ and the world product or GSP is now about 15 T$. More honestly, the concept should be modified to NSP (Net System Product), where allowance is made for the costs of harming or depleting vital resources (thus making later starships more costly to produce). The NSP is presently some undetermined factor smaller than GSP (1/10 to 1/2). If total starship program costs are 0.1% of GSP, few will mind. If total costs are 1% of GSP, few would mind as long as greater than 1% of GSP is spent on a number of other more pressing local needs. In other words, the starship should not be taking significant sums away from the ability to pay for education, health, local research, pensions, and the like. If some exciting event occurs such as a message from other beings, or if society develops a high interest in interstellar work, then relatively rich space colonies might commit 10% of GSP to starship work. They could not commit much more because, after all, people need to survive here and now. The money spent on the starship is somewhat like money spent on the military: it is spent with no immediate return. The effort by people and the funds spent on hardware are encumbered in plans and hardware that no one will be buying (if only because no one could afford to buy it in the near future). [See also a-Martin; a-Parkinson; a-Bond]

When and how large might GSP become? Space colonies will come about when, at least in the long run, their estimated 100 G$ to 1 T$ cost is recovered by the products they make and sell to each other and to Earth, including energy [see also b-Johnson; b-Heppenheimer; a-Criswell; a-O'Leary; b-O'Neill; a-Glaser; b-DOE & NASA]. Value is produced when human population is able to convert raw material (lunar, asteroidal, cometary) into products that people can trade for food (eating remaining the basis of most economic activity). A 1 T$ facility might expect in maturity to produce 10% of its value per year, but at best 1% per year for a speculative program such as interstellar travel. Space colonies would develop in number exponentially after some point. There may be 2 to 4 by 2100, 20 by 2200, 100 by 2300, 1000 by 2400, and so forth. If this scale of growth comes about, then 100 of these 1 T$ colonies would be able to generate 1 T$ per year for a starship program. Starships cannot and need not be built in a hurry. In a 100 year program, 100 T$ is made available, and really much more because over that time the space economy may have grown another factor of 10. This level is beginning to approach the amount needed, and already exceeds the cost of a small probe.

Another way to estimate future economic power is in terms of productive population. A 1 T$ colony might have a million people. If annual productivity of each person is valued at 100 k$, almost 10 times greater than the present average, and if 10% of value produced goes into a starship program, then again a space population of 100 million people can provide 100 T$ over 100 years. We would seem to have enough people and wealth on Earth now to begin a major program except that the work must be done in space to avoid another factor of 100 or more in launch costs (discussed later). There are other reasons why Earth cannot pay for a starship program. One is that Earth would be doing well to fund and launch part of the first small colony that would bootstrap itself into 10 and then 100 large colonies to support the main starship program. Another reason is that colonists cannot be hired by Earth. They cannot use cash or credit sent up from Earth. They need food, water, air, safe room to live, and other lesser goods which Earth cannot afford to send up.

At one extreme a civilization must have trillions of people to afford a starship program paid at a rate of a few dollars per person per year. If space colony residents do not want to devote 10% of their labor to a starship, then planners must simply wait for the space population to grow. At 1 billion or 10 billion, perhaps 500 years hence, a starship becomes more affordable. If all early starship programs flounder and perish because of high costs, there will come a time when a new set of program organizers can say: "We need 100 T$ over 100 years, and our population of 10 billion can afford this merely by everyone foregoing one new pair of shoes or the equivalent each year." Humans may continue to be stingy about idealistic programs whose results individual people will never see, but perhaps a few other civilizations take the longer view.

Rather than an interstellar mission being a spin-off from space economics, it may possibly become the main reason for living and working [a-Martin]. Space colonies may evolve to need more exciting projects than building more space colonies, exploring local planets and moons, and gathering astronomical data. If future people look for projects large enough to challenge their best cooperative efforts instead of floundering and fighting as now while society falls apart, then interstellar travel would fit the need. Such people would gladly put more than 1% of their output into interstellar missions. If messages from other civilizations were received, even with inconclusive content, the support would be even stronger, at least to begin sending probes. If the messages describe reachable planets which we would like and would be welcome at, the impetus to travel would become unstoppable.

Since interstellar travel is very expensive and risky, only a large peaceful well-nourished civilization of billions (or a titanic warlike one of trillions) can afford to try it. The discussion here continues, with some trepidation, to consider Earth-based society as one possible supporter of interstellar travel. If, as some indicators show, humankind is on the decline on Earth, unable to cope with energy, environment, and social breakdown, ceasing progress and eventually military activity too, the wealthiest period has already passed. We may not be able to afford to start space colonies, the first step to bootstrap at least part of civilization back to large-scale progress. Already over 10 T$ have been spent on military armaments, enough money to build 10 space colonies or send a small intersteller probe instead. Surely a modest mission to a star would provide more inspirational, philosophical, scientific, and economic benefits to many nations than preparing for war has.

Unfortunate systemic failures are possible in a starship program. If the program is publicly funded, it is controlled in part by voting citizens. Recent experience has shown that in complex high-tech societies such as US, commitment varies erratically, and the populace is difficult to convince to support a program for more than a month at a time. When space colonies were debated widely in 1970s, the general public reaction was mixed [b-Brand]. The starship program requires at least 100 years of commitment and possibly 1000 years. Only authoritarian societies in the past have carried on multi-hundred year programs involving large expenditures. If the public is presented with a choice between funding a feasible starship now and waiting for the results of research to produce a better one (faster, or cheaper) they may choose to wait since the do-it-now attitude seems to be dying out of the system if not the people. The public is easily scared by programs with enormous costs, even if the cost can be spread over a long period. At present the public does not seem able to comprehend a program requiring a century or more. Research might be supported at a low level like 0.0001% of GNP or GSP indefinitely, perhaps rising to 0.01% in a happier society, but never high enough to start the main project. Besides concern about near-future profits from such a venture, the leaders and bureaucrats of society who decide on and continue to support a starship project will look to their personal and political gain. Unless prestige value is high and politicians' constituents get some large subcontracts, leaders have little reason to support such a large program which might seem to have indefinite tap on public funds. Tangible return such as data would trickle in much too slowly to satisfy political time frames. [see also a-Molton]

An approach to try in unjamming the process is to start the program when the optimum feasible approach becomes affordable, perhaps the 0.01c 1000 year mission derived earlier. Political determination may revive, but only in space when space colonies are growing and prospering. A space-dwelling society would be more sympathetic to interstellar programs but could not afford one until the space population is very large as estimated above. Initiation and control of the program can be by groups other than government. Prosperous space colonies might consider themselves corporations rather than governments and join on this program. An alliance of many nations (either their military or civilian agencies) and business consortiums on Earth and/or in space might choose to fund an interstellar mission. Governments and other groups may have difficulty agreeing on such a complex project, but once it is started they must realize that they cannot keep much control after the starship departs. They would best begin a program for a further mission.

If a starship is sent on its way and depends on sunbeams or other form of propulsive or supporting power sent from space colonies for 100 or 1000 years, this is likely to be expensive and requires a very careful plan. The voyagers and local technical supporters must agree precisely on their plans in every conceivable situation, even cut-off of transmitted power. Perhaps the original mission plan should include an endowment invested to provide income perpetually for the project (if the concept of investment continues to exist). An organization must continue the plan despite being operated by many new generations. A substantial number of people here would oppose cutting off the vital needs of the mission, but only so long as people are educated to its existence, purpose, and needs. Over the centuries the political situation could evolve to one of cut-off and abandonment as less and less people learn about the mission and identify with it. Or perhaps an equipment failure will occur and the resources will be lacking to repair it. Or space colony war might break out, with the sunbeam equipment sabotaged or destroyed in spite.

One procedure for making political decisions about projects is cost-benefit analysis. The costs (sketched in the next section) can be compared to estimated economic benefits. The comparison must be in compatible terms; sometimes values other than money are used. On this basis the starship program might not win out. The main benefits to humankind would seem to be that Earth and colonies have acquired the technology to do more of the same kinds of missions, much cheaper the next times. The technology may be useful at home, for travel, for power, for yet unimagined applications. These could be major benefits but probably not equating to the 100 T$ or more spent until centuries have passed. Also a stream of new scientific data will be slowly coming in over the centuries, perhaps from all the way to the destination. Knowledge is difficult to price. The data would not be obtainable another way or that way would have been used already. While interstellar information is very expensive, the alternative may be not to obtain it. It is unlikely that those who begin the voyage would be economic beneficiaries. And no one is able to think in terms of the benefits to the generation that arrives at a new star after some 30 generations before that have made large sacrifices. Beyond the economic, we have vague ideas about perpetuating our species, and interstellar colonization may be the ultimate way.

10.6 COST ESTIMATES

Estimating the cost of interstellar travel is difficult but order of magnitude attempts can be made. The results are so large that they can indicate what era in future human history would be able to handle the projects. The Daedalus project has been estimated to cost 1 T$ to 10 T$. Others have tended to estimate starships low, perhaps extrapolating from the Apollo program which cost about 30 G$. Extrapolating in distance (a factor of several 100,000 more than a Mars mission), in speed (about 1000 more than chemical rockets would be nice), or in time (a factor of 1000 longer than a Mars mission) do not seem adequate. Can any of these factors be multiplied against current costs to predict future costs? Perhaps the million times greater kinetic energy would indicate interstellar mission costs. It is also not safe to argue that, since astronauts now travel a 1000 times faster than early runners, in the middle future star voyagers might go another 1000 times faster (requiring a million times more KE than now or a trillion times more than the prehistoric era).

Another cost approach is to choose the one or two most costly parts of a mission and consider what it would take to purchase those parts now. Or if a system has many identical units, the cost of one can be estimated, then multiplied accordingly. A method to apply to the cost of operations requiring local (solar system) travel by humans, such as acquiring fuel, is to equate it to the cost of a Mars mission, say 100 G$ after development. Still another is to extrapolate from present costs per kilogram of hardware after development (e.g. a building at 1 $/kg, a car at 5 $/kg, a computer at 100 $/kg, a space shuttle at 30 k$/kg). Launch costs from Earth are omitted here because at present rates via shuttle they raise the cost of hardware a factor of 100 or so (to a nearly intolerable 10 k$/kg to low-Earth orbit). Predictions of much cheaper launch, approaching the true low cost of the energy depth of Earth's gravity well, do not seem useful yet. (Recall that less than $2 of electric power could theoretically place 1 kg in orbit 1000 km up, moving at 8 km/s.) A starship must be built from materials already in space, or as transported from much shallower gravity wells such as the Moon. Space transportation to shift asteroids to near Earth orbit or to carry large equipment to the vicinity of Mercury will be substantial because of the large orbital speed change. For estimating the cost of such transportation, a conservative factor of 10 is multiplied onto the value of the material or equipment.

Economic estimates are given in current dollars. Future inflation or industrial efficiencies cannot be predicted and may not make much difference in grasping the magnitude of the human effort. While people's earning power has changed from the equivalent of say 100 $/yr in prehistoric times to 100 M$/yr for some entrepreneurs now, there is no reason to believe this trend will continue either. It also is difficult to see how interstellar travel would become immensely profitable in itself, especially if entrepreneurs must wait several millennia for a return shipment.

The KE of travel, and therefore the fuel energy, seems to be a likely economic limiting factor. A 10,000 tonne starship plus 100 times more mass in fuel traveling at 0.01c requires almost $(10)22$ joules of KE which, if purchased at present electric rates, would cost 100 T$. The situation is worse when rocket inefficiencies are included. It has been shown that anti-proton fuel would provide a reasonably fast journey (more like 0.1c than 0.9c) and a low mass ratio, but at a fuel cost up to 100 T$ for a 1 tonne 0.1c flyby mission. The anti-proton cost would be of the order of

(10)19 $ for a 100-person 0.1c colony mission that stops (about one million present US GNPs!). The formidable cost can be traced to literally making the fuel from other matter at the particle level, in a rather inefficient (0.1% to 1%) process. A society less than infinitely rich either must plan to devote many centuries to making the fuel or try a less desirable drive which uses fuel which nature has made and we need merely collect or purify. Automated space solar energy collectors and anti-matter factories probably could provide anti-protons much cheaper.

The Daedalus study has indicated that a medium-fast flyby probe could be sent with 50,000 tonnes of fusion fuel and cost up to 10 T$. The fuel as collected from Jupiter was estimated to cost about 1 T$ with some costs omitted [a-Parkinson]. While this resembles the shuttle launch costs for deuterium, the helium-3 could not be obtained on Earth at any cost. Fusion fuel should be cheaper to collect in space than to make. A one tonne probe might need only 100 tonnes of fusion fuel. A solar wind collection and separation station 100 km diameter near Earth can gather about 1 kg/yr [a-Parkinson]. If a frugal 1 B$ is needed to build one station and 100 M$/yr to operate it, then obtaining 100 tonnes in 100 years would require 1000 stations for a total fuel cost of about 10 T$. This is only somewhat better than anti-proton costs and about 100 times more than the cost which the Daedalus project finds for the same fuel from Jupiter. Moving the stations 10 times closer to the Sun would increase production density by 100, but travel costs would be much higher operating far from Earth and deep in the solar gravity well.

Separating and shipping deuterium from Earth would be much cheaper than collecting solar wind, but helium-3 would need to be made in a large expensive set of deuterium fusion reactors on the Moon or in space. The feasibility of this can be estimated another way, by energy economics. Suppose that fusion reactors instead of solar power are the power source for space colonies. With a basic deuterium-based energy economy for stationary use, the infrastructure would exist for making helium-3 for propulsion. If each 1-million person colony operates at 1 Gw of power, and 10% can be spared for producing fusion fuels, then more than 100 colonies would be needed to accumulate 100 tonnes of helium-3 in 100 years. Obtaining the deuterium from Earth or from a small comet would be a minor part of the problem, compared to this scale of operation for fueling just one small probe. And it is more likely that about 1000 colonies would be needed to support this operation.

A 100-person mission with 20 times more payload than Daedalus, using fusion drive, and needing to carry fuel to decelerate, would seem to require much more development and therefore perhaps 100 times more money than Daedalus, or (10)15 $. To provide the fuel and so forth to decelerate it would require at least 10 times more money, or at least (10)16 $ (more than 1000 US GNPs). To afford this mission over 100 years, about 10,000 hard-working space colonies would be needed as described earlier, or a space population of about 10 billion people.

Many other propulsion methods have been discussed. Another somewhat feasible one, pushing a "sail" with a lightbeam, is also chosen here for economic estimate. For feasibility the sail material must be less dense than (10)-5 kg/m^2, multiple-layered and reinforced. The manufacturing rather than the material cost is likely to dominate. Production energy cost is likely to be the cost of solar energy as collected by space colonies. Their costs will be higher than Earth costs, but without high launch costs. The costs of acquiring the materials in space will be substantial even after sail production is routine, as it will be for interplanetary transport of asteroidal material to make more sails. High-tech processes cost (now) of the order of 100 $/kg. On this basis a sail might cost about 1 k$ for 100,000 m^2 including all structure. A 100 km diameter sail might cost 100 M$. This seems very low for so large a system assembled in space, so let's raise it to 10 G$. NASA wants much more to build (and launch) a modest space station. Although NASA's hardware costs of 10 k$/kg and up seem far beyond reasonable, this estimate is already up to 100 k$/kg. Let the cable system of about 10,000 km of assorted size cables of exotic material be another 10 G$. This is a million dollars per kilometer, similar to what a freeway once cost, and far more than optical fiber costs. This sail system still seems to be a bargain considering what it will do. If it were not so massive the starship should carry several.

The collector-reflector would be similar in size and cost to the sail except for the need for a very accurate optical surface. And it must be transported near the Sun and held in place. As for all propulsion methods, we should not be surprised that this system costs 100 T$ or more. At 1 k$/$m^2$ and a factor of 10 for transportation, for a total of 100 T$, a permanent starship launcher (our end, anyway) might be purchased. However, the mirror must cope with the same reaction force that the starship sail receives. Counteracting this could cost energy and money, but much less than propelling the starship directly since there is no need to push an enormous fuel mass. Unless a big mirror can be anchored to Mercury, a large project of providing asteroids or of holding position with rockets or very long cables would be needed. Very accurate large-scale tracking will also be needed, to be maintained under intolerably hot conditions for 5 years per mission. Cost estimation should be continued to carrying and furling the sail and braking the starship at the other end. But the starship itself seems to cost of the order of 1 T$ to 10 T$ (to be shown), with the sail much less. 10 T$ should cover it easily. The whole program for a human sunbeam-pushed slow mission to a star looks like 100 T$, much less than a fusion-powered mission. A little more might be spent on sun-pushed probes and on test missions to establish the technology better.

Others have proposed that the lightbeam be produced by a giant bank of lasers, solar driven and therefore near the Sun. Not only is efficiency low (perhaps 10%), raising cost by 10 for larger initial solar collection, but cost is surely another 10 times greater because of complexity of the laser process. Now the total cost has risen to about (10)16 $. The cost of large lasers is difficult to estimate. Another way is to suppose that each laser costs a low 10 k$/$m^2$ of aperture and (10)10 m^2 of them are needed. Then the total cost is 100 T$, and another factor of ten is needed for placing it near the Sun. The solar power system to operate it may cost similarly. Giant fresnel lenses seem very expensive, especially because of their delicate adjustment, and are to be avoided in lieu of phase-locking the lasers.

Still another launch method is the 25 million km or longer magnetic launcher with 1/4 billion stations, each with solar power supply. At, say, 40 M$ per unit the cost seems to be about (10)16 $ or 100 times more than the sunbeam collector and about the same as the laserbeam and rocket systems. This system would surely cost much more to put into place, but it could be used many times. For comparison, the 10,000 superconducting magnets for the new SSC particle accelerator in Texas are expected to cost about 0.5 M$ each. Their size is approximately 1 meter across and 10 meters long.

Compared to the fuel or other propulsion method, the cost of all other starship components may be much smaller but still large (unless a part measuring thousands of kilometers is needed). Research and development would be substantial too. Boeing uses 4000 engineers to design a jetplane over several years at a billion dollars or so. A starship with complexity, if not size, at least 1000 times greater might require 1 T$ just for a prototype. NASA used several hundred thousand people for the Apollo program, paying an average of less than 100 k$ per person. A starship program might require a million people at times, 1/10 of the population of 10 large space colonies, each on the average engaged in 1 M$ of productivity over a lifetime. A 10,000 tonne starship (without fuel) is about 200 times more payload hardware than a shuttle (without fuel). On the basis of mass one might expect it to cost about 1000 times more, or of the order of 1 T$ (if NASA's costs are efficient). Making and testing many subsystems and several starships might require several more trillion dollars.

Another approach is to make a unit mass payload hardware estimate, again exclusive of drive or fuel. A 10,000 tonne starship built at the very high-tech rate of 1000 $/kg would cost only about 10 G$. This estimate seems very low and probably is less than assembly-line costs. NASA spent 30 times more per kilogram to manufacture a shuttle, and in view of reports of NASA inefficiency this might be taken as a high estimate. Dedicated starship builders could surely achieve higher efficiency. To improve the estimate in other ways, multiply by 10 for the difficulties of building in space and by another 10 for obtaining materials. This results in a 1 T$ starship, same as the much more massive and much larger colonies that build it. Regardless of the amount of R&D, uncertainty over the drive, pre-testing, mistakes, and setbacks, it does not seem possible to spend more than 10 T$ on the basic 100-person starship. As before, fueled propulsion would raise the cost another another 1000 or so.

One proposal for a 1 tonne probe found a 4 km diameter sail and laser beam necessary. Again propulsion would dominate costs. The 1000 km semi-precision fresnel lens proposed to focus the laser beam seems beyond the range of estimable structures but fortunately seems unnecessary. The laser system need produce less than 100 Gw, and at 10 times space power plant rates would cost about 1 T$. Establishing a 100 km or larger effective aperture for it might become costly. If the probe is instead sent by fusion rocket as a flyby at 0.1c, about 100 tonnes of fusion fuel is needed. This mission would be less than 1/100 of Daedalus, and might cost 100 B$ plus research, development, and testing. Only the fuel need be obtained or made in space; the rest would fit in one shuttle load. Overall a 1 tonne probe seems to cost in the trillion dollar range, affordable over 10 years to one very determined nation. It is difficult to imagine a 1 kg probe costing less or being as technically feasible.

For probes, a self-reproducing system has been proposed as the lowest cost method, after the enormous cost of developing and building the first one (estimated at a low 100 times larger than Daedalus or about (10)15 $) [a-Freitas]. This approach assumes that each probe makes 10 or more copies when it finds a resource base. The first one might be sent no farther than Jupiter's moons to begin work so that we might supervise the second generation and check that the system is working. But it was shown earlier that after a few generations of spherical expansion from Earth, at about 40 LY when they number in the hundreds, probes would rarely need to make more than one copy of themselves. It may work out cheaper for us to supervise all probe making here rather than make an automatic and more massive system. The probe program then reduces to producing "regular" probes, a major feat in itself. A Solar System civilization with colonies out to Jupiter and Saturn would be needed to support the program. Probes can boost all the same way (assuming Jupiter can provide the fuel) and coast the 10, 20, or more lightyears to their destinations. It is uncertain whether even a growing space civilization can afford to send probes indefinitely. Mission monitors may not hear from more than the first thousand anyway.

At the other end of the mission scale, the cost of sending a colony ship ("worldlet") should be considered. One way to estimate this is to assume that suitable propulsion such as fusion rocket can be provided and use a per mass estimate. If a space colony costing 1 T$ is to be converted to a starship (really, a pair of them), and the drive and fuel costs at the same per kilogram rate, then a 2-leg slow mission with mass ratio 1000:1 would cost 1000 times more, or about (10)15 $. This is an underestimate for several reasons: for example, advanced fuel costs more than most hardware per unit mass, and the space colony consists mostly of (10)12 kg of rock and dirt. When the Daedalus rate for the mass of the fuel is used, the cost is more like (10)19 $. A civilization that can raise 1 T$ per year would never prepare this mission, and we must wait for larger local space civilization, perhaps 1000 times larger. A population of a trillion would have real purchasing power and would be living in a million space colonies! If the first colony doubled itself in a fast 20 years and each succeeding colony reproduced likewise, only about 400 years would be needed to reach this number. Population itself must grow about 3.5% per year to keep up and fully populate these colonies. Space development is more likely to go slower. Gathering (10)12 tonnes of fusion fuel sounds far-fetched at this time, despite visions [a-Criswell] that later humans would become very ambitious in obtaining materials, taking apart Jupiter and even the Sun.

10.7 TECHNICAL PROGRAMS

Appearing formidably difficult now, interstellar travel may begin one step at a time. The task would be many orders of magnitude greater than Apollo or the Voyagers. Almost anyone can envision the task, but no one person will be able to study enough of the details of such a program to predict feasibility or establish design at a practical level. As teams were required for Apollo and Voyager, so will mega-teams of experts be needed for an interstellar program. Having to divide the work among perhaps a million people in many levels of organization reduces efficiency (and raises cost) but there seems to be no alternative. Coordinating such a program appears to be an ultimate challenge for human ingenuity, and almost all fields of study will be pertinent. Unprecedented long-term cooperation will be necessary, something humans achieve more readily on technical than social projects.

The departures of the Voyagers and the Pioneers have been pointed to as the first steps to the stars, but these have not caused any wave of public or other support for the next step. Sometime next century more probes to the outer planets will be affordable. New methods of propulsion, fission- or possibly fusion-powered ion drives and solar sails with gravity boosts will likely be tried, setting the stage for faster, more ambitious drives. A probe to and well beyond Pluto has been shown feasible now. Finding money, motives, and humans to take a long lonely roundtrip beyond Pluto might be a further step after closed ecosystem survival has been proven in several roundtrip voyages to Mars [b-Reiber] and Jupiter's moons. Meanwhile some asteroid prospecting will have occurred and possibly one or two asteroids might have been found worth deflecting into orbits that would bring them to Earth vicinity. As space colonies become ambitious, more asteroids would be sought to supplement the increasing flow of minerals mined and launched from our Moon. A 1 km metal-rich asteroid would mass about 2 billion tonnes, providing enough material for 10 large space colonies and thousands of starships (less fuel). [see also a-Martin; a-Hartmann] Meanwhile fusion drives may have been developed to the point that they can be used for interplanetary travel, providing many chances to improve the technology. Space colonies may earn money by separating deuterium from the solar wind with a low-cost process and selling it for stationary reactors and rocket drives. This capability can be scaled up for collecting starship fuel.

Who is doing all this? NASA (US) and/or Europe and Japan would work on the next local probes. Corporations and governments may join in permanent space research and manufacturing, requiring the establishment of small space colonies. Large consortiums of various national governments and/or large corporations and universities and possibly private foundations might support some missions by the 22nd Century. If Earth is having severe problems, the consortium might have a growing space colony component. Some new form of organization might appear that has a large people base and can provide funds and resources for space work. Assuming that launch remains chemical and costly, minimal support would be launched from Earth: very skilled people, very expensive low-mass equipment, and nitrogen, carbon, and hydrogen until asteroidal ices are located. As soon as people are sufficiently comfortable to raise families in space and operate schools and universities, the costly launching of people would also diminish. Later, space research and mission organization and planning would become a by-product of a growing network of prosperous space colonies. Mission research, hardware development, and assembly would be occurring in space near some suitable group of colonies. Simultaneously, astronomical work in space and from our Moon would be establishing which nearby stars have planets worth visiting. Much more unexpected would be receipt of a message from a nearby star or from any star at all.

As early as possible, interstellar enthusiasts would be assembling the smallest, lowest cost reliable probe to a likely star with planets, to obtain some hard data. If the feasibility looks good and the price is less than 1 T$, a long list of grants from every government, corporation, and foundation on Earth and in space might be collected. This would be a project worth doing at least once, and as soon as possible. If no better destination is found, a probe would be sent through the Alpha Centauri system. The wait for the data may be several hundred years. If the data support it, a second more ambitious probe might be sent, or possibly a mission with people who probably cannot return.

By now possibly five centuries would have passed, and human space civilization, if it exists at all, should be quite large. If some remarkable scientific breakthrough is possible to make interstellar travel easier, it will have occurred. Also the chances are good that signals from some other civilization will have been found, since the search would have been long and nearly exhaustive. But there may not have been time yet for interstellar dialogue. If support and technology to send a small colonizing starship at 0.1c has not been found, by the 30th Century the resources should exist to send a larger slower colony ship. If propulsion for a starship continues to be a formidable problem, space colonies and space mobility will have developed so that Solar System-wide resources can be used for a launching, with mirrors and sails or possibly a magnetic track. This may require techniques for surviving in small colonies closer to the Sun than Mercury and even enjoying safe stable life there.

It has been suggested [a-Lacey; a-Bloomfield] that starship departure be slow. That is, the selected voyagers might spend a few years in the starship while it is still near local resources, pretending to be isolated as they finish building, fine-tuning, and checking-out the starship and their compatibility with each other. If trouble occurs, they can get help. If people want to leave the mission and others join, trading can occur. If some supplies were forgotten, they can be obtained. Beyond the first interstellar mission by people would come a series, depending on reports of success and location of more destinations. It would be very important for the first missions (and probes) to transmit back to the Solar System full information on all failures and successes enroute and at the destination. These reports will trickle in over many centuries, but any early ones that provide guidance for improving starship design would save builders trillions of dollars at the least.

Other beings must discover and use the same sciences and technologies that we do, but their psychology, sociology, and philosophies could differ widely from ours, hence their goals and motives. Humans who travel to the stars must adjust themselves to the technical constraints more than adjust technology to their needs. Some other beings may have more flexibility and travel better, and some would consider such travel psychologically or socially impossible. Can humans adjust to each other and to the rigors of interstellar travel sufficiently well? Our descendants will find out.

Chapter 11

Interstellar Life and Civilizations

Our prospects for interstellar travel are closely linked with the incidence of civilizations of other intelligent beings, with the likelihood of our detecting messages from other civilizations, and with their engineering ability for communication and travel. At a fundamental level, the prospects depend on whether life can arise elsewhere, in connections not immediately obvious. The Search for Extra-terrestrial Intelligence (SETI) has become a vast subject in itself, with many papers, conferences, and experiments. The search incorporates astronomy (which stars might have planets), planetary studies (which planets might have conditions for life), biology (what sorts of biospheres can evolve on planets), a mix of biology, sociology, and philosophy (what kinds of intelligent beings can arise and how would they act), technology (what means would "they" use to send and receive messages, or themselves, across many lightyears), mathematics (expected to be a common language), and other fields (as we define them). SETI includes Communication with Extra-terrestrial Intelligence (CETI) and can include sending messages as well as receiving them.

Results in these areas bear on the central problem of interstellar travel, but only a survey of SETI topics with selected details can be provided here for background and to show those links. Along with a sampling of the many available references, some detail is given to convey the enormity of using interstellar travel to "contact" other civilizations. Earlier chapters have touched on questions of what stars can provide habitable planets and what technology is needed for communication. A very wide range of views on the relation of life, intelligence, and the Universe has been published. The work has been informed speculation due to lack of experimental or observational results. Most of the books listed for this chapter are collections of articles and provide general or specific discussions. While conference reports may sound arcane or too technical, such books provide reliable papers which are often also quite readable and lively. Conferences on this subject seem to have some of the most fascinating debate in all of science, plus glimpses of scientists' personalities. Many clever sf stories are also relevant, but, as with the rest of this book, there are too many good ones to cite.

Anthropocentrism, the pre-Copernican viewpoint which puts humans in a central role in the Universe, is an impediment to useful thinking about other life and civilizations and is very difficult to avoid. In an unavoidable bit of human-centeredness, all units of measure used here (such as Earth years) continue to be ours, even when applied to beings who cannot know we exist. Our base 10 number system is also continued, but our idea of an order of magnitude estimate, will differ from estimates made by beings with other number systems. Non-rhetorical "we" and "our" continue to be used here to denote our common humanity as distinguished from other supposed beings. References to "them" are often not put in quotes, to avoid excessive human chauvinism.

11.1 ESTIMATING OTHER CIVILIZATIONS

To discuss interstellar travel in regard to humans visiting other beings, hearing from other beings, receiving help from other beings (e.g. on methods of travel), having them visit us, or finding evidence that others have mastered travel, the existence of other beings must be ascertained. Interest in the existence of other beings has distant historical roots in human culture, back to ancient Greece at least [a-Beck; a-Papagiannis; b-Engdahl; b-Goldsmith]. In the 16th Century Giordano Bruno supposed an infinite number of worlds with life to exist. A tradition has been established in talking about other civilizations of examining the "Drake equation" which purports to predict their number. This general analysis of life in our galaxy presumes some context and prior understanding. A few common concepts are briefly described here first. Some are expanded upon in later sections.

"Extraterrestrial" defines "them" as other than us, setting us humans up as the standard by which all intelligent life might be measured (not a good start and not justifiable). The acronym "ET" was, for a while, in danger of becoming widespread after a touching film of that name. Many refer to "them" as being a different "species", which is certainly true but is a misuse of our definition of species. Earth itself has millions of different species in regard to types of creatures which can reproduce within their species but not outside it. The term "humanoid" sounds too limited and self-centered, like an inferior version of us with two legs, two arms, and a head. It has been meant to refer simply to an intelligent being. Some people have referred to other beings as "aliens", a prejudicial term evoking images of blood-thirsty bug-eyed monsters. If they exist, many may indeed appear gruesome or strange to our sensibilities. The term selected here is "being", implying creatures who have evolved to the level of recognizing their own existence and therefore likely to study their place in the Universe. Self-knowledge or self-awareness might be an important part of "intelligence", a characteristic commonly assumed for beings worth hearing from, seeing, or visiting. The term "sentient" has also been used to refer to being able to sense oneself and others at higher levels. (Biologists would enjoy traveling to see another world with abundant non-intelligent "life forms", playing out whatever version of evolution has evolved there.)

We assume intelligence also includes rational thinking (which some humans practice some of the time), problem solving, artistic creation, development of science, technical competence (tool use expanded to building, flying, computing, transmitting messages, etc.), complex societies with evolving cultures, and more. Other intelligent beings might not need to do all these to qualify, and might have some ways which we have not discovered. A computer-like brain with many data-processing elements, sensory input, and effective output would seem to be required.

But it has been pointed out that other species on Earth, for example insect societies, solve complex problems of survival with abilities evolved over long times, and that many species can adapt to bizarre changes in the environment, given enough time [a-Raup]. Being intelligent seems to have an evolutionary advantage, since an individual can solve complex problems fast that natural selection would require millions of years for a species to adapt to.

Discussion here assumes that beings have developed "civilization" in the sense of cities, nations, and space colonies, but this is not a requirement for intelligent behavior. Civilization could be think-tank villages in the forests, artists working from communal burrows, crystals communicating by radio, or something else. Other beings might live closer to their nature (the ecosystem on their planet), especially if it is more benign than ours, and they may not reach the stage of knowing to try to contact other civilizations of different types. Our anthropocentric assumption, that how things are for humans is the way things must be everywhere, is an ever-present obstacle to clear thinking about other civilizations. We probably can never know the extent of our prejudice in the definitions of life and intelligence, no matter now many other planets with life we learn about.

The field "exobiology" can be defined as the study of life as it might occur elsewhere, in different conditions, possibly but not necessarily leading up to intelligent beings [a-Lederberg]. Some call it a science without a subject. The primary term "biology" is usually assumed to refer to carbon-based life because, to the extent that our chemists can imagine, no other element provides such a versatile basis for life. But other chemistries and forms of life, including intelligent, cannot be ruled out. No basic proof has been found that carbon life is inevitable or that life can use another chemistry. Outside the particular examples of biology, biosphere, and ecology found on Earth, defining "life" and "biology" becomes difficult. Thus far humans have explored by remote control several worlds (Mars, Venus, and many outer planet moons) and found different chemistries but no life. Earth-type biochemical methods were used to test for micro-life on Mars [a-Horowitz]. The term "xenology" has also been used for exobiology, but the Greek root "xeno-" carries a meaning of "strange" and reminds us too quickly of xenophobia, fear of strangeness. Beyond exobiology is the field of "exosociology", the nebulous study of the lifestyle, culture, economics, and politics that other beings might have.

The Drake formula or equation reduces the question of the existence of other civilizations to a series of cascading factors, or probabilities that multiply together. It was invented by Frank Drake, a skeptic about interstellar travel, in 1962 to define better this problem [apparently unpublished but described in several forms and many places, e.g. b-Shklovskii; a-Sagan 1963; b-Sagan 1973; see Appendix D]. The equation is specifically intended to estimate the number of civilizations which achieve technology sufficient for radio communication, and its factors start with the astronomical, involve the biological, and climax at the sociological. The number of new stars per year, those with habitable planets, the probability of life on those planets, the probability of that life achieving intelligence, and the probability that those beings develop the ability to send and receive interstellar messages, are all combined with the average lifetime of a technological civilization to arrive at a number of civilizations existing at any given time in the galaxy. Most of these factors are very difficult to estimate, and the result may be no better than a wild guess made without considering the various subguesses. The most secure estimate is the number of new stars formed per year, about 1 to 10 in our galaxy.

Most published estimates have been optimistic about all Drake factors except the lifetime. Educated guesses for the factors have resulted in estimates of the number of civilizations from about 10 for short-lived ones up to millions for long-lived ones. The geometric mean is around 10,000 civilizations and pertains to high-tech civilizations which last just 100 times longer than our present 100 years. Since there are about 10 billion reasonably bright stars in our galaxy, there could be a billion chances for habitable planets, and the Drake estimates could even be low. On the other hand, the Drake equation assumes all factors are relatively unchanging over the billions of years for life to develop into civilizations, but various possibly disruptive processes in the galaxy (e.g. spiral arm motion) occur on a time scale of 100 million years and may cause an overestimate [a-Dunlop]. If the total time (several billion years) available in the galaxy for stars and life to develop is used in the Drake equation, it then gives the total number of civilizations ever to exist in our galaxy, from perhaps a thousand to a billion [a-Papagiannis 1982; a-Freeman].

The Drake equation is intended to be conservative, considering factors for which there is no known reason not to take as reasonably large (probabilities like 0.1 to 1). But scientists still differ widely on whether what has happened here with Earth is very unusual or very usual. Since the Drake result comes from a concatenation of probabilities, two kinds of results are possible. Probabilities mostly near 1 multiply together to give a probability near 1 and many civilizations. Probabilities mostly much less than 1 (0.1 illustrates this) multiply to give a very small result. If the probabilities of a habitable planet, of life developing, and of intelligence evolving are all independently taken at 1/100, then for ONE civilization to be present at any given time somewhere in our galaxy, it must last at least a million years. Estimating these probabilities is very difficult because human science thus far has little basis for determining the full range of habitable planets and then the full range of kinds of life that could develop on strange but habitable planets. The number of things that can go wrong is surely larger than realized, but also the number of ways that life can arise may be larger than we can imagine. No one can know now the overall trend of the complex mechanisms behind the Drake factors, but glimpses of the variety of fascinating active icy worlds found as moons around our outer planets increases optimism. Planets elsewhere will surely not be all cold airless dry cratered rocky worlds.

The lifetime of a civilization is crucial. Most suitable stars near us could have had, or might sometime have, a civilization, but if each lasts only a thousand years, the chances of one occurring at the same time as ours is very small. A breakpoint or hurdle may occur in civilization lifetime, with those lasting more than 100 to 1000 years entering a realm of extreme longevity [b-Shklovskii; a-Bracewell 1979]. The problem of synchrony has other facets. A civilization might be at a stage suitable for communication with us (have technical and cultural similarities) for only a short time [a-Beck]. The few civilizations, if any, within 100 LY of us (the closest likely spacing) would be unlikely to arise and endure during the lifetime of ours (unless we endure millions of years).

Given an estimate of the number of civilizations in the galaxy, the average density of them can be estimated, and then the average separation [Appendix D; nice graph in a-Oliver 1980; a-Bracewell 1974]. As an example at the geometric mean, suppose 10,000 are expected at any time. One must be cautious about simultaneity, but this number can pertain over spacetime as we observe it. They are less likely to be at the galactic core because of the shortage of heavier elements and the possible violent black hole at the center. For galactic volume about (10)13 CLY (cubic lightyears), excluding core, each civilization is surrounded by (10)9 CLY, and the average spacing is thus about 1000 LY, or about 1/6 the thickness of our galaxy. This large separation would make two-way communication very tedious. We cannot connect with all these civilizations either. If they endure for thousands to millions of years, then our signals can reach the oldest ones across the galaxy but the youngest only near us. However, other short-lived ones will be replacing the short-lived ones that fail to respond to our signals.

11.2 THE FERMI PARADOX OR DILEMMA

A story often told is that the nuclear scientist Enrico Fermi was in a free-wheeling conversation in 1948 that turned to the subject of other beings, and he suddenly asked "Where are they?" [b-Hart]. The premises are that there are many civilizations out there, that at least some must be adventurous, that some must last long enough to travel, and that interstellar travel is achievable, however slowly. Yet no one has come here. To avoid what has been called the "Fermi paradox", one or more of the premises should be false. Here is the key link between other beings and interstellar travel. Life has been possible in our galaxy for so long that some other civilizations should have been around a long time and engaging in widespread travel, if only by probe. Yet they are not known to have visited us.

A true paradox cannot be resolved. It is a collision of ideas, with both sides acceptable or irrefutable. Very few paradoxes have been left standing long in science because of their inconsistency. We should hope that the Fermi paradox is an apparent one and therefore has an explanation that removes the paradox. Many explanations have been sought and presented for why contact with other beings is much less frequent (zero thus far) than estimates of the likelihood indicate. The Fermi dilemma has also been called "the great silence" and could be extended to lack of messages thus far. It has been rephrased as "absence of evidence is evidence of absence". And it has been called simply the Fermi question with premises insufficiently supported to present a paradox [a-Seeger].

Since "they" are not here now and may never have been (nor have they sent obvious messages), many explanations have been proposed, spanning many fields. More headway will be made here if technical reasons (physical reasons, based on known laws of nature) are examined more seriously than the abundance of more speculative but possibly more important social or philosophical reasons (e.g. they don't feel like traveling). One far-fetched resolution is that no other beings exist, and proof is obtained from reverse reasoning such as the following: If "they" did exist, their probes (at least) would be everywhere. Since probes are supposedly easy to build and send, perhaps in von Neumann form, and are not here, other beings do not exist in our galaxy [a-Tipler; critiqued a-Sagan 1982].

Many psychological and social difficulties that interstellar voyagers would face have been described in the last chapter. There may also be biological reasons, some different from those making travel difficult for us. When we have no idea what sort of life might evolve elsewhere, such reasons are nearly impossible to guess. Then there is the possibility of bad timing (e.g. we're the earliest civilization, or we came along between two waves of galactic civilization). Some reasons are too anthropocentric to be credible when applied to the expected very broad range of types of worlds and civilizations out there. These put Earth at the statistical fringe as being unusual in some way. (Some postulate other beings at the fringe, but we cannot judge one way or the other.) And some speculators believe that others have already been here or are here now, contradicting the general observation that "they" are not here.

Many reasons for no contact via message or visit have been compiled and argued [a-Brin; a-Hart 1982; a-Deardorff; a-BIS; a-Martin; a-Tang; a-Drake 1985; a-Wolfe; a-Jones 1985; listed on a spectrum in a-Ball 1985; a-Kuiper; a-Cox; a-Ball 1973]. They include reasons why the incidence of intelligent life is not as high as suggested. Habitable planets like Earth may be rare. Life may not start easily on habitable planets. Higher intelligence might be a rare step in evolution, and technical civilizations still rarer. We are located where none have ventured or sent messages. We may have the only civilization ever. Or civilizations could be too content to venture out. Or they are busy using cheap radio contact only. Or Earth has been put in quarantine, under embargo, on probation, or in "zoo" exhibit status, to be observed but too dumb or dangerous to bother. Or life on Earth is a controlled experiment by "them". Or Earth and its life, fascinating though it is to us, may be considered too dull or bizarre by others (who presumably have seen it). Or the opposite situation may hold, we are nearly the cleverest beings ever to evolve, and others are still struggling to invent surface transportation.

Or some self-inflicted or natural catastrophe wipes out most civilizations before they can do much message-transmitting or probe-sending [see also b-Asimov 1979; b-Verschuur]. Perhaps Earth is in one of the least dangerous regions, and other planetary systems are too often devastated by dust clouds, supernovas, or comets to permit life to evolve to civilizations. Or the types of aggressive beings which would be able to master travel to stars also engage in war and mass destruction and eliminate themselves first. Or they cannot control overpopulation, mass disease, and famine. Or travel may be too difficult or expensive. Or too few individuals are willing to endure travel for long times. Or establishing a new civilization on an alien, barely habitable world might be too difficult. Or civilizations may develop rapidly to high levels uninterested in travel or in radio searches or even in technology, preferring mental, social, or spiritual pursuits. Or, if laws of nature permit, they may develop near-immortality and prefer very safe lives (even fearful of sending messages?). Or most beings are very gregarious and would not venture lightyears from their many nestmates or relatives. Or perhaps an armada is already quietly on its way to conquer or eliminate us. It might consist of robotic starships automatically sent to eliminate all life detected everywhere.

Responses to these reasons or excuses are given throughout this chapter. When a synergistic or feedback effect is considered [a-von Hoerner], it is more puzzling why we have not seen other beings. After a few civilizations leisurely discover by radio or travel that each other exists, they would surely increase their efforts to signal and to explore, and the rate of contact should accelerate. They will also try harder and learn more about having longer-lasting more stable civilizations. Unless we are the first, or costs of full-scale searches far exceed payoffs, again we should have already had visitors.

Evidence is absent that any other being has visited Earth and left a monument, artifact, or observation station here, or an obvious satellite in orbit, or just garbage (lost flashlight, dirty food containers) on the ground [a-Benford; a-Freitas 1983]. Searching solar orbit for a tiny station would be very difficult if it watches us passively and emits no signals toward Earth. If a 10 meter satellite is shiny, Earth telescopic surveys might not find it beyond a million kilometers. Large space colonies or active self-reproducing probes could be in the asteroid region and we might not find them for centuries unless perhaps by sensitive infra-red surveys that detect their waste heat [a-Papagiannis 1982, 1985]. Or they could have stopped or stayed at the outermost planets where they could replenish on the lightest elements [a-Stephenson]. We cannot search our Moon for hidden stations ("2001 monoliths") without going there, and a signal from station to elsewhere would be difficult to detect. If a monitoring satellite has marked our entry into the nuclear or space age, we are unaware. Observations have been made of the stable Lagrange points L4 and L5 near the Moon where equipment or space junk would remain for long periods, with no positive findings of large shiny objects yet [a-Valdes]. Making, bringing, placing, and hiding their long-term reliable robotic stations here [a-Freitas 1980, 1983] would involve them in much more severe technical difficulties than we face making a first starship or probe that operates a mere thousand years.

Many have supposed that visitors might be wary or ethically cautious about visiting Earth [a-Deardorff]. Probes might be avoided because of the possibility of error or disclosure. Any recent observers with ultra-high-tech means could probably glean the essentials of what is going on here from space without letting any detectable evidence reveal their presence. If Earth were a tourist attraction, the general run of beings might be sloppy and leave trash in space or on the ground. If we are being watched by professionals, then we would be within some jurisdiction of a civilization or a galactic organization. Some bureaucrat could have responsibility for the observations of Earth and must ensure that a "codex galactica" is followed in regard to visiting or colonizing here. Over long times, stars drift apart in galactic arms, and such a jurisdiction is not stable. Many different civilizations may have passed on the responsibility for us to the next nearest one. All this is very fanciful thus far because it depends on easy interstellar travel.

Once some hypothetical long-ago visitor has noted that Earth had an active biosphere producing ever higher lifeforms, a message might have been put out to leave Earth alone, or just watch and see. Perhaps a small distant observation package was left among the asteroids or near the Moon that might signal us (or them) when we reach a certain level of ability. Their advanced and experienced scientists might have predicted from earlier stages of life on Earth that intelligence would soon arise here, an event that might cause caution, jubilation, widespread interest, or who knows what around the galaxy. Earth may already be in a "galactic encyclopedia", although surely not with an up-to-date entry. We would like to believe that our worst habits--incessant warfare, ever larger weapons, atrocious treatment of biosphere, petty greedy squabbling instead of major space projects--have not been observed and recorded for all to see. We may still have time to clean up our act.

Military considerations occur on many sides of the question of communication and travel among the stars. Some humans fear that other beings carry on interstellar war and might invade, conquer, control, destroy, plunder, or eat us. Such plans would seem to require campaigns of very long duration with stable government and organization the likes of which we have never seen. Even if some beings have the patience or near-immortality to carry on such programs, no continuity or coordination is possible. The time scale is very long and the economic scale very large to send out starships to attack and plunder, to learn of their success, and to bring results home. Beings might need 10,000 year lifetimes to carry on in this way. The galactic-scale conflict shown in Star Wars films and many novels is much more restricted by technology than indicated. Undoubtedly some beings might have a generally hostile attitude and would do this if they could. Some who master the technology of travel would likely also carry formidable weapons.

Consideration of travel between galaxies has been used in a far-fetched extension of the Fermi dilemma to show that other civilizations are very rare else one or more would have spread among and through galaxies in the available time [a-Fogg]. But the conclusion from the Fermi dilemma that seems least acceptable is that interstellar travel is too difficult to permit visiting. Even if there are many civilizations out there not too far away, very few might have the resources, ambition, or longevity to attempt it. Of those few, perhaps almost none have beings which can survive a physically and socially hazardous journey. If there are very few explorers out there at any given time, Earth may be overlooked for millions of years. But explorers who know they are rare might leave signs at all planets visited that they were there, signs that would last millions of years and come to the attention of any emerging intelligence. The present status of the Fermi paradox is uncomfortable and indicates that the field is open to new ideas and to new discoveries of flaws in its statement or premises.

11.3 RADIO SEARCH AND COMMUNICATION

SETI has been primarily a radio search (CETI), listening with assorted large "dishes" (parabolic reflectors) and sensitive receivers for meaningful signals buried in galactic noise. Some of the technology has been described in Chapter 7 as needed for a starship, including requirements for bandwidth, information capacity, power, and the like. The million directions in space (or million stars) at which to point a receiving dish, the billions of one-cycle-wide channels to be examined, the short time available to "listen", and the rapid decrease of power with distance are some of the factors making search very difficult. Power is not the main problem. Presently we have a receiver (using the Arecibo 300 m dish) which could detect messages sent with a similar dish from 15,000 LY away, and further if they use more power than Arecibo engineers have. Our receivers are already working near the noise limit, and great improvement in sensitivity by us or anyone is not possible. A starship mission would surely do an ongoing search for intelligent signals during travel. Likewise, "they" might be transmitting from starships or remote stations not associated with stars. Thus a search should not be limited to likely stars, and a 100 m dish provides a billion directions to look at 30 GHz. [a-Tarter 1985, 1985, 1980; b-Sagan 1973; a-Sagan 1975; b-Shklovskii; a-von Hoerner; a-Morrison 1985; a-Oliver 1981; b-Morrison; a-Mallove; a-Murray; a-Papagiannis 1985; a-Drake 1976; classic paper a-Cocconi; classic general articles a-Morrison 1963, a-Purcell; b-Swift].

Presently only about $(10)-17$ of the multi-dimensional "Cosmic Haystack" has been searched [a-Tarter]. Especially well covered have been several hundred near stars at logically likely frequencies, those for hydrogen atoms (21 cm, 1420 MHz), for hydroxyl molecules (18 cm, 1666 MHz), and others for which a rationale has been found. This common atom and molecule happen to emit radiation at each side of the quietest band in the spectrum. Since they combine symbolically to form water, a common substance needed for life, this band has been called the "water-hole", a place where many beings might choose to congregate in the radio spectrum. Discriminating intelligent signals from natural molecular radiation very near the same frequency is not a major problem. The search at 21 cm would have found signals if they had used at least a megawatt spread no more in bandwidth than 7 kHz [a-Verschuur]. It is slightly discouraging not to find signals from the most likely near stars at the most likely frequencies, and this is some indication that the number of civilizations is less than 10,000 in our galaxy. If civilizations are about 1000 LY apart, then there are about a million bright (F, G, K) stars to check. We can see most of them optically, although many have not been cataloged yet.

A major proposed NASA SETI program slowly starting would cover only a tiny fraction more of the haystack using existing dishes [a-Papagiannis 1985]. Requiring 10 years, it would check 1000 stars closer than 80 LY, too close if civilizations are somewhat rare. The general survey would use only about 1 second to listen to each discrete direction in space at each narrow frequency band. The search could go on for centuries even if many new big dishes on Earth and in space are dedicated to it. If the search takes too long, some near civilizations may abandon their transmission programs or go extinct before their signals are found, or a new one may come on after the search looked in that direction. The lack of messages received thus far is more likely due to the size of the search than a lack of senders. Several intelligent signals might be reaching us now, within our ability to detect, if only someone checked those particular directions and frequencies. SETI has been thought of as a cheap alternative to interstellar travel (still requiring billions of dollars and much donated effort for a semi-thorough search) [a-Finney 1985]. But SETI would be useful in finding destinations for travel, and could be a program parallel to a starship program.

Occasional past SETI programs have achieved credibility for SETI on modest funding. Now a wider range of scientists recommend a search for what some might consider impossible or incredible. Each year US Congressional commitment to a modest program vacillates as public support varies from the derisive to the serious. As much thought, calculation, experiment, and scientific dialogue has occurred as could be done, given that no results have been obtained thus far. Those devoted to the search feel that the work is worth doing and worth long-term support, because of the immense possible payoffs. As indicated throughout this chapter, positive or negative results bear on the prospects for our achieving interstellar travel. SETI has been called a no-lose endeavor: if messages are detected, that's wonderful; if not after an arduous search, we would know that we are essentially alone and might try harder to have a better civilization. Negative results could never be final, especially since the size of a thorough search is very large.

Several kinds of signals can be counted as communication. As Earth has for some 60 years, other civilizations might use radio, radar, and other transmissions at even higher power for their own purposes and we could detect the leakage [a-Sullivan 1978, 1985; a-Woodruff]. Military radar effectively about 1 Gw should be detectable more than 10 LY away by a 100 m dish. A TV signal is transmitted at 10 Mw at best, but a civilization with larger dishes or a large array could detect TV carrier signals to 10 or 100 LY. Because of Earth's rotation and many transmitters, a large amount of nearby space is covered daily. Much less power is in the picture, and others watching Earth soap operas is (fortunately) unlikely. However, more general characteristics of Earth such as rotation, orbit, size, and details of our radio technology could be learned by analysis of our leakage. We can do the same. However, narrow-beam messages from one star system to another (not our own), or to a starship, occupy so little space and align with our Solar System so rarely that we cannot expect to find signals this way.

Other civilizations might transmit simple repetitive messages as beacons for their interstellar use or as announcements of their existence or desire to communicate. They might aim these at particular stars, surely selecting our Sun as one likely place. Or they might use very high power in very narrow bandwidth and cover much of space at once, or use high power in a scan so that each direction in space receives a short message repeated at regular intervals. It would not be wise of them to use a very narrow bandwidth to save power, or to use an improbable or naturally noisy frequency, or to use a complex modulation method, as they might prevent even sophisticated civilizations from detecting or recognizing the beacon. If they believe that civilizations are rare and widely separated, then they may include substantial messages, such as an encyclopedia mainly "for your information", in their longer and necessarily more powerful transmissions. They may send to selected stars or to a whole galaxy with one beam and not expect two-way dialogue. To save power they may transmit so very slowly in very narrow bands that present searches are not prepared to find such messages. They might send a beacon message which tells exactly how to find a simultaneous long message. A simultaneous "dictionary" message and other more complex protocols have been supposed [e.g. a-von Hoerner]. More far-fetched is the transmission of instructions for making copies of themselves along with encyclopedias for their education, a novel way to travel bodily among the stars.

Presumably, when we or other beings detect a beacon, we would know where to send a longer message as the first statement for an ongoing dialogue. To understand a message we expect that other beings have also discovered the landmarks of knowledge that we have: mechanics, relativity, spacetime, quantum physics, chromodynamics (quarks), many kinds of mathematics (including Godel's theorem), information theory, Turing "machines", robotics, engineering on a genetic code, multi-dimensional descriptions, and much more. Most of these are necessary on the path to building a large transmitter or a starship, and most represent a compression of a large set of ideas into core concepts (like deriving all laws of nature from U = 0). There seems to be only one way to do mathematics, and perhaps science and technology [a-Minsky]. A culture should also find the simplest ideas to be the most useful, assuming that simplicity is a universal idea itself. Maybe we have already found all the general ways to organize a society. However, a persuasive and opposite case has been made that others would develop different sciences (as well as incomprehensible cultures) despite the physical universe behaving the same everywhere [a-Rescher]. Their science may not be just a more advanced version of our science, nor would their science necessarily lead to technology.

Once an interstellar message is received, it will profoundly affect our world. Principles and protocol have already been prepared for initial action upon confirmed receipt [a-Michaud]. In brief, scientists have agreed to cooperate in confirmation, then announce the evidence to the public fully and openly. Presumably this includes all data on the content of the message even though deciphering it early is unlikely. Indeed, wide cooperation may be needed to understand the message. If a reply is at all feasible, international effort is planned on that too, with all local political manipulations discouraged. It is assumed that the message will begin with basic axioms of math and basic laws of physics which "everyone" out there could know. Quite complex text and pictures could be sent in a supposedly "universal" language [b-Freudenthal; a-Sagan 1975]. Search for more messages would increase while reply is made to the first. Some will forget that signals are difficult to find, and continue to wonder why there are so few (just one). Perhaps just one other civilization has survived in this neighborhood out of hundreds that began. It would be sad to receive a message from a distant civilization that is either so far, or so short-lived, that that civilization has deceased before our reply can reach them. We would never know why a further message did not come. If the message comes from a near star, serious study of traveling there would begin while many exchanges of messages are planned in preparation.

The effects of successful SETI could be paradoxical or oversold, and a foretaste of the effects of travel to meet other beings [a-Regis]. Would we try harder to exist longer and better, knowing that others are out there surviving and calling to us, or would we feel inferior or become more reckless, knowing that we are an insignificant species among many? If SETI is not successful after a long search and we deduce that we are alone in the galaxy, will we improve our ways here and become self-centered again? Will we survive against the odds, knowing theoretically that few or no other civilizations have? When humans thought they were the center of the Universe earlier, they still carried on mayhem and would have destroyed their world if they had had the technical ability. Yet it has been claimed that we would improve our ways if we discover other beings. Either way we are said to be going to improve, yet what are we waiting for? Maybe we have no intention of improving regardless and should not be surprised that the possible existence of others many lightyears away would have no effect on us. And if we receive an invitation to visit and help in developing starship technology, we should wonder why they are not traveling here instead. They might want a group of us to travel because they are afraid to take the risks.

We have rightly assumed that other richer more advanced beings would do most of the transmitting. The question of why anyone would want to send us messages and information, presumably for our benefit, is less asked. If civilizations want to communicate with other more advanced ones, to learn from them, then they might send very sophisticated messages (and we have failed the test for sophistication by not finding them). General galactic friendliness is an anthropocentric assumption. We are not all that friendly either. Will our society come to spend millions of dollars per year (billions in a serious program) to send out our best knowledge to unknown recipients? And if a dialog turns to threatening or embarrassing issues, humans are sure to become suspicious.

Experimental work in SETI has been primarily observational on our part, and only one message has been sent intentionally from Earth. That was in 1974 using Arecibo and beamed at a whole star cluster M13 in Hercules 25,000 LY away with effectively 3 Tw [a-Sagan 1975; a-staff]. This distance keeps our existence secret for a long time, unless radio astronomers at a nearby star or on a starship intercept the message. It consisted of 1679 bits that form a picture if arranged in a two-dimensional array 23 by 73. This method is a simple illustration of the many possibilities for sophisticated communication.

First attempts at sending information in physical form into interstellar space have been done with plaques on the Pioneers and with recordings of sounds and pictures representative of Earth and its cultures on the Voyagers [a-Sagan 1972; b-Sagan]. Although other beings are very unlikely to find these information packages, they were designed for easy reading or playback and interpretation. Further transmissions (or packages) from Earth are unlikely unless evidence is found that many stars near or far have civilizations to receive them, or if society deems such activity much cheaper or more practical than our traveling. If we set up a beacon, it would need to use 100 Mw to reach about 100 LY with a 100 m dish, covering only a millionth of possible nearby space [a-Oliver 1981]. In 100 years $(10)11$ kwh would be used, costing 10 G\$ for the energy and much more for the operation. Considering the time and cost of a full search in the radio spectrum, not to mention other possible channels and sending our own messages, interstellar travel is not necessarily slower or more costly in the long run. [Chapter 6 compared radio contact with contact by physical travel.]

Communication is not limited to radio waves but can be done with light (photons) over a wide range of wavelengths [a-Schwartz] and with neutrinos, gravitons, or other neutral stable particles that we have yet to discover. Laser signals from others seem quite feasible, requiring only 10 kJ pulses to outshine stars [recall Chapter 7; a-Ross]. Gamma or x-ray photons are difficult to generate in quantity but not easily absorbed in interstellar matter. Neutrinos are much more difficult to control or detect. Gravitons are the yet unproven quanta of gravitational radiation and therefore very difficult to control or detect. We can produce massive neutral particles such as hydrogen atoms in modulated beams, but a trillion times more energy is required to send the beam at speed near c across space than is needed to send radio waves carrying the same information. Even more might be needed to send biological spores in the "panspermia" idea of spreading life among the stars [covered in Chapter 6]. This is in the microgram mass range, and more massive forms of "communication" are properly called probes and discussed in other chapters. Watching for unexpected methods of communication would be important in SETI.

11.4 IF THEY TRAVEL AND VISIT

Human reasons for travel would be a small, possibly irrelevant, subset of the reasons other beings might attempt the long lonely travel [a-Wolfe]. Some would be better equipped biologically: more tolerant of radiation or zero gee, much longer-lived (and patient), much smaller-massed, less metabolism or inherently hibernating. They might travel for scientific reasons (organized curiosity), political plans (spread galactic government), economic trade (of very high-value low-mass items), colonization, daring and adventure, or wanting to meet other beings for dialogue, sharing arts, play, physical contact (new ways to love), teasing, fighting, cruelty, or many other purposes. If galactic-scale dominance or hostility to life were a common motive, life on Earth would already have been conquered or eliminated. While tools and rationality are needed to build a starship, other beings might then use starships for emotional, artistic, or other softer or non-rational "reasons".

To resolve the Fermi dilemma many reasons have been proposed for why they would not travel; for example, the technology is too difficult or requires too much wealth and energy, or they are content to stay "home", or just plain lazy [a-Drake]. They might limit themselves to sending out probes [classic paper a-Bracewell 1960]. Some have suggested that beings commonly evolve in oceans where electronics and rocket travel are much more difficult to conceive. But Earth has assorted electric sea creatures, and humans have underwater missiles. Much general speculation on the possible philosophies and sociologies of other civilizations has also been published [e.g. a-Beck; b-Christian].

Although it would not cause humans to migrate to other stars until some 5 billion years hence, the fact that nice well-behaved long-lived but bright stars like our Sun become red giants might be making other beings move their civilizations to save them from being incinerated. Civilizations that reach this predicament are likely to be very old and rare. They could use planets or colonies farther from their expanding sun, but eventually the star flares up more and then shrinks to a white dwarf. A more pressing reason why some civilizations must move is a supernova near them. As discussed in the next section, a supernova a few tens of lightyears away is likely every hundred million years, and more often and closer in a star-forming region. The radiation at this range could be fatal to life like ours.

If other beings have a biosphere similar to ours with oxygen atmosphere and plants that convert CO_2 to carbohydrate energy, then they may find, as we are about to, that a planet's worth of fossil fuel (liquid and solid), methane gas, and even uranium is not enough to supply a space colony empire or to start interstellar travel. They may have found that energy is the key to long-term civilization when used carefully from the beginning, and the end of civilization if misused. Energy is closely linked to population, with uncontrolled population likely to expand energy usage (but a small population could also waste much energy) [a-Soule]. Inability to control population can be the doom of a civilization for other reasons too. In evolving to intelligence we have become very good at keeping our offspring alive, unlike, for example, insects which reproduce copiously with virtually all perishing. Some of our other big problems could be listed as reasons why other civilizations may fail--ignorance, greed, technology for mass destruction, environmental damage--but other poorly succeeding beings might have other problems at the top of their lists. It seems likely from our experience that intelligence alone is not sufficient for long-term survival.

This book has stressed the difficulties humans face in attempting interstellar travel. Other beings should surely face many of the same problems, especially those with a common physical basis such as limitations in energy density and energy supply, using materials to their limits, laws of motion, and natural force laws. They would face the problem of where to go: that is, how to find a place like Earth in the billions of cubic lightyears around them. If we do not signal ourselves (beyond local radio leakage), they cannot find us. They might need to approach within 10 LY with very large telescopes just to detect that our Sun has an inhabitable planet and to check for unusual radio and other signals. They may have difficulty stopping here and pass by at high speed with a short time for observations and radio signals (which we may miss). If someone has spotted our civilization, they made need centuries or longer to accomplish a return visit and stop here. These are other possible reasons why we have not had visitors.

Thus far no observations have been made of the travels of others, in the form of lightbeams and their reflections, more energetic radiation and particles, and radiation from disturbances of interstellar media, as they use focused sunbeams, giant lasers, energetic particle drives, or braking methods [a-Viewing; a-Harris; also Chapter 7]. Electro-magnetic pulses from launchers and radar pulses from starships also have not been detected, but these would be too weak lightyears away. Most methods of travel involve large effects with large doppler shifts and sudden un-natural changes. But these may not be detectable from beyond our Solar System or as far as 1 LY. Neutrinos or gamma photons from a very large anti-matter drive should be detectable farther. Beamed emissions for any purpose, such as particle exhaust, laser, radar, and radio, should be detectable from much farther if pointed near us. There is one reason why a beam would be pointed precisely at Earth and Sun: a message or visit. The exhaust of a starship braking to visit us may travel at 0.2c or greater and be detected here years before the starship arrival (an effect apparently not investigated yet). Occultation of a star would be a very rare event. We should include in observation programs a systematic search for this evidence, which if found could prove two vital facts at once: other beings exist and interstellar travel is possible.

However they travel, it will not be by "flying saucer" [a-Sheaffer; a-Markowitz]. This unproven mode of transport seems technically unfeasible, whether an interstellar vehicle or just a landing car. It is not surprising that none of the many claimed sightings of and encounters with saucers have been verified. The magnitudes of magnetic field, currents, material strength, mass flow, and so on, needed in the various proposed magnetic, anti-gravity, and other vaguely described methods are much larger than those needed for the very difficult but more conventional travel methods discussed earlier. The typical alleged size of "observed" flying saucers was used in Chapter 4 to estimate the magnetic field needed and comes out closer to neutron star strength than to presently achieved strengths.

If visitors come here, humankind would go into a tizzy at the least [see also a-Michaud 1990, b-Ridpath]. The meeting between humans and other beings has often been called "contact", meant in a stronger sense than a dialogue by radio but with an unnecessary implication of physical touch. People would be thrilled and scared over a personal visit. Present levels of xenophobia indicate that a general mood of fear and panic may prevail. Many would hope that visitors would have good intentions. We would have much to learn from them (assuming effective translated dialogue occurs somehow) as we learn from ourselves that we need a new attitude toward our role in the Universe. We might learn to think bigger and longer-term, with many factions drawing closer together. We might come to see ourselves as we imagine them seeing us, perhaps as a struggling, fussy, overbusy clever young bunch of creatures crawling over the planet. How they would really view us could be quite different and multi-dimensional, ranging over thousands of descriptors from grand or mysterious or beautiful to petty or contemptible or ugly.

The suggestion has been made that many humans who live in guilt, justified or not, might look toward superior visitors as godlike, dispensing judgment. Many more might trust everything that "they say" without question instead of being cautious (not suspicious). And others would feel hopelessly inferior regardless of the actual facts provided by the visitors. Many might forget that other beings could have as many faults as we do (a possible universal law), perhaps just handle them better. Many sf and other writers, in showing how we are not as free or as successful as we could be, have implied that visitors might have the more fulfilling lives that we are afraid to have. Scientists (and artists) would bypass most hysteria as they realize the scope of new work they can engage in. Almost every field of study would be affected deeply. Visits with the visitors would need to be leisurely to treat them kindly and respectfully, and new complex programs and procedures would be needed to allow many people fair and full access to the visitors and their knowledge. On top of all this, they may bring information about still other civilizations.

Contact would start in space, with messages which we may not understand before they reach Earth orbit. A delegation of humans would probably meet them first in Earth orbit or a space colony. We may need to help them land on Earth, after much study, or perhaps they would not want to reach planetary surfaces (or be able to, biophysically). Difficulties of communication are likely to arise, and major problems of trust on either part could occur. We would hope they had superior but kindly tools for handling these situations. We should be prepared to provide them a safe place to live during and after getting acquainted, exchanging information, and providing many tours of Earth. Their material requirements may be rather different and it may cost much money to provide suitable atmosphere and nutrients. From our knowledge of the difficulties of travel, they may also become permanent guests (perhaps at our mercy) and not be able to travel on. Negotiations should begin to find them a place, perhaps in space, where they can expand to a colony and to a civilization without conflicts. If they want to resume traveling, they may need much technical help (fuel etc.) and perhaps "buy" it with extensive information. We may not be able to provide that help for some centuries.

Their arrival should have a lasting effect, unlike most major events on Earth these days (here today, forgotten tomorrow). The excitement of knowing beings able to make such an interstellar voyage would be rekindled often, even if the media did not continually stimulate it. The prospects are exciting and foreboding, knowing past human indulgence in territorial and other wars. On the positive side, if visitors arrived soon, they would be an inspiration and aid for humans to begin serious space development, as well as improving our social, environmental, and other philosophies. If they stay and develop their own civilization, ours will surely change drastically, possibly better in some ways and worse in others. Overall, we do not seem prepared for visitors, but it is unlikely we would know how to prepare, at least the first time. Their technology will interact with our economics, and every economic disruption will be blamed on "them". Maybe some of the problems we have will be "shared" by them on a larger scale--e.g. intelligence in the Universe is absolutely corrupting and unstable, especially when technically superior. Perhaps they will meddle in all our affairs and take advantage of our fears. What we are unlikely to need is a unification of the human species as a force against them. Perhaps they will be masters of the ability to relate well. And of course they would bring proven technology for interstellar travel. [more exuberant and scary possibilities in a-Michaud]

11.5 HABITABLE PLANETS

Habitable planets are defined (somewhat anthropocentrically) as planets in the pleasantly warm biozone around a star with adequate water, breathable (to us) air, stable orbit, low eccentricity, reasonable rotation and axis tilt, absence of very violent weather and high radiation (including UV), and other factors [a-Cohen; a-Archer]. The interstellar environment must be relatively stable, with minimal passage through dust clouds and infrequent supernovas. A supernova is expected within 30 LY of any spiral arm star every 100 million years or so [a-Clark 1977; a-Tucker]. Since it can increase radiation levels by 1000 to 10,000 at that distance, a supernova is very disruptive of the evolution of life. It is not known if evolution on

Earth as been affected by several less close ones, but apparently Earth has escaped exposure to a major one.

A long-term stable (billions of years) star is also required, and these types F, G, K, and M comprise over 90% of all stars. The rarer and short-lived ones (A, B, O types) are much more massive than our Sun, rotate fast, and can be fickle, changing luminosity rapidly and having other unstable habits. As an example of stability requirements, the few percent variation in Earth's orbital distance and orientation seem to have been sufficient to cause ice ages, and a star whose luminosity varies more than a few percent in a billion years can strongly affect the biosphere. Our typical Sun has increased brightness about 30% in about 4 billion years. This section focuses on habitable planets which could acquire biospheres and repeats as little of Chapter 6 as possible, where the general relevant characteristics of stars were presented. [a-Oliver 1980]

Present theories of stellar formation [e.g. a-Cameron 1975] indicate that planets are more likely to be formed when the condensing nebula does not form pairs of stars [b-Ridpath; a-Harrington 1982]. Thus searching binary stars (about half of all stars) for planets might be futile, but arguments to the contrary have also been presented [a-Oliver 1980; a-Harrington 1981; a-Huang]. A system with two or more stars would not permit stable orbits for planets unless the stars are far apart, at least four times the distance to the best place for habitable planets. Jupiter is as close to our Sun as another star could be and permit a planet like Earth. Possibly stars are formed in clusters from large nebula rather than in close pairs or triples, presumably with planets around most. Our Sun may since have carried its planets away from a group birth. Planets would take up the rotational (angular) momentum that the central star cannot without splitting apart [b-Sagan 1973]. The star's magnetic field provides one way to transfer angular momentum outward to the nebular gas disk. Angular momentum is a conserved quantity, and a set of planets can acquire about 99.9% of it because of their distance, leaving a slowly rotating star, another clue. Whatever small amount the original gas had appears to be concentrated as the cloud shrinks until gas, star, and planets are rotating and/or revolving at higher speeds, all in the same direction and in the same plane.

The kinds of planets that may form depend on the temperature and composition of the contracting, condensing interstellar material and on random events in the motion of that material. Beginning contraction may have been caused by arrival of shock waves from nearby supernovas. Since the distribution of elements seems broad and everywhere about the same in spiral arms of galaxies, away from the core region where the oldest stars are, the same sorts of planets with the same capacity for life might appear with almost any star [a-Trimble]. The detailed process of condensation to a disk of gas and dust and a proto-star is yet unknown [a-Stahler]. Some separation of elements occurs during nebular contraction because the gas and dust disk grows hotter near its center from gravitational PE converted to heat and possibly from a proto-star. The inner region is too hot for the lightest volatile elements (the bulk of the material) to condense or form compounds, and these accumulate farther out to form large gaseous planets. The vital inner planets form from the small residue of heavier nonvolatile elements, especially silicon, sulfur, phosphorus, iron, and other transition metals. A trace of volatiles (carbon, nitrogen, oxygen, combined hydrogen) stays with them or arrives by bombardment. Planet formation is thought to require less than a hundred million years of collisions of small objects to produce larger objects which melt, mix, and then congeal in layers during the process. When meteors and planetoids are nearly exhausted, a few large, widely-spaced planets remain which cannot be broken up and which continue to collect remaining debris until planetary space is almost as empty as interstellar space.

Other vital variables are planetary mass and distance from the star also forming (which forms first is uncertain). Low-mass planets do not have the gravity to hold the lightest gaseous elements hydrogen, helium, nitrogen, and oxygen, especially if they are warm. A small difference in gravity or temperature has a large effect on the time over which gas molecules can escape [recall molecular speed in Appendix B for 1.2]. While the first two are by far the most abundant elements, the star gets nearly all of them. We can observe in our Solar System that a planet as small as Mars at 1/9 of Earth mass is too small to hold nitrogen well and that Uranus, 14 times more massive can hold any gas very well. The least massive planets also cool most quickly from the hot accretion stage and they have less internal radioactive heat to dissipate. There is a distribution of planet types from rocky to gaseous. More specifically, planets might range from rapidly-cooled rocky and/or icy (Mercury, large moons) to warm rocky core with thin wet atmosphere (terrestrial) to warm rocky core with thick liquid and gaseous overlayers (jovian). Habitable planets might range from half to several times Earth mass.

Many models have been calculated to find what range of planetary masses and distances might result from random accretion of primordial material [b-Sagan 1973; a-Sagan 1975]. Approximately 10 planets are formed as gas and dust condense in the proto-stellar disk, with planetary sizes ranging from greater than Jupiter-size to small moon-size. Typically large gaseous ones are in the outer regions and smaller rockier ones in the interior, in accord with heat distribution, but with considerable variation. These model results (not unexpectedly) resemble our Solar System. A recent study even predicts an asteroid belt like ours where an intermediate planet could not form due to Jupiter's gravity [a-Wetherill]. But if Solar System planets and moons are any indication, variation in size, location, and composition can lead to an even wider variety of surfaces on planets and moons at other stars.

Distance from the star is crucial for having some warmth but not too much. Once a planet is formed it may have an orbital radius that is nearly constant (nearly circular) for billions of years in a stable system. Symmetry apparently dominates during planet formation so that orbits end up nearly circular, a requirement for habitability. An mildly elliptical orbit might be acceptable if its ends remain in the biozone. Its star's luminosity also should not vary much over time. Our Sun seems to have become about 30% brighter over billions of years, and CO_2 content of Earth's atmosphere seems to have diminished to hold a more constant temperature. Any such natural regulation cycles are rather uncertain and of great interest. That CO_2 content varies along with solar luminosity to hold planet temperature steady may be a rare coincidence. Taking Earth's position as a standard, the biozone for our Sun seems to be about 100 to 200 million km radius. For cooler stars (or hotter) the biozone would be closer and smaller (or farther and wider) in proportion to the square-root of the star's luminosity [see Figure 11.1 and Appendix D]. The biozone seems sufficiently wide that at least one planet is likely to form in it, except for M stars which may be unlikely to have a planet form in the close and narrow biozone. It has been argued that our biozone is much narrower, therefore making Earth an unusual case and eliminating biozones from many stars [a-Hart 1979].

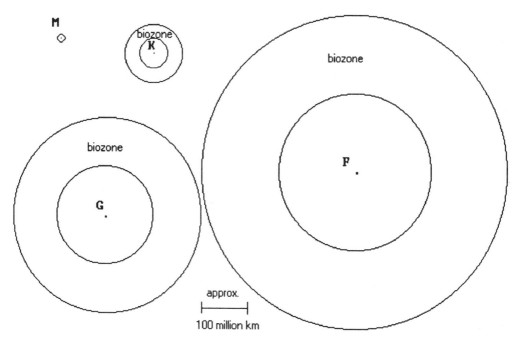

Figure 11.1 **Biozones for stars of various luminosities, shown in cross-section. Zone for Sun-like G star about 100 to 200 million km. M zone smaller than shown.**

At the bottom of the biozone scale an M star could have luminosity about 0.0001 of our Sun and would require a warm planet to be about 100 times closer or about 1.5 million km from the tiny star. Such a star would have mass as small as 1/10 of our Sun. The main concern with this extreme situation is that this example planet would orbit about 300 times faster, going around in little over an Earth day [see Kepler law in Appendix B for 1.4]. The small orbital radius has a large effect on orbital speed; the smaller stellar mass moderates this effect only a little. This hypothetical planet would be much deeper in the star's gravity well than Mercury is with respect to our Sun. If the planet is Earth-size, the gravity due to its star would be about 1/3 of the planet's own surface gravity, an unusual situation that would result in planetary rotation permanently locked to planetary orbit, with one side of the planet toward the star. This may not prohibit life if the planet is farther away so that its illuminated side does not become too hot.

The effect of seasons, if any, would be very different for a rapidly orbiting planet. Other fluctuations in the planetary system might bring about longer-term variations in sun intensity at the planet to stimulate the biosphere. But for these and other reasons, such as the predominant long spectral wavelengths of the star's light (much red, negligible blue), M stars, the most common type, should not be assumed to have habitable planets. Stars brighter than M type with mass from 0.4 to 1.4 solar mass and still long-lived have been suggested as having suitable luminosities [a-Cohen]. These types F, G, and K comprise about 25% of all stars. Planetary atmospheres can have a strong effect on the habitable zone since CO_2 and other "greenhouse" gases can raise the planet's surface temperature well above what it otherwise would have at any given distance from a star. A dense atmosphere or the presence of other gases such as ozone can protect the surface from the severe ultraviolet radiation that F and G stars emit.

The range of possible day lengths (rotation periods) does not seem to have received much study, and judging from our Solar System, almost any day length is possible. A day between 10 and 50 Earth hours seems offhand to be necessary for relatively stable temperatures and weather, unless the planet does not rotate at all (one side toward the star). Moons should not be overlooked. Since Jupiter has moons with much water (ice) and almost large enough to hold atmosphere, larger moons whose planet is closer to the star could be warm, habitable, and impressive places to live. Moons may also affect the evolution of life through tidal effects and illumination. When all the above is considered in reference to the Drake factors, the number of stars with planets could be as high as 1/2, but the number of habitable planets per such star (discounting M stars) could be less than 1/10 instead of 1 as some maintain. A factor in the range of 0.001 has also been estimated [a-Pollard].

Detection of planets at other stars is attempted by measuring the wobble in the star's transverse path due to a large orbiting planet and by doppler shifts in stellar spectra. Large orbiting planets wobble the star back and forth in the line of sight, and where jovian planets occur, smaller ones might follow. Large jovian planets may require a decade or more to orbit (although they could be closer and faster), so that the wobbles are very slow and a long time is needed to accumulate evidence. At present large planets are only suspected at several stars. Direct, interferometric, or spectral images might be obtained of smaller planets in visible or infrared light with sets of accurate space telescopes (less than 0.00001"arc). Detection of infrared radiation from a dust or asteroid disk around the star would indicate that planets are also present or will be formed [a-Black 1991, 1981; a-Worden; a-Aumann; a-Bracewell 1979]. As with SETI, a wide range of scientists recommend a long program of interstellar planet searching with present and new instruments, culminating in a 10 km array of telescopes on our Moon which can detect life-related ozone at terrestrial planets [a-Burns]. [Chapter 6 covered more of the science of detecting planets.]

11.6 ORIGIN AND EVOLUTION OF LIFE

How life begins on a planet is very important to whether intelligent beings arise and are plentiful. How life is to be defined (or fundamentally described) is also an open question. Some approaches have included life's anti-entropic behavior (converting free energy into complex structure and information), its open system behavior (with special semi-permeable barriers between organism and environment), its self-reproduction, its program or genetic code, its ability to change (at least at the species level) in response to environmental change, and other biophysical factors. Life might also be defined more broadly to include the whole planetary surface involved (biosphere) [a-Shapiro; b-Feinberg; b-Lovelock]. To work counter to entropy, life needs to take in high-grade energy and exhaust low-grade energy, but not as a heat engine. The principal energy input to life on Earth is sunlight, with energetic blue light playing an important role. But even on Earth life has evolved to use other energy sources than light, such as superheated sulfurous water deep in the ocean, a natural chemical source. The far-fetched idea of life spreading as spores through space, expelled somehow from terrestrial planets, is not sufficiently convincing to discuss here as an alternative to the spontaneous origin of life (and spores must arise first somewhere). But life, even human, could adapt to space, given time and technical help [b-Dyson].

Science has difficulty on the general problem of life because Earth provides just one proven example still poorly understood [a-Horgan; b-Dyson]. All life on Earth can be viewed as one fundamental kind and may be unified at high levels of interaction with our planet's surface and atmosphere (the Gaia hypothesis) [a-Margulis 1981, 1974; b-Lovelock]. The self-regulation of the biosphere would be an important explanation of how a planet can remain habitable over long times and extend the chances for life and intelligence. Indeed, the dominant form of life on Earth, by mass and by global effect, is micro-organisms, mainly bacteria. Bacteria are a large source of oxygen. Plankton in the ocean surface remove much CO_2 and release dimethyl sulfide that helps nucleate clouds (which help reflect excess solar input).

The most elementary forms of life may remain dominant on all planets where life begins. Division of life into micro-organisms, plants, and animals is also too readily assumed, as is partitioning into cells. Although plants can be defined as that which converts oxidized carbon to stored hydrocarbon energy, and animals that which harvests this energy and oxidizes it back to CO_2, there may be other ways of arranging things (e.g. the plants may be mobile and bring the food to the animals). How life began here is still uncertain in detail although much work has been done, beginning with an experiment that produced amino acids and other biochemicals with UV radiation in a simulated primordial Earth ocean and atmosphere [classic a-Miller 1955; a-Miller 1959; a-Ponnamperuma; a-Dickerson]. Organic molecules may also arrive via comet collisions.

While there is only the one case of Earth to study now, hope exists for breakthroughs in several directions in the next few decades. One is high-precision optical observation of not only the existence of other planets but the spectra of life-produced gases in their atmospheres. An atmosphere with oxygen, methane, ammonia, nitrous oxide, dimethyl sulfide, terpene vapors, chloroflurocarbons, and other life-produced gases is an atmosphere far from equilibrium [a-Margulis]. Another direction is to use many modern fields to study more of the origin and evolution of life here with the goal of finding general principles applicable anywhere [list in a-Ball 1985]. NASA's exobiology program is addressing these questions as well as study of possible life on Mars, Titan, and perhaps other local places.

Besides the still-disputed Gaia hypothesis, other interconnections of life or biosphere and general planetary resources have been suggested. Oceans and/or life may be required for concentrating some mineral deposits, affecting the potential to support technical civilizations [a-Smith]. In turn, the large water supply must remain stable for billions of years, requiring a stable climate and negligible breakdown of water to release hydrogen that is lost [a-Papagiannis 1989]. Estimates of the Drake factor for the appearance of life on an already habitable planet in the biozone have been as high as 1/2 to 1, but even 0.1 would not limit the number of potential civilizations very much. One study estimates the probability of life to be extremely low, making us nearly unique [a-Hart 1982]

The potential number of inhabited worlds is increased if life can develop in other environments than lukewarm water. A few other liquids besides water (e.g. ammonia, oils, silicates) might be good solvents for the complex chemistry needed for life at other temperatures from cold (-50°C) to very hot (1000°C) [a-Clement; a-Shapiro]. For several reasons such as slowed chemical reactions, life might work (therefore evolve) very slowly at much lower temperatures. Conversely, life might evolve quickly at high temperatures. Over a wide range of temperatures, the common elements with small atoms are among the few which can form hydrogen bonds that are more facile than the bonds that hold most inorganic solids together. Silicon, sulfur [a-Clark 1981], germanium (rare), and possibly other multi-bonding elements besides carbon might provide the basis for complex "organic" structures at higher or lower temperatures. Laboratory studies with other chemical bases for life have not produced encouraging results; that is, not only has no life been created, but also no working chemistry has been established, certainly nothing like silicon DNA in ammonia! Some planets and moons in our Solar System have liquid methane, ammonia, nitrogen, or sulfur environments, but communicating civilizations have not readily developed there or we would know it by now. Carbon-based chemistry still appears the most versatile, and Earth life has utilized only a small fraction of the possible building blocks and structures possible with it.

Beyond the spontaneous origin of life on a planet comes a Drake factor of great interest, the evolution of intelligent life. Evolutionary biologists tend to see intelligence, particularly in humanoid bodies, as very improbable [a-Simpson], while physical scientists tend to believe that once the conditions for life are available, life and intelligence will appear about as fast as possible, in a few billion years at most. An information theoretic approach compares the 200 bits of organization that can arise randomly in a fertile chemistry with the 100 million or more bits that represent an intelligent being [a-Argyle]. Explaining the enormous increase in organization is a biophysical problem. Progress has been slow, as evidence undermining a definitive 1964 argument against an evolution of humanoids (really, any intelligence) elsewhere under the best conditions has yet to be presented [a-Simpson]. More generously, at present no one can say how likely a repeat evolution of humanoid form and/or abilities is under similar conditions. Biologists are familiar with the nearly endless number of variations and branchings in evolution and the lack of ultimate purpose or design (teleology) in biological nature. Thus they are inclined to a view of uniqueness for any life form. And almost any form may appear as time goes on and the environment changes.

Despite its obvious (perhaps temporary) high survival value, intelligence must first arise by chance (and in degrees), then have its value tested in the environment. High intelligence is not automatic, else the cat family (and some other animal lines) with excellent senses and brains highly developed to process sense data would have evolved to still more sophisticated levels with language, tool use, control of fire and water, etc. Some biologists and others point to the concept of "convergence", the repeated independent appearance of the same traits and physical forms in evolution on Earth, and propose that intelligence would follow this pattern despite it not having appeared twice on Earth. There seem to be biophysical limits on the size of intelligent beings, and a sampling of factors is mentioned here. Much smaller beings might have too few molecules per neuron (or other information processing unit), might live too fast (high metabolism), might cool too fast, and might be unable to supply a brain with sufficient energy. Much larger beings cannot move their bodies quickly, need too much food (for energy input), and cannot cool down easily. All this assumes a gravity near one gee, and the mass of the planet would affect the abilities of beings there.

Given the presence of well-evolved life, the appearance of intelligence which achieves communication technology has been estimated at a chance of 1/100 to 1/10 for the Drake equation, a high probability in the view of some and low in the view of others. The final factor is the expected lifetime of the civilization that arises. Again, perhaps hastily taking from Earth experience, a technological civilization would seem to either solve its problems in 100 years or so and go on to a long life, or reach a catastrophic or dismal end within 100 to 1000 years. The result is a fork in the Drake result, predicting either a low number of successful civilizations (10 per galaxy?) or a high number (millions) of temporary ones. A mixture seems likely, with the mix depending on how pessimistic one feels about the ability of ambitious intelligence to cope with energy, nuclear weapons, pollution, overpopulation, greed, and the like. One philosopher (who is not alone) feels that intelligence carries the seeds of its own destruction [a-Beck, but with many other more positive observations on the nature of other possible beings]. But surely some, perhaps many, other forms of beings have different kinds of intelligence and different sets of instincts and drives and might avoid many of these traps, or fall into others we are spared.

11.7 GALACTIC CIVILIZATION (OURS AND THEIRS)

Some studies of the possible spread of civilization throughout the galaxy have found that one population of beings can, by a spreading wave of colonization, occupy the many habitable planets in our galaxy in a few million years [a-Hart 1985]. As before, the meaning of "colonization" is limited to a process where the home planet sends out a few colony ships with small populations, they establish colonies that grow into civilizations at several nearby stars, the new population centers send out small colony ships, and so on. Each new colony ship travels outward, or at least sideways, to stars not previously visited. It is assumed (hastily) that expeditions would not want to compete with each other on the same habitable planets. Starting from a single source, beings, but not likely large populations, move in a slow wave spherically outward in a diffusive process much like heat at the center of a planet moves outward. The time to spread involves the speed of travel and the civilization recovery or rebuilding time. At speeds like 0.1c rebuilding time would dominate, at low speeds travel time might dominate. The probability of encountering a civilization can be estimated by comparing the volume that a spreading civilization can fill with the volume of the galaxy [see Figure 11.2 and Appendix D]. If a high estimate of 10,000 civilizations are actively spreading in our galaxy in an epoch of 10,000 years and typically advance 100 LY (or 0.01 LY/yr), then the chance of encountering one in a given region is only about 0.001, rather low.

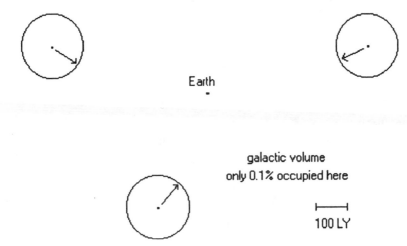

Earth

galactic volume
only 0.1% occupied here

⊢————⊣
100 LY

Figure 11.2   Expansion of several civilizations into galactic space over 10,000 years at rapid rate of 0.01 LY per year (0.01c), shown in cross-section. Volume of galaxy may be less than 1% filled, and Earth may be in an unexplored interstice.

A crucial part of the process is the control of population growth, possibly reaching a high rate after colonization to establish a new civilization, then reducing to a low rate for a stable civilization there that can start space colonies and starship programs. Beings must have very good control of themselves to thwart original nature and control reproduction this well. Those who have rebuilt themselves for near-immortality (or been so gifted) must especially avoid much reproduction. All this assumes that advanced civilizations find reasons to send out colony ships. Comets have also been suggested as a means of very slow diffuse spreading [a-Jones 1985; a-Stephenson]. Since spreading by comets can be continuous and more pervasive, the effect may be large despite the low speeds. The reason here for considering the speed of colonizing the galaxy is to estimate whether other beings could have spread so well by now as to be visiting us. Since they do not seem to have, one premise or another of the Fermi dilemma is at stake.

Other studies of population diffusion (by us; can "they" do better?) find different slower results and longer times of 10 to 100 million years [a-Jones 1978, 1982]. At the extreme, a study showed that if the nearest civilization arises only 200 LY away, it may need to last 20 million years to visit more than 100,000 stars before reaching us [a-Newman 1981, 1985]. At this speed (literally some 3 km/s, chemical rocket speed) most of the age of the Universe would be needed to fill a galaxy. Thus there may not have been enough time for other beings to arrive here, an explanation of why no one has visited us. Indeed, this spread is so slow that stellar motion in the galaxy disrupts any empires that form. Perhaps we should wait for a star system with beings to pass near us, which could happen in a million years or so. Reasons offered for the slowness include near zero population growth habits and concern about the dangers of travel, but not any difficulty in traveling. If we are the first (or only) civilization contemplating the galaxy, then it is widely believed that we should colonize it as quickly as feasible, perhaps pre-emptively.

An analysis in Chapter 6 showed that after the wave has advanced a few steps, most new centers do not find more than one new place at best to send a colony ship (or robotic probe). Also, the emigration of large numbers of beings does not seem feasible (to us), so that visions of migrations through the galaxy, with populations of millions or billions fleeing catastrophe or oppression are incredible. And habitable but unoccupied planets could be rather rare. Another theory is that there are many long-lived civilizations interacting (competing) in many kinds of ways during colonization [a-Turner]. A few or many centers of civilization may each start a spherical wave of population. These waves may collide and interfere, with two different civilizations encountering each other in cautious, peaceful, warlike, or other ways as they resolve who gets what territory. When two or more civilizations interact, their colonization is slowed, perhaps left incomplete, a phenomenon observed in Earth ecology. Our Solar System could be in one of the interstices [recall Figure 11.2]. But when expansion is nearly complete, the total overlooked volume may be only about 10% of the galaxy. This would put Earth in a somewhat unlikely situation and never an impressive argument. A related reason given for why no one has visited Earth is that we are in a "backwater" of uninteresting stars, at the fringes of several nearby expansions.

As indicated before, population pressure is one reason that does not apply to a program of galactic colonization, regardless of who does it. Mass migration seems very unlikely unless the beings are very small and their technology much better than we can envision. Those who cannot control reproduction at home may not be able to in their starships. Because of the time and effort, they are probably also not expanding in space to gather up more resources to send home, or to trade, or to conquer and pillage. Every star system already has approximately the same share of chemical elements. Those that do not will not have life in the first place. Moreover, the energy cost of shipping is far greater than costs of local mining and manufacturing. Interstellar trade is not a likely reason for travel or colonization [a-Tang].

Contrary to science fiction stories, home planets may not become the capital worlds of large interstellar empires. Reasons include long time delays for radio signals and even longer times to travel, the feasibility of travel, and the need to use the home planet carefully just to begin space civilization. The development of galactic "empires" would be much hindered by enormous distances, but if they exist, we should observe much radio signaling. Control of events at a star 10 or 100 LY away would be very difficult for the government of a local empire if administrators or police forces must travel 1000 years to reach the site of a problem. Only very long-lived beings might bother to try such far-flung control. We may already be in a galactic jurisdiction, with a starship of bureaucrats and tax collectors due to arrive anytime. They may not expect any resistance if they assume that all beings would feel honored to pay homage to galactic royalty. Applying uniform political rules to many star systems seems infeasible. Local customs and rules are bound to evolve faster than interstellar communication can influence them. Languages will change. The spread of civilization to nearby stars may take longer than the time in which beings can learn to alter themselves with genetic engineering (or equivalent) but less time than natural evolution requires for changing species (a million years or so) [a-Finney]. Thus the beings will change themselves as they spread, perhaps making themselves live longer and be more compatible with space. Although an empire can have different beings, the purpose of empire itself is in doubt. [a-Hart 1985]

If civilization does not or cannot spread to other stars, much can be done near the home star with O'Neill colonies, hollow asteroids, Alderson disks, Ringworlds, Dyson spheres, and more [a-Niven]. A "Dyson sphere" has been proposed as an ultimate construction project, a spherical assemblage all around the star to capture all its energy and provide enormous living surface [a-Dyson 1960]. There would be no gee on the inner surface, but beings can self-evolve to such an environment. The difficulties of constructing a Dyson sphere might be greater than appreciated. A solid one is unstable, hence Dyson's clarification of it as a "swarm" of orbiting objects. The concept is mentioned here because the effect is observable. If the "sphere" is partially constructed, or less ambitious space structures occlude part of a star, astronomers can detect that the stellar luminosity does not match its type and has excess infra-red. If the star is completely covered, astronomers would observe lukewarm infra-red radiation from the waste heat rejected from the back, detectable out to about 3000 LY [a-Sagan 1966; a-Slysh]. All the star's power cannot be bottled up and must flow somewhere. Such observations would be strong clues that other civilizations exist. Over time a gradual change or cutoff of a star's light might be observed as space engineers expand their work.

Much larger-scale engineering has been proposed for super-civilizations, such as moving stars around, dissembling them, rearranging galaxies, and the like [a-Kardashev 1985; a-Criswell]. Our not-too-humble scientists do not hesitate to predict the types, evolution, and astro-engineering of advanced civilizations, based on our knowledge of energy, information theory, and mathematics (e.g. Godel's theorem) [a-Kardashev 1964; a-Sagan 1973; b-Sagan 1973, 1973]. Three types (levels) of civilization are based on energy usage: I Earth scale, II stellar scale, and III galactic scale. Other classifications of levels of civilization have been described [a-Stephenson]. Technical growth of civilizations may require the solution of ever increasing problems, not a tapering off to an optimum existence, and Godel's theorem seems to imply an increasing number of unsolved problems. If civilizations spread and grow, there may be few limits to their growth. In our galaxy, or at least in some nearby galaxy, a civilization may be carrying on activity so large as to produce other observable and unmistakable evidence besides waste heat. We might find it with optical spectra of unusual elements, or unusual motion, radio emissions, gamma photons, neutrinos, gravitons, or exotic stable particles. If they are dumping matter into black holes, it may be difficult to detect from the radiation that the process was artificial.

Going from the super-technological to the cosmological and philosophical aspects of civilizations, one postulate about all civilizations is that if just a few more are found besides us in this galaxy, there are likely to be many more. Advanced life would have been proven to follow from suitable conditions. If not many in this galaxy, then many in others or very many among all galaxies. This is a possible consequence of what is called the "cosmological principle", that the Universe must be physically (therefore biologically) about the same everywhere, that nothing special occurs in one place that cannot occur elsewhere. This principle is a modern version of the Copernican view in astronomy, that Earth is not the center of everything.

A second related postulate is that the variety of intelligent beings and civilizations would be as wide as nature permits. Even if many others are like us in many ways because of some underlying regularity we do not yet understand, they will differ in some ways, and some will differ drastically. This postulate bears on explanations of the Fermi dilemma in that the wide range of explanations cannot be as wide as the kinds of civilizations. Thus if there are enough civilizations, some should find it feasible to visit us while most find it infeasible. If intelligent beings generally sabotage themselves and have short histories, a few somewhere are certain to be exceptions and find by intention or by accident ways to survive nearly forever. If some or many do not bother communicating or traveling, a few certainly will. If humans have difficulty controlling "free will" with ethics and laws, some beings might have rules for social behavior internalized while retaining enough initiative and creativity to travel to the stars [a-Ruse]. Our Universe is not known to be infinite now, so that the idea that we are exactly repeated elsewhere is neither likely nor relevant. Even if it becomes infinite, it will be running down entropically, and the same holds.

A different sort of principle for the Universe is the anthropic principle, named by Brandon Carter in 1974 [a-Rothman; a-Gale; b-Barrow; a-Carr]. In one form called "weak" it states that the Universe appears the way it does because we are here to observe it. This sounds absurd, like anthropocentrism again, but our presence is closely linked to some coincidences and precise balances in physical laws. Some cosmologists are interested in a possible intimate connection between life and the physical evolution of the Universe, and some disdain this view as nonscientific. Often overlooked is that such a principle is surely not limited to humans but should apply to all intelligent life (or to all forms of life). Some proponents [b-Barrow] maintain that we humans are the central cause of the Universe and that there can be no other life or beings. But if the misnamed "anthropic" principle were firmly established, it would constitute a general proof that life must arise nearly everywhere to justify the Universe existing.

As if the anthropic principle were not sufficiently grandiose on a self-serving level, some have claimed that the Universe is certain to be dominated by life, not just us but many beings, who are expected to continually remake it on ever larger scales until they insure their perpetual survival [a-Michaud 1982]. Talk about fear of death! A slightly different view is that, if the Universe is open, never to contract, then the available energy can supply a very large set of civilizations indefinitely if they gradually reduce their needs while expanding their domains [a-Dyson]. This includes reversing entropy as needed for conscious information processing, the most lasting feature of life. If somehow it is proven that life is an important part of the Universe's evolution, then the domination of the Universe by life would require mastery of intergalactic travel and handling huge amounts of energy, perhaps ever more slowly as the scale increases. This all would occur in the far, far distant future and does not bear on whether beings are traveling about now.

11.8 VISITING OTHER WORLDS

Reasons for developing and using interstellar travel are to visit or colonize a found habitable planet or to visit a found civilization, some of the possible missions of Chapter 6. If the habitable planet has life of supposedly lower types, a human mission would start a colony there for studying the new biology found and possibly for establishing a new human civilization. If there is no life, the planet would still be of much scientific interest, and an Earth-like ecology would be started if at all possible. If a civilization of any sort is already present, the human mission would try communication. Agreement would be attempted on the timing, method, and purpose of a personal visit rather than just "dropping by". If they cannot communicate into space, then only a cautious landing would allow further contact. More detailed reasons for visiting include exchange and learning about science, arts, cultures, philosophies, and perhaps other fields yet undefined. All information would be transmitted back to Earth. Maybe Earth/space civilization will reach a middle age, burned out on just human imagination and activities, and yearn for the exotic freshness of a new biology, new arts, and the new culture of another civilization.

A vision about interstellar travel must be questioned. It is surely mythical that voyagers will find a pristine planet ready and unoccupied. If the planet is at all habitable, it may be teeming with strange if unintelligent lifeforms. At best humans might arrive at a planet in an early stage after acquiring sufficient oxygen atmosphere. Most likely they must compete with and adapt to local life. To subdue or eliminate local life seems very unethical, but some groups might do this, or be forced to. Should any sort of life found elsewhere be respected or should people begin exploiting it? Certainly voyagers would have ample time to study the historical record on the human attitude toward Earth. Just by their presence humans would begin altering local life. After locating breathable air and adequate water (only water is easily purified), a reliable food supply becomes crucial. If local life is inedible, humans must find places to produce their own food without conflicts between the micro-organisms and plants from Earth and the local life of all sizes and types. Humans would surely attempt to start and raise Earth animals unless local ones are suitable for food.

Another myth is that landing is easy. The uncertain technology has already been discussed. There will be no shuttle or spaceplane landing and launch facility, maybe not even a convenient level solid field. Building one would be a separate massive project if the voyagers manage to land. If the planet has an early civilization, help may be available. If they are a technical space-faring civilization, reception might be in space, with planetary landings deferred. Perhaps our voyagers can no longer live in gravity. Our hosts might be able to refuel our spaceplanes and landers, or they may have a cable elevator system.

If the planet is known or found to be deficient in habitability, a common idea is that humans will transform it. To the extent that our science understands ecosystems and that planets follow a Gaia principle, there are some clever ways to use micro-organisms and trace gases to prepare a planet--making oxygen, warming or cooling it, removing methane and $CO2$, and so on. The process has been called "ecopoiesis" and raises exo-ethical problems [a-McKay; b-Lovelock]. Humans desperate for a new home will not be stopped by any "right" that planets might have to be left alone. Nor will they refrain from guiding a new biosphere at every step to insure that it evolves in ways most suitable for humans. If water is very scarce, it cannot be added without dropping in thousands of comets. Changing a planet might require a million years rather than a thousand with the best micro-organisms and other tools. Humans might live thousands of years in their orbiting starship or worldlet, building more if material is obtainable, while the planet improves. If transportation is good, and it must be, they can replenish many supplies from the planet, but more cheaply from comets and moons.

If intelligent beings are present, higher ethical questions arise. The mission may not know that they are there beforehand. As the starship approaches and orbits the planet, any technical civilization would make itself known by radio, radar pulses, lighting, unusual chemicals in the air, spacecraft, or visible patterns. At this point an intensive radio effort would attempt contact. Assuming that landing equipment is available, a landing among a pre-radio civilization could be like other beings having landed here near an 11th Century French castle and trying to make useful contact.

Prior arrangements for visiting are much less troublesome. If messages from elsewhere are received here and deciphered to be an invitation to dialogue, or to visit, a human mission is likely to be sent if at all possible. Many more messages can be exchanged during development and travel of a starship. An interstellar mission can be much easier if friendly hosts live at the other end. If lightbeam propulsion is chosen, perhaps upon their advice, then they will surely provide braking at the end. If a fusion drive is used and a sail carried, they could again provide the braking. Any such cooperation greatly reduces the total cost of the mission. An arrangement where both ends provide the propulsion has a total cost to both that is only double a one-leg mission, not greater by another mass ratio factor. Other beings should be glad to pay the cost of stopping visitors. What will we do if we receive a message asking us to put together specified technology in a certain number of years to halt an approaching starship for a visit?

Unlike visitors coming to Earth, a human mission to another civilization could not be widely shared with humanity. This event might be decoupled from Earthly affairs by a thousand years, although a video recording of the meeting would be sent to Earth. The other beings would learn of Earth only through recordings and messages from its distant past. But some of the joys, cautions, principles, and protocols of receiving visitors here would apply to a first meeting by a human mission with another civilization [recall a-Michaud]. We tend to assume that we can develop or participate in a universal etiquette, but maybe "contact" will not be so simple.

The hosts must provide many other services such as local transportation, translation, tours, cultural education, and permanent guest quarters. If the human expedition has endured many generations of travel, the arriving generation must be well-schooled in the history of the mission, the planet Earth they represent, and the general problems of search and contact with other civilizations. Perhaps along with a profound anxiety they will acquire the exuberance for the historic meeting that the mission founders had. If congeniality can be a universal quality, then the host civilization must do much to make the humans permanently comfortable. This includes providing a place where humans can begin a large colony which will not conflict with the hosts' own lifestyles. Anything less would be dismal for the humans. An indefinitely long collaboration would be possible with much mutual benefit, but one can see also here the setting for local space war if the two quite different civilizations begin to conflict. If peace reigns, then further interstellar missions, perhaps joint, might occur. SETI would continue vigorously if this contact was fruitful.

Chapter 12

Long-Term Prospects

Many requirements for interstellar travel have been reviewed in broad outline and in crucial details. Needed advances in technology for propulsion (Chapters 2 through 4), in design and operation (Chapters 5 through 8), and in biology for survival (Chapter 9) have been discussed. Chapters 10 has shown the need for Earth to develop local space civilization while it can. Chapter 11 has covered the question, why, if travel is feasible, other beings have not been here. If prospects for mastering the ultra-technical problem of interstellar travel in the near future seem somewhat discouraging, this chapter tries to show where we must try harder.

The possibilities for pushing technology well beyond current limits are examined here, along with basic scientific break-throughs needed. A model human mission has evolved throughout the book and should be kept in mind for this concluding discussion on the general problem of feasibility of human travel to a near star. Other probe and human mission options should not be excluded. This is a more speculative chapter, but also an integrative one, with a summary of what has been learned and recommended. Again, when "we" is used, it refers to our human civilization, in comparison to some other. Few references are repeated here as results of preceding chapters are summarized.

12.1 ADVANCES IN SCIENCE AND TECHNOLOGY

One more device in the field of interstellar and other futuristic studies deserves mention: Arthur C. Clarke's Third Law [b-Clarke]. This states that the ordinary science and technology of an advanced civilization would appear as magic to us. It must be understood that magic is not mystical or a-natural (also inaccurately called "super-natural"). Instead of being outside of nature, magic consists of tricks done within strict and well-understood laws of nature that confound an audience less acquainted with those laws.

With this said, nature and scientific progress remain difficult to outguess, especially in physics. Scientists are often trying to anticipate nature so as to find new laws and theories. Very few stones have been left unturned in the general search for energy sources. Science consists partly of revolutions in understanding (e.g. relativity). More revolutions are needed to make interstellar travel easy, although laws of nature are not so much overthrown as refined. Many of the findings in this book are based on 19th Century classical physics, with the refinements of relativity appearing mainly as limits on travel, the limits of lightspeed and of mass-energy for propulsion. Nuclear physics, which provides the most likely energy sources, will be refined by the quark-based standard model of particles (discussed later). However, no startling new energy methods are to be expected at the nuclear level.

Other kinds of refinement are prediction of possible new materials and new uses of electro-magnetic fields. At present technology gropes to find better materials (e.g. warm superconductors) and tinkers to assemble reliable, stable, high-power systems (e.g. fusion reactors). The laws of nature are already at hand for precise calculation of what given atoms and fields can do, but sufficiently powerful computation continues to be sought. Only the real devices themselves, found perhaps by intuition or luck, serve as the most advanced "computer" models and tell us what nature will do in given complex situations. Very sophisticated computer-calculated models seem feasible for doing almost the same, provided that a person or a model creates the new idea or system. The computer is one of the great tools of the 20th Century for solving complex problems. Two others are "seeing" and identifying the atomic structure of matter atom by atom, and measuring and manipulating those atoms one by one.

Whether our science is slowing for lack of startling new evidence, or for lack of deeper, more clever human thinking, or from reaching final barriers inherent in nature cannot be determined. The situation is perplexing. Some would say that scientific progress continues to accelerate, not just applied science but basic science. Our science is at the beginning of understanding nature, and advanced beings, if any, must have advanced much farther. Yet "they" do not seem to have made breakthroughs that enable them to spread everywhere including visiting or contacting us. If other beings do not exist, at least in our galaxy, one of the many explanations or consequences is that humans have a special role in the Universe. Humans may be the first to find some remarkable way to travel, and may need to if we are to be important in the limited yet vast realm of this "our" galaxy. For a balancing view, consider also another explanation, that life such as ours is both rare and insignificant.

Interstellar travel is partly an engineering problem, and building a starship is assembling materials of the right size, reliability, performance, and cost to produce a material system. Often, designing starship sub-systems has called for pushing known materials to their limits. Summaries of the state of knowledge are given here, along with recommendations, for scientific and technical work on assorted starship and travel sub-problems. Not to be overlooked are large and vital support problems such as programming, library preparation, materials storage (including biological), and people preparation. Such work should have many important spinoffs for more mundane science and technology. Thus far military interests have driven work in science and technology relevant to space travel more than under-funded curiosity has.

Some of the technical sub-problems of interstellar travel (e.g. stronger thinner materials, warm high-current superconductors, denser and longer-lasting computers and information storage, denser energy storage, efficient conversion of light to electricity) may be solved with the improving ability to manipulate atoms one by one and build materials atom by atom [a-Whitman]. [See also a-ScAm on ways to find new materials.] This is expensive on the laboratory scale, but robotics and other technology might make automatic assembly of very complex materials routine. The entropy-energy costs of such high-information materials would be prohibitive in a closed starship, and the work must be done mostly in an open, less energy-limited construction facility. Atom by atom sorting might be necessary for ultra-efficient recycling of materials in a starship, although the entropy costs may get out of hand. A related materials problem is stability over time, especially with radiation present. Reliable computers, information storage, and strong materials require elimination of most impurities and that component atoms not change states nor move or diffuse in very long time periods.

Superconductors are often thought of as solutions to nearly all problems. The discovery of superconduction at "cool" temperatures (about 100°K) has raised hope of "warm" superconductors (about 300°K), a very large jump in nature. These could solve the important problem of making equipment work here on Earth and in warm space-labs, then having it work later in the deep cold of space at very low power. Probes that can depart with functioning computers at 300°K should work later at 3°K with no significant changes. Having two sets of computers and other equipment, one for each temperature range, is undesirable and less reliable. While small, fine equipment using warm superconductors seems remotely possible, obtaining also high current density, high fields, and high strength from warm superconductors for use in propulsion is unknown territory. While present cold superconductors have reached only about 35 tesla steady-state, a few starship ideas such as the ramjet and shielding can be more effective at much higher fields. How far can physics go in this regard?

The limits on magnetic field are strength and exposure. Magnetic field might be limited to a strength which would not rip atomic bonds apart with the Lorentz force. On this basis, field can be 1000 to 10,000 tesla, enough for the drives seriously discussed but not enough for the most fantastic applications. Neutron stars produce much higher fields, apparently without damage, but that is irrelevant unless our technology of neutrons advances to material applications (see later section). Biological exposure to high fields must be avoided, and high-field devices can only be ones located far from starship habitats. Ironically, organic superconductors are another promising avenue, although they can handle only modest temperatures (less than 30°K), fields, and currents at present.

Very thin "sails" and very strong cables seem almost within reach now, more a problem of large-scale, precise space-manufacturing than of need for new materials or physical theories. Structures only 100 or 1000 atoms thick, of many layers or just metal, likely can be made sufficiently strong. Composite materials with pure crystals in a durable matrix for high strength are a maturing technology. Very thin or low-mass methods of converting light energy to electricity seem more elusive. But given the relative simplicity of the problem and the great complexity that can be built into materials at the atomic level, new solutions seem possible. Efficient electric energy storage for short-term operating power is another case where precise and complex atomic structure could be a solution (and very important for problems here on Earth). If tanks of hydrogen and oxygen can store energy to be released by combustion or in massive fuel cells, surely similar mass and volume could store electrical energy more directly. (For long-term power much denser energy storage is needed, such as in the form of fusion fuel.)

Technical sub-system problems that seem difficult are, in approximately decreasing order of importance: shielding and prerequisite ionization of interstellar gas, artificial gee, landing (see later section), power supplies for operation, 100 to 1000 year reliability of everything, conservation and recycling of all materials, total and compact medical care, and reliable growing, harvesting, and preparing of a tonne of vegetable food per day (and liking it). Many shielding ideas have been presented, primarily against penetrating cosmic particles. The gas the starship runs into (at speeds less than 0.1c) comprises a much larger but weakly penetrating dose of radiation. A mere millimeter of metal would shield from forward radiation, except for the large but uncertain amount of erosion from hitting dust. The choice seems to be between carrying the extra mass for thick (1 to 2 m) shielding of the habitat volume or developing a large electric and/or magnetic ion repeller. Preionizing the gas may be the major hurdle despite hasty claims that lasers can solve such problems.

Artificial gee can only be obtained by a 1 km rotating habitat with large inherent mass, strong structure, large area facing forward, and very large non-rotating shield mass. If such a wheel is made, first for local use, many other problems must be solved, such as low-friction very smooth (magnetic?) bearings, spinning up, entering and exiting, and complex interactions with its rotating inertia. The problems of gee and shielding need a drastically different solution. In a starship without a high power drive, it is unclear where about 1 Mw of operating power will come from. Development of reliable small fusion reactors seems mandatory even if large ones for propulsion are hopelessly massive or tricky. Starting a drive might require additional or larger power supplies temporarily. Nuclear fission power plants would not last very long and would be very hazardous and massive.

At a more basic level, the use of large electric and magnetic fields, charges, and currents is a general area needing more study before the possibilities of shielding, braking, maneuvering, and propulsion can be fully explored. Every opportunity should be taken to develop electro-magnetic equipment for use in space near Earth to stimulate development (e.g. collect fusion fuel from solar wind and shield space colonies). Sustaining the large structural stresses that result from high fields is at odds with keeping large electric and magnetic equipment low in mass. The unlikely ramjet idea would require a magnetic field very large in size and strength, and any fusion or anti-matter drive or launching track has similar requirements.

Starship technology may need to operate as long as 1000 years, less for faster probes. Humans may be able to oversee, repair, and replace everything to insure such long reliability, but it is difficult to imagine robotic systems accounting for every contingency and every part. The prospects for artificial intelligence are still uncertain in regard to long-term reliable robotic control of an interstellar mission. And in the essentially closed system of the starship, accounting for nearly every atom of material seems very difficult. If recovering elements from household dust is not worth the effort, then effort must be put into preventing most materials from being worn down to dust. There will be limited power and human labor to recover, recycle, and remake all parts.

The idea of molecule-by-molecule and cell-by-cell imaging, identification, examination, manipulation, construction, and repair would also help in biological problems such as precise and reliable medical treatment and creation of plants and other organisms best suited for a starship ecosystem. Medical care should be automated with quite complex yet compact diagnostic and repair equipment. Only one doctor can be available at a time, and he/she must be computer-assisted to cope with the millions of variations in body parts, diseases, treatments, and medicines that follow from the millions of human genes. The medical resources of a large city must be carried in one room. Going further with our best technical tools, detailed computer models of human physiology should point the way toward long lives (near-immortality) and/or controlled hibernation. Perhaps these problems would be solved during a voyage itself.

Biological solutions to gravity (tolerating zero gee) and to radiation (tolerating high levels) are less likely to be found without total rebuilding of the human organism. At least bio-technology seems likely to solve the problem of carrying a broad genetic pool for humans, animals, and plants. And the no-longer far-fetched "bionic" idea of combining physical electronics and systems with biological systems has already begun, with work on sensors and even neural nets being done. These techniques may help with extending human life and capabilities and help solve the problem of reliable computation. Cells already "know" how to make and maintain micro-structures, and humans are learning how to control and maintain cells, and soon how to make them.

A different kind of combination of biological and physical sciences began partly with SETI. Radio searches for other civilizations have been complemented with studying how life began and evolved here, studying the environments on accessible worlds and moons in our Solar System, and looking for evidence of planets at other stars. A more general and necessary astronomical project is a nearly complete catalog of every star within 1000 LY. And biological work on stable ecologies for large space colonies might complement, if not enhance, essential work on re-stabilizing Earth ecology.

Scientifically, the crucial advances are needed first in physics. If powerful propulsion is developed, along with solutions to some of the operations problems, then biological, social, and economic problems of travel would eventually be solved. Those who study interstellar travel in a fascinating blend of fantasy and real physics have found a large number of different ways of getting there, almost all meriting careful theoretical study, none yet leading to affordable hardware even if scientists had ready access to space now. Unfortunately, there is very little indication in general relativity or elementary particle physics of new higher energy sources, forces, or effects that could be used in propulsion. If there were, motion in the context of general relativity (curved space-time) would also merit more study and be part of this book.

Many reasonably well-established findings in elementary particle physics and in gravity and cosmology were thought preposterous one or two decades ago. Predictions of fundamental particles called quarks were greeted with dismay; now they are confirmed despite the impossibility of isolating them. Quantum chromo-dynamics with its quarks and gluons (components of the nuclear force) has recently been combined with electro-magnetism and the weak force to form the so-called Standard Model with families of quarks and leptons (electrons and relatives) as truly elementary particles [e.g. b-Lederman]. These theories and models are partially confirmed and explored, and much work remains to be done. Other more sophisticated (probably not simpler) models may replace them. Physicists are attempting to join together gravity and the Standard Model into a unified field theory, one that might keep them busy a century studying its ramifications and finding new wonders. New ways to catalyze fusion of hydrogen or to control nuclear and weak forces may be found as a result. But experiments at the unified level are out of the question, needing about $(10)16$ GeV per particle (Big Bang level), almost a trillion times the energy of the planned SSC accelerator.

Working with anti-matter has a well-understood theoretical basis already, and the practical details of making, storing, and using anti-matter should be studied on a very small scale [b-Forward]. "Very small" because the best estimate with known technology of the cost of anti-protons as a commodity is about ten trillion dollars per kilogram. Perhaps a way will be found to make them more efficiently, but anti-matter factories must ultimately be located off Earth, using "free" sunlight, away from Earth's limited economy and ecosphere. Full understanding of the Standard Model above or its successor theory would seem to be the route for finding ways to make anti-matter more easily. If some sort of self-catalyzed reaction is found, solar energy must still be consumed literally by the kilogram or tonne. If anti-protons become abundantly available, their best use would be in dilution with hydrogen, for lower exhaust velocities.

High speeds are an unexplored area for applications other than starship motion. Between 10 km/s spacecraft and heavy nuclei moving near c is a large unknown realm of applied physics. Electro-magnetic drivers may be needed to send fuel pellets across space at speeds near c. The collision of interstellar gas and dust with a fast starship needs further study to determine damage rate and effective shielding. The collision of dust with dust would be of interest. Further study of high-speed collisions could aid development of propulsion and shielding. Heavy nuclei collisions or collision of manufactured dust or pellets moving near c could be keys to igniting fusion or other reactions. Shooting clusters of deuterium oxide (heavy water) at deuterium has recently been tried at about 100 km/s to cause fusion [a-Pool; a-Amato]. When collision energy far exceeds rest-mass energy (e.g. nuclei moving with 100 GeV per nucleon), the higher realm of Standard Model (partly unified field) physics can be entered. This is not in itself an energy source, but new physical phenomena may lead to a new source.

Much earlier mention was made that traveling from star to star is essentially a zero net energy change problem, hence the frustration of needing to expend 100 or more US GNPs on energy to make the transition to another star and another such expenditure to stop. All this energy would be dissipated as radiation and particles all over our part of the galaxy and benefit no one else. If Fundamental Problem #1 is that too much energy is needed for timely travel, Fundamental Problem #2 is that most of it is lost. Propulsion methods such as streams of particles or pellets are conceivable which keep this energy in play, converted from storage to KE, then recovered, stored, and used again. Practical implementation of this philosophy needs further study.

12.2 MORE ENERGY

The ideas covered here are a little less wild than those later. Basic models and theories in physics support these, but details for practical applications are almost nonexistent. In Chapter 4 an estimate was made of how much energy must be drawn from each kilogram of matter to make interstellar travel fast and easy. Einstein has set the limit available from matter at about (10)17 J/kg, widely confirmed. The best studied anti-matter propulsion idea would put less than 1/3 of this into starship KE. Moderate gamma travel (2 to 10) would require 10 to 100 times more energy per unit mass if carried, an amount that should be called "ultra-energy" to distinguish it from presently known sources. Very ambitious travel would use 1 gee steadily for several proper years, covering tens of LY, reaching gammas in the tens, or might use several gees to cover thousands of LY, reaching gammas in the thousands. In general, the size of gamma reached indicates what multiple of mass-energy must be packed into each kilogram of mass to achieve that gamma. Or such travel requires an energy source not carried in the starship as mass. Used in a momentum drive, this ultra-energy source would leave a very dangerous trail across the galaxy, destroying any vehicle or life that happened through it. Regardless of the titanic and unimaginable machinery required, we must wonder whether there is some way to obtain more energy. Feasibility can be studied later.

High-energy physics is now studying particles (e.g. the Z0, unfortunately very short-lived) which carry 90 times more mass-energy than protons. But these and more massive predicted particles all obey Einstein and have proportionately higher inertial masses, making them difficult to transport. All known and suspected particles carry about (10)17 J/kg, and none carry more. Black holes (discussed for other reasons in next section) are also no panacea, since the smallest conceivably useful ones would be very massive to transport and would obey Einstein in every way. They can contain or release no more energy than any other form of known matter. The fundamental energy problem remains: how to get more energy from less inertial mass.

A basic law of "uncertainty" in elementary quantum physics states that variations in energy and time are limited and inversely related, at least at the quantum level. Ordinary "empty" space (essentially vacuum) is restless and seething at small scales. Varying amounts of energy wink in and out of existence in very short times, usually as a particle anti-particle pair. Few ideas have been put forth for harnessing such quanta of energy. Nevertheless, space is one place to look for nearly unlimited energy. Not just space but anywhere, even inside the electronic devices and other materials that run our high-tech society, particles are borrowing energy momentarily to move from one place to another. We would like to do that on a much larger scale. Energy can be borrowed from nature without penalty if the time of the loan is short. Planck's constant is involved and permits only very tiny effects [see Appendix D]. One could borrow (10)24 J, the whole KE requirement of a star journey, from space itself for about (10)-58 second, an inconveniently short-term loan. One could borrow portions of this for longer times, except that a piddly joule can be borrowed for only (10)-34 second. Clearly ordinary quantum borrowing will not provide significant energy. One may think, why not borrow the joule for (10)-34 second, then return it and borrow another, and so on and so on? But the energy is returned and not put into starship KE. A way to cheat nature on this law is yet to be found. Perhaps it will be overthrown in some subtle, unforeseen, dramatic way.

An energy supply need not be carried if it can be gathered along the way from "empty" space. At present the Standard Model seems to leave another related remote possibility open, the vacuum energy of space. A pervasive "false" vacuum field or Higgs field is conjectured as left over from the Big Bang and provides a net energy in all space [a-Abbott]. This field may also produce the masses of all particles by "resisting" their motion, at least until they reach a high-energy range (perhaps TeV) [b-Lederman; b-Pagels; a-Veltman]. The Higgs field would manifest itself as Higgs "bosons", particles thought to be have perhaps 1 TeV, not too massive to be made with a present accelerator and verified and studied. As the field pervades space, so do its bosons. The mathematical physics of the Standard Model is not accessible to many, but should gradually become more familiar, especially if practical applications are found.

The Higgs field arises also in a different context (a good sign in physics), and the amount of energy in it per cubic meter of space can be very roughly estimated from cosmological considerations. [See preceding references for background on cosmology, but the author bears responsibility for the wild estimate of energy here.] The numbers used here may seem extreme but are part of present theories at the smallest and largest scales and are supported by observations. At an early age from (10)-35 to (10)-32 second the Universe is thought to have "inflated" rapidly by a factor of about (10)50 from about (10)-27 m in size to (10)23 m (to about 0.1% of its present size) to achieve its uniformity. Much energy seems to have been added to space during expansion, as the Higgs field or "false vacuum". (An ideal vacuum, which probably cannot exist, would contain nothing, no mass, no fields, no energy.) The presence of vacuum energy is equivalent to Einstein's long disputed "cosmological constant". This Higgs energy perhaps can be deduced from the later energy content of the Universe.

During inflation most or all of the Higgs field energy is conjectured to have converted to particle mass-energy. After inflation there were about (10)87 baryons (protons and anti-protons), for a total of about (10)78 J in terms of mass-energy (including an extra 10 times more energy as a conservative estimate of the mysterious "dark matter"). Somewhat later all but 1 part in a billion of baryons are thought to have annihilated, leaving today's much lower count of protons. Suppose that some fraction of that Higgs field energy remained in space to affect all particles now, perhaps 1/10 of the total particle mass-energy. (Some say it is all still here, or it could be only a billionth of the original amount. Errors of factors of 10 here are small.) Or possibly another pervasive field with energy exists so that space is still a false vacuum [b-Pagels]. If, say, $(10)77$ J is distributed in the $(10)78$ m^3 volume of the known Universe, then the energy density is about 0.1 J/m^3. This is 10,000 times larger than the typical mass-energy density of interstellar hydrogen. Perhaps the Higgs field is "denser" within galaxies than between, or perhaps the contrary.

Collection of Higgs energy for use could be quite difficult. Almost nothing is known about this field, and no spectacular effect involving it is known. Ordinary matter does not seem to interact with it in extraordinary ways. A starship moving at 0.1c needs to collect about $(10)15$ J/s. It would need to scoop an area of about 300 million m^2, or about 20 km across, if all the Higgs energy guessed above were collected. Somehow the energy must be converted to particles or radiation that can be exhausted for the usual momentum drive. The possibility of substantial vacuum field energy in the form of the Higgs field or some other field seems worth further investigation.

To make progress in this direction, elementary particle physics should be well-supported, including the next generation of accelerators such as the Superconducting Super Collider (SSC) which will reach for the 40 TeV realm. Answers about the Higgs field and boson and more answers about quark interactions are expected somewhere below 40 TeV [a-Lederman; b-Lederman; a-Jackson]. The Higgs field is needed for both particle-field theory and for cosmological theories, and it has been called a focus for physics of the 1990s. After the SSC has explored this realm awhile, the need for still higher energy in a space-based accelerator might be established. Perhaps a solution for interstellar drives will appear then.

Some theoretical studies have been published on using vacuum energy in another way, quantum fluctuations in space in the ordinary vacuum [a-Froning 1980, 1981, 1986 all similar, note cit. Wheeler]. If the fluctuations occur on the scale of $(10)-35$ m (structure of space at this Planck size discussed in next section), then the typical energy in a fluctuation is said to be about $(10)9$ J and the equivalent mass is about $(10)-8$ kg. Again, the smaller the scale, the larger the energy fluctuations. It has been further argued that the possible energy density is huge since this much energy can "materialize" in a volume of only about $(10)-105$ m^3. The catch is that an extremely small-scale method is needed to capture these tiny, fast fluctuations. On the average the energy of the vacuum may be virtually zero. Extrapolating from these ideas, an entirely different quantum view of the interstellar travel problem has been suggested that might indicate a new realm of solutions [a-Froning].

If energy is to be captured from space somehow on the fly during travel, ramjet analysis and technology of some sort seems appropriate [a-Froning]. Practical problems to ponder include obtaining stable reaction mass to employ captured energy in a momentum drive. Somehow the rapidly fluctuating mass from space must be used as exhaust despite its rapid fluctuation during action-reaction. Since abundant vacuum fluctuation energy seems available, very low efficiencies of utilization can be tolerated. If all the energy one could want were available, the minimum needed density for capture would be about 10,000 J/m^3 for a 0.1c starship with reasonable 100 m diameter collection area.

Space has many "vacuum" fields: gravitational, electromagnetic, hypothesized false vacuum (Higgs), nuclear, and others that may be discovered. All should be ruled by the law of quantum physics in that the amount of energy that temporarily appears is determined by Planck's constant. One more of these fields that has been investigated is the electro-magnetic field in vacuum. A Casimir force develops as a result of electro-magnetic fluctuations in the vacuum between two very close conducting sheets [a-Forward 1984; a-Boyer]. The sheets limit the fluctuations to standing waves with definite numbers of wavelengths between the sheets. The force can be made to do work, i.e. yield energy. The force has been calculated. Although relatively small, it becomes stronger as separation decreases. Experimental verification has been difficult and therefore with present materials and methods not much useful energy is expected to be extracted this way.

Then there is the mysterious missing or "dark" matter of the Universe, the matter which has a strong gravitational effect on the motion of galaxies yet seems to be undetectable here and has no measurable effect on our activities or on Solar System motions. Some 90% to 95% or more of the Universe may consist of matter of unknown type. Perhaps ways to detect it and eventually to scoop it up and use it will be found. However it does not represent much energy density. If ten times the observed baryons, or (10)53 kg, is "dark" matter, its mass-energy is about (10)70 J in the (10)78 m^3 volume of the known Universe, or a density of about (10)-8 J/m^3. For comparison, the energy density of interstellar hydrogen is about (10)-5 J/m^3, rather meager if collected or used by a ramjet. Little gain would occur if the mysterious dark matter could be collected and used. More discouraging is that the bulk of the dark matter, if it exists, seems to be present not within galaxies but in intergalactic space where we cannot use it.

If any of these possible sources of ultra-energy exist and can be harnessed, other beings do not seem to have done it on a significant scale. No evidence of ultra-relativistic starships or titanic beams of energy has been seen. Earth has not passed recently through a trail of exotic particles or holes or other disturbance of space by a long-ago passing starship. The existence of ultra-high power drives would have a strong effect on the spread of civilizations through the galaxy and indeed would have brought us many curious visitors. If a transportable ultra-energy source were found, we must hope that its economic cost is not proportional to its energy content as happens for anti-matter. If the energy content of ultra-fuels had to be purchased, then civilization would choose to go the slow route instead. A reactor method seems needed to produce ultra-fuel, one that bootstraps the fuel out of nature from some huge energy reserve in nature, using a very small technical input by human engineers.

12.3 DISTORTING SPACE-TIME

Space is thought to be strange, distorted or "warped" (to use a favorite sf term), full of holes like a sponge or foam, but at the Planck scale (pico-pico-pico-scale) of about (10)-35 meter. If space could be made to act this way (or inherently does) on a scale of meters to lightyears, a popular conjecture is that a starship could be dropped into a suitable hole here and emerge "moments later" from a hole near another star. Black holes (BHs) have been proposed often as appropriate holes for travel, but their name is misleading. The simplest kinds (Schwarzschild "singularities") cannot be holes to anywhere because everything is squeezed to zero in them. BHs are one of a family of interesting phenomena that follow from general relativity and greatly distort, disconnect, or connect space-time [b-Hawking; b-Kaufmann; a-Forward 1989; a-Penrose; a-Matzner twice]. The idea has also caused some distorted literature. Some kinds of BHs are thought to form "wormholes" in space-time at the Planck scale and perhaps at larger scales as "space bridges", "tunnels", or "subways". Wormholes are speculated to have exits called "whiteholes" in some universe, not necessarily ours, but no evidence has been found for such energy emitters.

Warped space-time is not just a physicist's fantasy. There is already much evidence for Einstein's curved space-time on a large scale. The gravity wells of planets and stars can be thought of as shallow "dents" in space-time in which objects and photons follow curved paths, thus explaining gravity in terms of curved geometry. Our Universe is thought to be a closed "bubble" of space-time so that all straight paths are really great circle paths. If special relativity (in Chapter 3) was found marginally important for practical interstellar travel at speeds less than 0.2c, general relativity seems even less relevant at present because no tremendously dense masses seem available to us to warp space artificially and severely. However, the possibility deserves mention because general relativity deals with the essence of space-time, gravity, and mass-energy.

Black holes have not been placed in the later "wilder idea" section for discussion because of a theoretical basis, much disputed, for feasible travel when the BH has net electric charge and/or rotation [a-Forward]. However, rotation must approach lightspeed (unlikely) to eliminate the "horizon" of no return and permit objects to leave again. The rotating mass is said to be distributed as a toroidal singularity in a rotating Kerr BH, and a starship which travels through the hole of the torus might enter a different space or time. There is no evidence yet of practical usefulness, and some theoreticians find that no BHs can be used for travel [a-DeWitt; a-Birrell; a-Matzner]. Some say that travel inside could take longer than the equivalent travel without the BH, and no one really knows where BHs might connect to normal space. Worse, any disturbance such as a starship passing through could cause additional singularities inside that prevent safe travel. Besides trouble inside, just approaching a BH is dangerous because of intense radiation from infalling material. The gravity field itself is titanic, although not dangerous because of its strength. Any orbit would feel like freefall as usual. But the variation of gravity with distance (gradient, or tide) near a BH of stellar mass is so large as to pull any material objects apart. Perhaps kilogram probes, or surely atoms, could pass through a charged or rotating BH "tunnel" without damage if starships cannot. Much larger BHs of many stellar masses are much less likely to be available to us but do not have a dangerous gravity gradient.

Black holes warp space with a vengeance, pinching off a piece completely so that not even light can escape. The zone of no escape for a simple BH is a spherical "event horizon" whose radius depends on the mass [see Appendix D]. BHs are an ultimate destruction of space, time, and matter, although nothing essential is lost. They retain all the mass-energy, net charge, and angular momentum that went into them, and the mass-energy sits there with infinite density in the singularity, gravitating. The matter in a BH might be described as a sixth or seventh state of matter, a more extreme state than found in a neutron star. BHs are probably real although finding sufficient evidence to be convincing has been a major task. The nearest partly-confirmed one Cygnus X-1 [a-Thorne] is thousands of lightyears away. If one were big and much closer, we would know it by now.

If BHs were otherwise useful for travel, they are too scarce. Humans cannot travel hundreds of lightyears just to use a BH. No starships are popping into space nearby and visiting us. Perhaps some other former civilization had a BH nearby. If they lived in a binary star system, one of their stars could have gone supernova, wiping out all life and planets, then sinking into BH oblivion. BHs are expected to come from very massive bright short-lived stars. One's own star would never become one, since life would not have time to evolve. If colonists from such a deceased civilization returned to their former home, they would then own a bonafide dangerously radiating BH to use if they could.

Cosmologists speculate that "small" primordial BHs are sprinkled about, left over from the Big Bang. Very small ones would have already "decayed" and ended their lives with smaller but impressive bangs. Sizes up to proton size with a billion tonnes mass might last about ten billion years and could be useful [b-Hawking]. They could be dangerous to drop matter into, but tiny ones are also thought to be very slow to swallow matter [a-Matzner]. They seem unlikely to consume a gas giant planet fast enough to make a gravitational booster for starships. Indeed, these "small" BHs are not really black but hot and radiating, emitting about 10 million kw [a-Hawking; b-Hawking]. A much bigger BH would be needed to power a starship drive, but the idea is unthinkable because of its nearly uncontrollable radiation and nearly unmovable mass. Carrying a billion tonnes just for a portable power source is quite unrealistic.

Two methods which involve gravitationally-warped space, but without holes or connections, are sometimes called gravitational machines [a-Dyson; a-Forward 1963]. One, mentioned much earlier, uses two rapidly revolving massive stars for gravitational boosting or braking. Along with the usual attractive effect of each star, close and fast stars also generate substantial tangential gravitational force as predicted in general relativity. Some BHs and neutron stars are expected to exist as pairs revolving at high speed around each other, providing two compact large masses that one can travel close to. It would not be necessary to enter the BH, just fly by it for a large boost if the severe radiation and tidal force are not deterrents.

The other method handles mass at high density like electric current to produce new forms of gravitational force predicted by general relativity. For example, increasing mass flow around the minor circumference of a torus would produce a gravitational field through the center which could be repelled or attracted by ordinary gravity for propulsion (fabled "anti-gravity"). Ten billion tonnes of mass starting to flow in one second around a 100 m torus would produce a field of only about $(10)-15$ gee [see Appendix D]. The scale for effective operation for such devices clearly must be much larger, and is about c-squared larger than electro-magnetic methods. Matter densities should approach neutron star densities $(10)18$ kg/m^3, material not accessible for reasons given below. The requirement of increasing mass makes steady-state generation impossible. Anti-gravity machines also do not seem feasible for interstellar travel because gravitational fields far from stars in interstellar space are too weak to provide much interaction. However, a strong boost might be achieved from the home star and good braking from the destination star. In this case the extremely high (star-size?) mass of the starship drive would not be a disadvantage; the more mass, the more force obtained for acceleration. But the helping star might be displaced as much as the starship!

It is difficult to pick one's way through the various sf inventions and distinguish them from physical possibilities. "Hyperdrives" or "spacewarp drives" seem to have no physical basis, being proposed machines which would cause a starship to leave its space-time continuum and move to another or to leap over a long piece of its world-line. [If this is not clear, some reading of introductory books on general relativity such as cited above is suggested.] Scientists have no knowledge of how to begin to design such machines since warping space-time is not sufficient. A BH might be able to tear a hole in space-time, but a controllable way to depart from our familiar four dimensions is sought. Our space-time is only gently curved, near masses to produce the effect of gravity and overall about the Universe to "close" it. It is difficult to believe that "outside" our space-time the Sun is very "near" Epsilon Eridani or some other near star. Or that our Sun is "near" any other star in any other galaxy in this or in any other Universe. Again, if warp travel were practical, someone's star should connect near ours and they should be visiting us.

Stranger, but still with a theoretical basis, is the possibility of faster-than-light (FTL) or "super-luminal" travel. As presently understood, Einstein's theories do not prohibit particles moving FTL, all the way up to infinite speed [a-Bilaniuk; a-Feinberg]. But they cannot move slower than light, which remains a barrier. Thus a new class of particles is required, called "tachyons" (making ordinary particles "tardyons"). It seems impossible that we slow people could build a starship using hypothetical tachyons as material. Attempts to verify the existence of tachyons have been extensive with null results. If they did exist, it is not clear how interstellar travel would use them. If nearly-instant tachyon communication were possible, it would help tie together a galactic empire, but it also would violate causality as we know it. The answer could arrive before the question was sent. Energy and therefore communication seem prohibited from traveling FTL, regardless of tachyons.

12.4 SPACE-GROUND TRANSPORT

A little-appreciated problem related to interstellar travel is lifting from a planet and landing back on it. As discussed, chemical rockets are well-matched to lifting mass from an Earth-size planet but are expensive and polluting for large projects such as building starships or space colonies from planetary materials. At least Earth can provide weather information, space stations, landing fields, runways, and refueling. The problem would be much more severe at the end of an interstellar voyage where no such space and planetary technical resources are available (unless a friendly advanced space-traveling civilization dwells there). Discussion here focuses on transport problems at arrival and assumes Earth/space civilization can prepare a starship for departure. Whatever the transport, another problem is carrying the methods there, or making them upon arrival. Unloading a starship would require many landing vehicles in reusable form, as well as the option of removing everyone and their supplies from an unexpectedly hostile planet.

Chapter 8 reviewed some methods of ground-to-space and space-to-ground transportation and more are mentioned here. If the destination planet is Earth-size, transport is difficult. Smaller planets, less likely to be habitable, would be easier to land on, especially if they have air. Chemical rocket methods require a large ground fuel source (preferably hydrogen and oxygen) for taking off, and the same source in space to use for retro-rocket landings (as done by Apollo on our Moon). Chemical fuels are too massive to carry across interstellar space, and must be found at the destination. Very many landings would be needed. A rocket or shuttle could descend with fuel that is used for braking, a procedure never tried on Earth yet, but it cannot be landed full of fuel. A source must be developed on the ground. Well-proven air friction braking of heat-resistant capsules, followed by parachute landing (on planets with atmospheres), is a method more for emergencies than routine landings because capsules carry small loads and cannot be relaunched. Some destinations of interest or necessity might be airless, requiring rocket landing, and they would probably be deficient in fuels. Even if anti-matter provided the energy, abundant hydrogen would be needed.

Air braking is preferable where feasible since no fuel is needed for landing. The US NASA shuttle is a somewhat fragile large vehicle which can land using only air friction and air lift. The proposed US space-plane (NASP) would use mostly air (requiring an Earth-like oxygen content) to burn hydrogen or hydrocarbon fuel for takeoff, thus reducing the fuel-oxidizer mass that must be lifted, but perhaps not eliminating the need for auxiliary tanks. It would land like the shuttle, easier but still needing a long smooth runway. Smaller versions are being developed by France (Hermes) and England (robotic HOTOL). It may be difficult to design for a range of unknown atmospheres, especially since the design process appears more difficult than for the shuttle. Small differences in air composition and pressure may affect space-plane performance more critically. Still, the space-plane is an idea worth much study and development effort, not only as one kind of lander to be carried with a starship, but also in support of a space colony program here.

Other more futuristic ideas are laser-heated rockets [see Chapter 2], anti-matter heated rockets [Chapter 4], launch tracks, high-altitude balloon and airplane launches, balloon-supported launch tubes, elevators, and orbiting rings. Some provide initial speed with large ground installations (e.g. electro-magnetic tracks or large conventional jet planes). Some provide high-altitude launches to bypass most air friction (and oxygen), such as balloons, jet planes, mountains, and metal towers. The launch tube would provide speed and altitude on a track into orbit. Some transmit all energy from a ground power station to a rocket (e.g. laser-heating). Many ideas have been proposed for elevator-like methods to lift people and supplies from Earth to space and descend again, using stationary or rotating cables or "fountains" of pellets [see e.g. b-Forward for these and more]. The orbiting ring establishes a "stationary" series of platforms in low orbit, making elevators easier to build. Some of these transport methods permit recovery of gravitational energy from descent, to be stored and used to lift a next load. Net energy usage would be positive if more mass is taken up than brought back overall in time. As with spaceplanes, most of these methods seem to use materials to their limits in strength, toughness, and/or thermal integrity. At an interstellar destination, most of these involve massive structures which cannot easily be carried or built on site.

Material for building transportation or colonies in space can be obtained from passing comets and asteroids, moon-mining, and distant asteroids. Some planets may have closer smaller moons. Mass drivers and rotating mass flingers seem feasible for throwing mined material from moons and asteroids into space at low cost. Experience with these methods will accumulate as local space civilization develops. An interstellar mission must be ambitious to roam the new planetary system for material before people can begin developing a larger habitat in space and/or land on a planet. Experience at more exotic transport methods would be valuable here to aid local space colonization. At present massive emigration from Earth to local space seems impossible due to the cost and complexity of lifting just the people, never mind their equipment, into space. Transport using cables and the like would change that.

12.5 STILL WILDER IDEAS

If stronger materials are needed for starship technology, nature has much stronger materials. They are formed only in extremely strong gravitational fields as degenerate and neutron star matter, and it does not seem feasible to make, obtain, or use them in the near future. However, unexpected tricks may be found, given the basic possibility of such matter. Ordinary matter is held together by "clouds" of electrons among the positively charged nuclei of atoms, with the tiny but massive nuclei rather far apart. The theoretical strength of covalently-bound whiskers of diamond and similar materials is near the theoretical strength of positive and negative charge attraction at typical atomic distances. For more strength, either the charges must be closer together or a stronger force used. Both are available.

Being about 100,000 times smaller than atoms, nuclei can approach 100,000 times closer than atoms if electron clouds around nuclei are removed. No negatively charged stable nuclei (forget anti-matter) are known for placement next to positive nuclei. Instead electrons must be packed in to provide an electric glue to overcome proton repulsion. (Unstable muons mentioned earlier as fusion catalysts would pack in much more easily.) Closely-packed matter occurs in the collapsed "degenerate" state of white dwarf stars, where a kind of quantum mechanical electron pressure is all that prevents further collapse to put the nuclei cheek by jowl [a-Ruderman]. In neutron stars collapse has gone further and the "star" has become a ball of seething neutrons (electrons having combined with the protons of nuclei).

Even here nuclei or neutrons do not approach closely enough (except perhaps in the core) to feel the effects of the nuclear force, about 100 times stronger than the electric force for the same distance. If neutrons did "touch" and bond, nuclear matter would be about a trillion times stronger than ordinary everyday matter. If neutrons can be kept from decaying (as in most nuclei) and spun out like thread, cables of incredible strength and low mass could be made for holding sails, for hoisting loads from ground to synchronous orbit, and for many other uses. (Gripping onto such cables is a further problem, remindful of the "shadow square wire" in the Ringworld novel.) The possibilities with large quantities of neutrons could be explored now that physicists are able to collect very cold (slow) neutrons in bottles.

Matter transmission as described in sf seems impossible. This mysterious process is supposed to involve stripping down an object to its atoms, then sending them as waves at lightspeed. But matter waves do not seem to travel over much distance. However, coded instructions for building machines, living cells, perhaps people, could be transmitted by ordinary radio over interstellar distances. With growing knowledge of fractals, all essential details might be reduced to a reasonable size of code. If this is an acceptable form of travel (at lightspeed!), then science need merely solve how to record and transmit the personality to go with the body (a feat that may be proven impossible) and hope that someone out there puts it all together. Even so, the original person here could never know anything about what his/her alter ego is experiencing out there, and continuity of travel is broken.

Although the simplest living organisms (in favorable environments with large genetic codes working at the molecular level) can reproduce and can evolve to higher complexity, the prospects for von Neumann self-reproducing probes as an easy way to solve our technical problems does not seem much better than a "wild idea" at present. But any sort of probe or peopled starship would benefit if equipment could diagnose, repair, and update itself. This subject bears and will get more study, if only for practical applications on Earth and in local space. More study of how living organisms accomplish these tasks might help. Full-fledged von Neumann machines which collect fusion fuel and reproduce on thousand-year time scales seem to require a much more complex starship development project than the difficult missions described in this book. Justifying a von Neumann program is difficult because after a few generations each one will usually need to reproduce itself only once to reach the next stars. If reproducing machinery could readily adapt to interstellar travel by machine evolution, perhaps incorporating the bionic "space caterpillar and butterfly" idea [b-Dyson], lack of observation of successful "species" in local space is telling.

A much-studied propulsion method that has turned out to be one of the wilder ideas is the interstellar ramjet. Its attractive use of interstellar material, has not lived up to initial expectations. Hydrogen, deuterium, and helium-3 are too scarce in space and often unionized. Required magnetic fields may need to be much larger and stronger than examined in Chapter 4. If abundant hydrogen could be collected (scooped), making it fuse for energy remains a fundamental problem because a different kind of weak nuclear force is involved. Breakthroughs on any of the major problems of the ramjet would justify increased interest in it. Another resource of space, interstellar magnetic field, is also too weak and unpredictable to use for maneuvering starships.

12.6 CONCLUSIONS ON STARSHIPS AND MISSIONS

Once the population of a human mission is determined (estimated 100 minimum, fluctuating as high as 200), many other design parameters are set. Required payload mass (everything but fuel and propulsion equipment) would be closer to 10,000 than 1000 tonnes. Biology, sociology, economics, politics, and much more figure in the problem of interstellar travel, but fundamentally it is a hardware problem to build such a large precision machine to accomplish such a titanic goal. Like all technology it can lead to triumphs and to tragedies. A starship could enslave voyagers to artificial conditions more restricting than any technology before while giving them the freedom to visit other stars.

The key practical step toward the stars is propulsion, but technology is inseparable from purpose. Hence possible missions or expeditions must be considered along with technical means. Disregarding the more speculative breakthroughs hoped for, an attempt is made here to assess the state of the art of propulsion as it might be based on early 21st Century science and technology. The means to build it may not be available for several more centuries. Many and varied scientists have worked on theoretical details of many imaginative forms of starship propulsion. No indication has been found yet that more precise models would make a difference in estimating feasibility. Until experimental work starts in earnest, physical estimates as featured in this book seem sufficient to continue to explore the general problem of interstellar travel.

A trend in the literature on starship propulsion is toward lower and lower speeds. From early prophecies of high gammas, with 100 LY journeys requiring just 10 years of voyagers' lives, projected speed has fallen to a barely relativistic 0.9c, then to 0.1c for Daedalus on pulsed fusion power, then to 0.01c or less for "worldships" on sails or fusion drive. That trend has occurred in this book while ever more powerful propulsion methods were studied. The fantastically expensive anti-matter drive has been shown to put less than 1/3 of rest-mass into motion, thus limiting speed to well below c. As planned speed decreases, travel time increases and required starship mass increases to support larger populations more completely. The need for very long-lived mechanical, electronic, and ecological systems increases. But required fuel mass decreases very favorably. A baseline 100-person interstellar mission has evolved, for purposes of discussion, to cover about 10 LY at about 0.01c and require 1000 years and 10,000 tonnes of vital mass (plus fuel if any). If fuel is not carried and greater speed is possible, payload mass also can be reduced. The best guesses here for propulsion at present are sunbeam-pushed sail or pulsed fusion, with a separate braking system. Farther down the list because they are still slower are an ion drive powered by fusion reactor, or a solar sail using direct local sunlight.

Lower speeds do not reduce the cosmic radiation problem but may permit more mass to be carried to shield a starship with sufficiently wide frontal area to carry a rotating habitat. In regard to mass, starship shape should be compact (near spherical) to minimize shielding mass. But the need to minimize interstellar erosion dictates a long, slender, streamlined shape. These three conflicting requirements on shape need resolution, perhaps toward a compact but wide cylindrical form. To obtain more structural strength from less material, the fragile starship would best have a smooth simple regular round compact shape. A habitat with artificial gee could be accommodated as a central wheel in such a design. Design of an elongated cylindrical stack of living levels at zero gee could be discarded, but a probe would be long and thin. For humans the indicated design is the large compact cylinder with rounded end, consisting mostly of empty volume to keep total mass reasonable. The starship could not approach or enter an atmosphere, or land on the smallest of moons, or withstand high gee from any means. In contrast to speculation elsewhere, fragile low-mass extended structures such as sails, screens, or coils have not seemed feasible in sizes larger than about 100 km.

Other design difficulties need resolution. Radio dishes and other instruments cannot face forward due to impinging dust and, if a drive is operating, may not face back toward Earth either. Dishes should be as large as 100 m, else much operating power is needed to transmit messages to Earth. On a 10 LY human voyage, our Sun and Earth should remain easy to track. But a probe would require very sophisticated astrogation to ensure successful data transmission to Earth after hundreds of years. If a sail is used, a blinding light reflected from forward must be contended with. Astrogation is hindered by all these restrictions. If instruments could look forward, there is the possibility of detecting and zapping large grains or avoiding substantial dust clouds (for speeds near 0.01c). Astrogation also requires having data on most of the stars out to 1000 LY, however dim, and a computer that can store, calculate, and display any view of them. The required measurements will surely be made by the time a starship is built.

The scale of operation of a starship with fusion drive should be kept in mind. A 10,000 tonne peopled starship may need to carry 100 times that in fuel, for an initial mass of a million tonnes. Total energy carried might be about $(10)23$ J, somewhat more than in all the fossil fuel deposits on Earth. The power of the drive might be of the order of a trillion kw, and the mass of the drive would be at least 100 tonnes. Such a power-to-mass ratio of 10 million kw/kg is far greater than any present power equipment handles. The power handled by one starship would be at least ten times larger than the total power (not just electric) used by all of present human civilization. The volume in which this power is developed might be about 1 million m^3, for a power density not much different from present nuclear fission reactor cores. About 1 to 10 kg/s of mass would be exhausted, at a speed of about 0.05c. The "engine" throat size might be about 10 times Daedalus size (preferably 10 or more engines in parallel). Then the power area density would be about 10 million kw/m^2, about 10 million times greater than the solar radiation we feel here and deadly to anything in its path. Controlling such a torrent of power would be very challenging, since a slight leakage that reaches structure or people would be devastating. Radiation shielding must be very thorough and durable. The propulsion must be highly stable and finely tuned. Leakage, instability, or breakdown could involve catastrophic energy releases.

If this scale of energy use proves too formidable, a small 1 tonne probe sent with fusion power would still be a very large project, still requiring a power level larger than US national usage, since it should be sent faster than a colony mission to be of value. An option that increases chances of success for human or probe missions is to send as many as 10 smaller starships. Aggregate mass, fuel requirements, and costs would be larger, but not 10 times larger. "Production line" savings might make the program cheaper overall.

Propulsion methods, carrying no fuel mass but with overall results not so much different from fusion, require development of low-density high-strength materials for use as very large solar "sails", very long cables, high-current coils, and other components. Basic limits to sailing size were shown, indicating that large missions must break into 10 or 100 smaller duplicate sailing starships, a redundancy that much increases mission success. If sails are used, Solar System-based sunbeam mirrors or laser beam generators might be needed. A 100 km sunbeam could push a mission to well over 0.1c, but halting it later is a problem. A fusion drive could be carried with fuel collected along the way, or, more likely, large structures for braking. Although low gee, three braking methods merit study in themselves and as integrated with propulsion: solar sails used for drag near stars, a charged drag screen that interacts with interstellar material, and magnetic interaction with interstellar plasma. A more difficult method to study, if gigantic space structures are favored, is a solar-powered magnetic launching track 25 million km or more long.

Large sub-system problems such as shielding, artificial gee, operating power (especially if sails are used), and landing remain to be solved as discussed earlier. Many suggestions for these have been reported, but none seem ready to be integrated into a starship design. As an indication of problems beyond propulsion, the estimated minimum habitat volume for 100 people was the size of a 15-level medium apartment building, including one or two levels for growing food, or the equivalent spread out in a rotating wheel. This may seem fine until one considers living in this for a lifetime plus as many as 30 more generations to be raised there. Worse would be living without gee, surely committing the population to permanent inability to land on a planet. Obtaining artificial gee by rotating the habitat seems difficult but feasible once shielding is assured.

A program of interstellar exploration and colonization will not happen quickly. The time scale seems to be a thousand years at the minimum. Robotic probes should be sent first to nearby stars to scout for destinations and to prove the performance of propulsion and other systems. Probes could travel faster than human starships but the wait for data could still be hundreds of years. Several successive probes might be needed. Making very small ones massing 1 to 10 kg would lower the cost and permit higher speed but problems of power supply, transmitting antenna, and shielding make these unlikely even if micro-miniaturization is successful. One tonne seems a minimum feasible size.

Mission planning raises many quandaries. When people travel, many must go, yet never as many as such a voyage needs. They would want to be certain of a place to establish a colony and civilization upon arrival. Only one habitable planet might be partly proven within 10 LY. Return is unlikely for reasons of energy and time, and contact with Earth is essentially lost. The first interstellar probe and first mission with people are the hard parts. Once a colony is established at another star, they can, over time, provide technical support for braking later starships and can make two-way travel feasible (if needed). And a growing colony would set the stage for farther interstellar missions.

If solar-pushed and interstellar-braked starships, for example, seem expensive but achievable, we must worry that other beings are not already traveling this way. It appears that humans must begin interstellar exploration on their own, and a contemporary technical civilization of other beings is unlikely to be found at a near star. Humans cannot count on assistance by radio nor a welcome with full technical and cultural support for their first interstellar travel.

12.7 CONCLUSIONS IN SOCIAL AND POLITICAL CONTEXT

Our space age began with great enthusiasm. After humans reached our Moon in 1969, Mars was thought to follow before 1980, and interstellar missions early next century. Progress has slowed considerably, even on the local robotic missions [but see a-Norman]. Space development such as nearby self-sufficient colonies was shown feasible in 1970s and thought certain to be started in 1980s. Public support and money seem to have diminished, although US NASA's budget increases year by year. Less and less is done, while some billions of dollars are spent on mistakes. A simple space station is now priced higher than the Apollo program, with several billion dollars spent already on designs without agreement [a-Marshall; a-Broad two]. Political problems aside, is this the trend of our future in space, less and less over longer and longer time, with perhaps bigger confusion and failures? Have we reached technical limits already? Delays on new propulsion would seem to corroborate this. Faster propulsion such as solar electric (ion) and sail have been under study for about 30 years, yet chemical rockets are relied upon for all missions thus far. We are still dreaming about the stars, but we seem too cautious about reaching for local planets. The general passion for high-tech projects cannot continue if we do not actually carry some out.

Considering the difficulties faced for a simple robotic mission to a near star, the vision of human empire across the galaxy is a very large leap. Many people believe that space and the stars are the next steps for us, as if part of our "destiny". From what is known about evolution at this point, neither life nor intelligence can have purpose, goal, or destiny beyond those self-imposed. Life could be a mere smudge on a few worlds rather than the ultimate result of the Big Bang. Nevertheless, there is a slight chance that the Universe in some metaphysical way "expects" us and other beings to populate it. If so, the Universe has made interstellar travel unaccountably difficult. Whether achieving travel might be a test of our (and others') abilities seems unanswerable. Some hope that messages from elsewhere will have instructions on how to build advanced propulsion and solve the other technical problems, as well as on learning to live peacefully with ourselves here. A yet unimagined drive would drastically change the situation.

The idea that humans will or must colonize the stars requires more thought later. After a few steps of colonization, most new colonies would find only one new star to send a mission to. Keeping contact with Earth to form an empire would be difficult, for the same reasons other beings might not have formed widespread empires among the stars. While much study has resulted in conclusions either way, a good guess on the spread of civilizations is that they spread very slowly. This is a possible reason why we have seen no visitors scouting our sector of space. Without hard evidence, careful intuitive estimates of the number of advanced civilizations out there at present settle on a low number. The nearest might be in the range of 100 to 1000 LY away, well beyond projected travel capability. Humans could place a colony at a near star before a message and response have occurred with another civilization.

Some preceding chapters may have been discouraging on the prospects for successful interstellar missions because of identified problems such as these: the difficulty of determining a habitable destination before going there, the difficulty of returning, the long ordeal of many generations traveling to a near star, the hazards of breakdowns, radiation, and zero gee, the long time commitment to find out if a mission yields results, and the lack of visitors here who have accomplished such travel. But it is too early in our social and technological history to draw final conclusions about this futuristic goal, and no single paramount conclusion can be presented here. To reach the expected scientific and technical adventures and achievements, we simply may need to try harder. Interstellar travel might follow in its own time after vigorous local space development occurs, and our civilization may be privileged as the first one able to "develop" our region of the galaxy. (The idea of "develop" as a human technical activity should be taken cautiously, since over-development of Earth may help prevent beginning vital colonization of local space.)

Achieving interstellar travel may require a sequence of vital steps at unique opportunities. At present Earth societies have just enough fossil fuel, uranium energy, economic flexibility, and expertise to keep civilization going while starting a space colony program. Cooperative space development could even be a cure for economic recession and benefit nearly everyone. Scientific studies on biospheres and propulsion might spin off applications to help problems on Earth too. Unless the first colonies become self-sufficient and sell energy and special products back to Earth, or unless deuterium fusion energy becomes easy and widespread, there may be no more chances for local space civilization. Without space civilization, full-scale interstellar travel seems impossible to support. The crucial link is space development, done at a pace that is self-catalyzing without limit. It should leap up the scale from 10 person space station to 100 person Moon mining and test colony, to 1000 person almost self-sufficient community in space, to 10,000 people living and working together and traveling for material, and to more powers of 10 in space population until billions of people are up there enjoying life and ready to support major interstellar missions. Concurrently, to preserve resources and keep a hand in the future, Earth population must stabilize.

Low-mass probe missions would be affordable in the next century, supported by a small space population and/or by a group of determined Earth nations and institutions. If money and effort spent on military behavior were halted, the money saved in a few decades would fund some space colonies and an interstellar probe. Some nations now toss trillions of dollars around where a century ago they balked over millions. This gain by a million indicates that societal wealth grows rapidly and may grow more. In a lesson from the past, when civilization becomes rich enough, almost anything becomes politically and technically possible. A 100 person mission appears to be in the cost range of 100 trillion present dollars and up. SETI programs by radio and by probe do not seem in conflict and could be complementary. Over centuries a major SETI program could be as expensive as a probe. If a message is received from nearby (an unlikely event), the impetus for human travel would be greater.

Along with noble goals and deep interests, humans have certain nearly immutable aggressive and destructive tendencies. The latter are presently being turned increasingly inward toward violence in crowded cities and toward more self-abuse, and are still used in military actions among crowded nations and in fights among ethnic groups around the world. Neither nearby space colonies nor interstellar travel can relieve the population and economic pressures. The energy, environmental, and monetary costs of launch from Earth already seem almost prohibitive. Some say that we can no longer afford space, that we must work on problems of population control, environment, energy, housing, education, equality, crime, and health at home.

Even if these problems were fully and cooperatively addressed, one should still ask, toward what end? Is this all a young civilization can do for the rest of its time, try to hold onto a high quality of life for some and establish a mediocre but adequate quality for the rest? Restricted to one planet, humans at most can try for an ecologically stable or "sustainable" set of "lifestyles", more complex, based on different economics, probably more austere than present varied styles. At best, a garden planet with a lower stable population would result. New frontiers in space can provide direct and symbolic outlets for untamable instincts, places for adventurers to fulfill themselves and initiatives for homebound populations to support at modest levels.

Sf has made travel among the stars seem very appealing, as if any determined person backed with high technology can get his/her hands on the starship controls and in a few short (proper) years personally visit and see several different suns, gaze upon nebula and other galaxies across space, dart past a close binary star system, walk upon many varied worlds and their moons. This tourist vision is many orders of magnitude beyond the already difficult task of traveling to one near star and landing on one planet to remain there. Another fascinating but technically incredible route would be for space colonies, whole communities, to acquire sails or fusion drives and fuel and become worldlets accountable to no one else, traveling without caring where they arrive or when, making travel among the stars their destiny. Contrary to historical political and social visions, very few people would be privileged to "escape" to the stars. Nor can interstellar travel "save" Earth. Practical results from a mission would require centuries or more to reach Earth and local space.

For those fractions of humanity with the "outward urge", or the desire to find (or escape to) a better place to live for themselves and their descendants, or the need to build and use bigger more powerful technology, or a general seeking of the infinite and the eternal "out there", the news from consideration of interstellar travel at a fundamental level is that Earth and vicinity are likely to be all that is available to most for a long time ahead. Thus does an examination of the problem of interstellar travel bring out broader concerns. An era when all people have enough food and other basic needs and are truly content with life on Earth remains unforeseeable. Yet an implication from interstellar studies (outer spaces) is that more caring and respect for each other (inner spaces) and for the planet (ecospace) is necessary and urgent. As a recommendation, this holds regardless of when interstellar travel may begin and is a prerequisite to insure that interstellar programs can ever begin.

Persuading Earth societies and/or local space civilization to commit to an interstellar program, to an ongoing series of missions with long-term results, to a thousand years before learning of a starship success if at all, seems politically difficult in light of our history. In the past some societies have worked on visible projects for centuries, but recently democratic public support wavers on a time scale of months. The changing populace is not likely to keep in mind that a small group of people is out there on a thousand-year voyage. Getting a series of generations in a starship to commit to the mission and to train in the specialties that support it seems almost as difficult. These conclusions are based on human nature as it appears now. But we can do differently, and more effort should be spent on why and how we can do differently. Predicting what directions biological and cultural evolution will take our species, whether guided or sporadic, does not seem possible at present. An altered human disposition would change the prospects for interstellar travel, perhaps favorably, perhaps not.

Studying the feasibility of travel and of humans becoming an interstellar or galactic civilization leads into nearly every aspect of human capacity and culture. Studying all prerequisites, requirements, and reforms for interstellar travel seems an unending process. Many general essays and studies have been published on planning the human future on Earth [e.g. b-Brown]. Few mention the role of local space and none the significance of farther travel. None are being taken seriously enough to cause widespread response and change. While the problem of interstellar travel is closely intermeshed with our future, it is an open question whether we can expect a study of interstellar travel to make a difference in the general human condition.

Regardless of scope, study of interstellar travel shows the problem to be very complex and the laws of nature unusually severe. Scientists would be surprised now for many reasons if detailed solutions were at hand, however costly. Different readers will obtain different impressions of the technical and philosophical aspects of the problem. A goal of this study and review has been to leave the case in the hands of present and future readers with a wide range of interests and let them judge what the prospects are now and later. All options should be set out for contemplation and planning. It may seem that every conceivable fact and idea has been brought forth, with little hope of new avenues of progress. The feasibility of travel may depend on consideration of all factors together or it may depend on a new person discovering what has been overlooked. Over time, further progress in science, society, and space should show which parts of the problem are moving toward definite feasibility. The immensity of the problem should be clear now, so that premature hopes are not raised again and time is allowed for resolution of major questions.

Interstellar travel is likely to be achieved on some modest level some time, if only a low-mass probe to the nearest star system. If further, more massive travel (with people) is not achieved in the next few centuries, the problem is still well worth studying. It stretches a wide range of our scientific, technical, and social abilities. People who learn about astronomy and cosmology would wonder about the prospects for seeing the stars and galaxies firsthand and may contemplate the extent of more direct human participation in the Universe. Those who study the science and technology of space travel would be interested in what travel is realizable. Whether our progeny are on Earth or in local space, we already have the scientific, technical, biological, and social means to survive or fail over the next thousands of years. More distant space opens the possibility of reaching and surviving at other stars and eventually populating some part of our galaxy. Thus does the problem of interstellar travel link human evolution to the deepest constraints of physics and technology.

Interstellar studies as a complex branch of science and technology could serve as one of many inspiring vehicles to teach the spectrum of sciences and applied math, to introduce systems and design, to teach integrated and creative thinking, to show the interaction of technology and society, and much more. Grasping and exploring the problem can engross a student for years. Education in the broadest ways possible is one key to a future that permits interstellar missions. Without better scientific and social education we may not keep a world where such endeavors are possible. An interstellar project makes quite clear the need for common human purpose and efficient cooperation, the need to take the long view and to use one or more lifetimes to prepare each major step, the need to account for the billions of people who would be supporting such a project, and the need to keep planet Earth in good condition.

Students, leaders, and anyone who want to contribute to making interstellar travel possible 100 to 1000 years hence could do so in many ways. Directly, one could study elementary particles and related physics, plasma and fusion physics and technology, and space travel, development, and colonization. Many more kinds of educations would contribute, since very many fields pertain directly or indirectly. Many relevant sub-fields of biology, astronomy, technology, and social studies are on the list. And all people can work for restoration of intellectual respect and integrity and against further resources wasted on weapons and violent conflicts.

When the capability of travel to the stars is near to hand, then humans may ask: If we can go to the stars, will we and should we? Enthusiasts by definition will think so, but the billions of unidentified people who support the effort must be prepared to help make the decision. The long haul starts here at home now.

—JHM, Sep. 1991, Pueblo, Colorado, US, Earth.

APPENDICES

Unit System, Powers of Ten, Conversions, Constants

SYSTEM INTERNATIONALE (METRIC) AND OTHER UNITS USED, WITH SYMBOLS:

Dimension	Basic	Symbols	Subsidiary	Symbol		
length	meter	m	kilometer	km	lightyear*	LY
mass	kilogram	kg	tonne	(Mg)		
time	second	s	year*	Y		
force	newton	N				
charge	coulomb	(C)				
current	ampere	A				
potential	volt	V				
magnetic field	tesla	T				
temperature	kelvin	°K	degree	°C		
absorbed dose	gray	Gy	rad*	rad		
biological dose	sievert	Sv	rem*	rem	roentgen	
energy	joule	J	electron volt*	eV		
power	watt	w				
frequency	hertz	Hz	cycles/second*	cps		
intensity*		w/m^2	luminosity			
spectral flux		w/m^2 Hz				
speed		m/s	lightspeed*	c		
pressure	pascal	Pa	atmosphere*	atm		
plane angle	radian	rad				
solid angle	steradian	sr				

Notes: * denotes non-SI unit. Units listed as "basic" above are not necessarily official SI "base" units. Symbols not used here to avoid confusion are in ( ).
The degree symbol ° is omitted from official SI use of K. Despite its prefix k, kg is the official SI base unit of mass. Square and cubic units are shown by superscript 2 or 3 thus: m^2 for area, m^3 for volume.

POWERS OF TEN:

Common	SI prefix	Symbol	Decimal	Power of 10
trillionth	pico-	(p)		$(10)-12$
billionth	nano-	(n)		$(10)-9$
millionth	micro-	(μ)	0.000001	$(10)-6$
thousandth	milli-	m	0.001	$(10)-3$
thousand	kilo-	k	1000.0	$(10)3$
million	mega-	M	1,000,000.0	$(10)6$
billion	giga-	G	1,000,000,000.0	$(10)9$
trillion	tera-	T		$(10)12$

Notes: ( ) denotes symbols not used in book to avoid confusion.
Powers are more commonly shown as superscripts on 10; computers use the form: E-6 or EE-6. SI prefixes have been named to two steps smaller and two steps larger than shown here but are not used in this book.

UNITS CONVERSIONS,
Usually Shown From Common/English To SI/Metric:

Distance

1 mile = 1.609 km
1 parsec = 3.26 lightyear
1 lightyear = 9.4608954 $(10)15$ m $\cong$ $(10)16$ m
1 astronomical unit AU $\cong$ 150 million km = 1.5 $(10)11$ m

Speed

1 mile/second = 1.609 km/s
1 mph = 0.447 m/s

Acceleration

1 gee = 9.80665 m/s^2 $\cong$ 10 m/s^2

Mass

1 tonne = 1.102 tons

Time

1 standard Earth year = 31,558,150 seconds $\cong$ 3 $(10)7$ s

Circular/Spherical Angles

1 "arc = 7.716 $(10)-7$ of circle = 4.848 $(10)-6$ radian $\cong$ 1 millionth of circle
1° (1 °arc) = 0.0175 radian
360° = 1 circle = 2 π radians
1 radian = 1/(2 π) of circle $\cong$ 57.3°
1 steradian = 1/(4 π) of whole sphere

Energy

1 electron volt = 1.602 $(10)-19$ J
1 MeV $\cong$ 1.6 $(10)-13$ J
1 GeV $\cong$ 1.6 $(10)-10$ J
1 kwh = 3.6 $(10)6$ J

Pressure

1 atm = 1 bar $\cong$ $(10)5$ pascals

Magnetic Field

1 gauss = 0.0001 tesla

Spectral Flux

1 jansky = $(10)-26$ w/m^2 Hz

Radiation

absorbed dose 1 rad = 0.01 gray = 0.01 J/kg
biological dose 1 rem = 0.01 sievert = 0.01 J/kg

Notes: commonly used rounded values are shown also.

PHYSICAL CONSTANTS:

lightspeed c = 2.99792458 (10)8 m/s ≅ 3 (10)8 m/s

gravitational constant G = 6.673 (10)-11 in SI units

electric charge e = 1.60 (10)-19 coulomb

Planck's constant h = 6.63 (10)-34 J s

electric permittivity of vacuum ε_o = 8.85 (10)-12 in SI units

magnetic permeability of vacuum μ_o = 4 π (10)-7 in SI units

Boltzmann's constant k = 1.38 (10)-23 J/°K

Stefan-Boltzmann constant σ = 5.67 (10)-8 w/m^2 °K^4

Wien's constant a = 2.90 (10)-3 m °K

Avogadro's number N = 6.02 (10)23 atoms per mole

proton mass = 1.67 (10)-27 kg ≅ 931 MeV ≅ 1 GeV (as mass-energy)

Notes: When working entirely with SI units, it is not necessary to know the units of constants to do compatible calculations. The accuracy given here depends on the importance of that accuracy for interstellar studies and does not necessarily reflect full known accuracy.

Classical Physics

This mathematical physics supplement for Chapters 1 & 2 is keyed by section number and written for brevity. Subjects are primarily Newtonian mechanics and electro-magnetism for the low-speed regime $v \lesssim 0.2c$. Calculus is used as needed. Where the order of mathematical operations seems in doubt, use standard calculator or computer format (space instead of * denotes multiplication). For more detail see any introductory college physics textbook. The advanced reader can solve almost any type of propulsion condition starting from Newton's law, and in more than one dimension, using numerical methods. Since there are not enough alphabet characters to go around, the same letter is used for different quantities where unavoidable and in conformation with the widest use in physics books. For units, conversions, and physical constants see Appendix A. Powers of 10 are shown e.g. (10)-13, without superscript. All additive terms of equations should have the same units of measure.

Throughout the book estimates of area and volume often ignore pi (π), thus assuming circles have about same area as squares (actually over-estimating by about 4/3), and spheres have about same area as cubes (actually over-estimating by about 2).

1.1 NEWTONIAN (CLASSICAL) MECHANICS

Newton's law: $F = dp/dt$, where F = force on system of mass m, momentum $p = m\,v$. If m constant (often not with rockets), $F = m\,a$. In general m, a, and v vary in time t and position x. Motion is found by solving for v and x (or r) as functions of time. When more than one dimension is involved, F and dp/dt or a are in same direction, v and x not necessarily.

When a is constant (rarely with rockets but sometimes for other propulsion), the solutions for motion are: $v(t) = a\,t$ and $x(t) = (1/2)\,a\,t^2$. Time can be eliminated and v found by $v^2 = v_0^2 + 2\,a\,x$, where v_0 = initial speed.

1.2 ROCKET WITH CONSTANT THRUST AND NONLINEAR ACCELERATION; ROCKET EQUATION; MOLECULAR SPEED

Newton's law becomes: $M\,dv/dt = -v_e\,dM/dt$, where M is mass of rocket, v_e is constant exhaust velocity, and dM/dt is rate of reaction mass ejection. Note dM/dt must be negative and is commonly constant, although this derivation works regardless of how mass varies in time. Eliminate time and solve $dv = -v_e\,dM/M$ to get Rocket Equation: $v/v_e = \ln(M_0/M')$, where ln is natural logarithm, M_0 is initial mass, and M' is final mass (usually payload). Note $M_0 = M_f + M'$, where M_f is mass of fuel consumed. For exponential form, convert solution to $M_0/M' = \exp(v/v_e)$. The natural logarithm and exponential appear in any problem where the change in something is proportional to the something. Apply to stages by adding speeds and multiplying mass ratios: $v_1 + v_2 + v_3 = v_e \ln[(M_0/M_1)(M_1/M_2)(M_2/M_3)] = v_e \ln(M_0/M_3)$. For rocket motion see 1.3

Speed of gas molecule $v_{rms} = \sqrt{(3kT/m)}$, where k is Boltzmann constant, T is temperature °K, and m is mass of molecule. v_{rms} is the kind of average speed that best represents molecular energy.

1.3 ROCKET MOTION WITH VARIOUS CONDITIONS

Constant acceleration and decreasing mass: Constant a_0 is an unusual condition in space travel and requires that mass not decrease linearly if mass changes. Newton's law gives M dv/dt

= M a = dM/dt v_e, where dM/dt is nonconstant mass ejection rate. The equation becomes a_o dt = v_e dM/M and has solution: M(t) = M_o exp(-a_o t/v_e). This works until burnout time t_b = (v_e/a_o) ln(M_o/M'). The motion is as before with constant a_o. Note time constant v_e/a_o. Note thrust F(t) = M(t) a = dM/dt v_e.

Constant thrust and linearly decreasing mass: Thrust is F_o = μ v_e, where μ is the constant rate of mass ejection and interpreted as a positive number. Rocket mass decreases linearly in time thus: M(t) = M_o - μ t until burnout time t_b = (M_o - M')/μ. Newton's law gives M(t) dv/dt = -v_e μ and has solution: v(t) = v_e ln[M_o/(M_o - μ t)]. Acceleration is: a = v_e μ/(M_o - μ t). Distance x(t) can be found analytically but is too complex to be useful. A computer program in Appendix E illustrates numerical solutions by giving one for motion with gravity included for vertical launches from planets.

1.4 GRAVITY AND CIRCULAR MOTION

Weight W is force on mass m due to gravitational acceleration g thus: W = m g. g is 9.8 m/s^2 or 1 gee at Earth's surface and can be calculated from gravity force law F = G M m/r^2, where G is gravitational constant. For g use r as radius of Earth, M as mass of Earth, and m drops out.

The Kepler problem is the general motion of mass m in the gravitational field of mass M, or both around their common center of mass. See [b-Goldstein] for analytic solutions for both orbit and position of m in orbit as function of time. The resulting Kepler laws include: orbits must be conics with M at one focus, and the period T is related to the average orbit size (semimajor axis) a by: T^2 = 4 π^2 a^3/G M. For more accuracy M must be the sum of central mass and orbiting mass (else a Jupiter-size planet brings approx. 0.1% error).

Circular orbit requires inward force m v^2/r equated to gravity force G M m/r^2. The speed that satisfies this equation is v_c = $\sqrt{(GM/r)}$. For low Earth orbit use r as radius of Earth. Centripetal acceleration v^2/r can be used for any fraction of a path that is circular, regardless of inward force.

1.5 ENERGY

Kinetic energy is defined: KE = (1/2) m v^2, where v is speed. Note that if v increases with time, then KE increases more strongly with time.

Gravitational potential energy GPE is found by integrating the work done to push mass out of gravity well or to find what KE it acquires when it falls into gravity well. For falling in, do $\int$F dr = $\int$(-G M m/r^2) dr with the limits of r from infinity to r. Result: GPE = -G M/r. Note GPE = 0 when r goes to infinity. Avoid considering GPE when r goes to 0, and stop at planet surface. For altitudes h just above planet surface GPE $\cong$ m g h, where h must be much less than planet radius.

For escape velocity (speed) v_e, equate KE needed at position r to escape to GPE at that position so that total energy TE = KE + GPE = 0. Result and relation to circular orbit speed: v_e = $\sqrt{(2 G M/r)}$ = $\sqrt{2}$ v_c.

Efficiency of rocket: Efficiency at any time is found by letting energy added be split between rocket and fuel: dE = dKER + dKEF. Fraction of energy received by rocket is f = dKER/(dKER + dKEF). Let M and v be rocket mass and speed, m and v_e be fuel mass and speed. Differentials give: f = M v dv/(M v dv + dm v_e^2/2). Use Newton's law dv = dm v_e/M to eliminate dv. Then f = 1/[1 + (v_e/2 v)]. f = 2/3 when v = v_e. f approaches 1 when v >> v_e, an absurdity since Rocket Equation prevents large v from given v_e. f and efficiency go to zero when v goes to zero.

Efficiency overall gets larger but does not approach 100% when v is large because so much fuel has been wasted getting there. There is a maximum in the overall efficiency η as shown in references [a-von Hoerner; a-Oliver]. This overall η is found by comparing final KE' with available KE0 in the fuel, where KE0 is defined as that part of fuel energy which could all be released into motion. The efficiency function becomes: $\eta = M' v^2/(M_0 - M')v_e^2 = (v^2/v_e^2)/[\exp(v/v_e) - 1]$ after the Rocket Equation is used to eliminate masses. This function has a maximum at $v \cong 1.6\ v_e$ or $v_e \cong 0.63\ v$, at which point it reaches about 65% efficiency. η goes to zero if speed is wanted $v >> v_e$. If v_e and final v can be chosen for the most efficient case, then the Rocket Equation gives a nice result for mass ratio: $M_0/M' \cong e^{1.6} \cong 5$.

1.6 GRAVITATIONAL SCATTERING

An object with mass m and speed v aimed distance b (impact parameter) to the side of the center of a star or planet of mass M, while far from it, will be deflected by angle θ after it has approached straight line travel after the encounter. The angle is given by: $\theta = 2 \tan^{-1}(G\ M/v^2\ b)$ [after b-Goldstein]. The angle is in degrees if the function used makes it so.

2.2 ROCKET/STARSHIP WITH THRUST PROPORTIONAL TO $1/r^2$:

Rocket mass M is assumed constant here for a situation where diminishing sunlight pushes on a sail (see 2.8) or negligible mass is expended in a reaction drive. Thrust decreases with r as power input decreases thus: $F(r) = +K/r^2$, where K is a constant to be found for the particular source (e.g. F is known at some r). This is very like gravity except F pushes outward. (The same math can be used for gravity when the starship falls directly toward the source of gravity without orbiting.) When a function of time is not available for Newton's law, convert it to space coordinate thus: $M\ dv/dt = M\ dv/dr\ dr/dt = M\ dv/dr\ v = K/r^2$. Solution is: $KE(r) = M\ v^2(r)/2 = M\ v_0^2/2 = -K\ M/r + K\ M/r_0$, from which $v(r)$ can be found. Initial condition is $v = v_0$ at $r = r_0$. Dependence on mass drops out. As r goes to infinity, KE approaches $M\ v_0^2/2 + K\ M/r_0$.

$r(t)$ can be found by integrating $dt = dr/v(r)$, but the result is not simple. $r(t)$ can also be guessed by assuming it to be a simple power of time such as $r = h\ t^n$. Newton's law gives $n = 2/3$. Therefore $v(t)$ is proportional to $t^{1/3}$, and $a(t)$ to $t^{-4/3}$. However, only partial solutions can be pieced together. The solution can be good only beyond some point r_0 after some time t_0.

2.3 THRU 2.5 POWER

Definition of power P is rate of flow of energy: $P = dE/dt$. Power to starship motion when no potentials are present and mass is constant is: $P = dKE/dt = M\ v\ a = F\ v$ at each instant, even when v, a, and thrust F are not constant. KE is that gained by the starship. Note that as v increases, more power is required to keep constant thrust. When mass decreases over time, the definition must expand to $P(t) = M(t)\ v\ a + v^2/2\ dM/dt$, where dM/dt is negative. Power, like energy, is conserved moment by moment. If P_0 is constant input power from onboard source (reactor) or received via solar collector, then: $P_0 = P + P_e + P'$, the division into starship, exhaust, and waste parts. A constant exhaust has mass flow but takes away constant power $P_e = F\ v_e - \mu\ v_e^2/2 = \mu\ v_e^2/2$. If the source provides more power than the starship can "accept" through its momentum drive, the difference goes into waste or exhaust, or cannot be drawn. Note: $P_0 > P_e$.

2.5 MAGNETIC FIELDS AND FORCES

Magnetic field B (in tesla) near center of circular loop of wire of radius R of current I is $B = \mu_0\ I/2\ R$, where μ_0 is the permeability of space. B is stronger near the wire. If loop has n turns, multiply result by n, as if n I amperes were flowing. If loop is spread out to be a solenoid with n turns in length L, then $B = \mu_0\ n\ I/L$. Field inside is uniform constant Bm. Far from loop or solenoid at distance r, field decreases as $1/r^3$.

Direction of B given by right-hand rule: If curled fingers showing direction of positive current (opposite to electron flow), then B points with thumb. Think of B as an array of lines in space, either straight at center of a large coil or at end of a magnet, or curved in a bunch around a wire.

Circular magnetic field around a straight wire carrying I is given by $B(r) = \mu_0 \, I/2\,\pi\,r$. If right-hand thumb points along wire, curled fingers tell direction of B. This field exerts a force on the flowing charges producing it so as to constrict them. Thus a current in a plasma self-pinches the plasma.

To find radius r of orbit of charged particle (charge q, mass m) moving at speed v at right angle to magnetic field B, equate force for circular orbit to (constant) Lorentz force thus: $m\,v^2/r = q\,v\,B$, where q is charge of particle. Result: $r = m\,v/q\,B$, also called gyration or Larmor radius. Positive charge follows right hand rule for motion, where thumb points in direction of B. Note charge:mass ratio (coulombs/kilogram) appears and is an alternative fundamental constant for a particle.

The gradient of magnetic field in direction z along the axis of a magnetic mirror (with increase in intensity of field dB/dz) is found [in advanced texts such as b-Jackson; b-Reitz] from $dB/dz \cong 2\,B0/v_t^2 \; dv_z/dt$, where v_t is typical speed transverse (perpendicular) to z axis, $dv_z/dt \cong 2\,v_z/t$, and B0 is background field in region. v_z is typical speed along z axis, same in all directions, and t is the time to travel along z axis from central region to end zone. $2\,v_z/t$ is approx. acceleration required to reverse particle motion. B must increase from B0 in central region to maximum Bm at end by amount dB in distance dz.

2.6 MOVING CHARGES (ELECTRODYNAMICS)

Current in amperes is a flow of charge $I = dq/dt$, where q is charge. Electric force F on charge is $F = q\,E$, where electric field E can be found from change in potential over distance, dV/dx, in volts per meter. Electric field points away from + charge, and potential V is higher nearer to + charge. Electric potential energy $EPE = q\,V$. If a charge is pulled by $F = q\,E$ through a potential V, it acquires $KE = EPE$ and its speed can be calculated.

Electric power P is: $P = I\,V$.

Lorentz force on charge q moving perpendicular to field B is: $F = q\,v\,B$ and can be rewritten as: $F = I\,L\,B$, where L is the length of wire carrying current I. (One cannot ignore how current gets to that piece of wire.)

2.7 MAGNETIC DRIVER

Two wires or coils carrying current in the same direction attract; opposite directions repel. Opposite sides of a loop tend to repel, expanding the loop or coil. Adjacent turns of a coil tend to attract, squashing the coil that way, turn to turn.

Magnetic force between two wires, loops, or coils carrying currents I and I' is given by: $F = \mu_0\,L\,I\,I'/2\,\pi\,d$, where L is the adjacent length, and d is the distance between the wires. The wires can be straight or curved (the same) or circular. For multiple turns, consider the aggregate current as if one turn. Also note that adjacent turns within one coil sharing same current attract. For circular coil, L is its circumference. Note that for two coils separated by distance approximately the same is their size, $L/2\,\pi\,d$ is approximately 1 and has little effect. Force still depends inversely on separation and is NOT an inverse square force.

2.8 LIGHT PRESSURE; MATERIAL STRENGTH; SAIL MOTION; SOLAR MILL

Momentum p of light is found from energy E of a photon by $p = E/c$ and from its wavelength by $p = h\,\lambda$, where h is Planck's constant. Energy of photon can be found from its frequency f by $E = h\,f$ (or from wavelength by using $c = \lambda\,f$ to convert.) Number of photons n in a light beam of intensity I (w/m^2) shining through or on area A is the total power divided by the energy per photon per second, or $n = I\,A/h\,f$.

Light intensity I is assumed proportional to stellar luminosity for stars (with Sun as reference). Light pressure (in N/m^2) from totally reflecting photons is $P = 2\,I/c$, and from totally absorbed ones is $P = I/c$. Force from pressure is $F = P A = 2 I A/c$. If reflectivity is ζ, a number between 0 and 1, preferably near 1, then: $F = (1 + \zeta) I A/c$. For total absorption, $\zeta = 0$ and F is half that for perfect reflection.

Material strength is expressed as the stress, or force per unit area, thus: $F/A = \sigma$, where σ is the ultimate tensile strength (preferably measured), or the lesser measured yield strength, characteristic of a given material in a given form, in N/m^2. σ should not be confused with Young's modulus, a larger value which expresses the elasticity of the material, the stress needed for a given proportion of elastic stretch [b-Popov].

Mass of a cable or other object can be found from volume times mass density ρ thus: $M = \rho L A$, where L is length and A is cross section area.

If the force on a sail or other object of mass m is proportional to inverse square of distance as $F = k/r^2$, then the speed acquired after this force acts over a distance is found by integrating Newton's law to get $v(r) = \sqrt{[(2/m) (k/r_o - k/r)]}$, where r_o is the starting position and r the position of interest. k can be found by knowing the force at the starting position r_o. Inverse square gravity could be added to this problem. Indeed, this result can be applied to gravity alone if $k = G M m$ where M is mass of sun or planet.

Angular speed ω of an orbiting or rotating object (e.g. solar windmill blades) is found from frequency f or period T as: $\omega = 2 \pi f = 2 \pi/T$, and from (tip) speed v and radius r as: $\omega = v/r$, and is measured in radians per second. Torque τ, or rotational force, is found from $\tau = F r$, where force F acts at right angle to radius r, and has units newton-meter (never joule). For sunlight pressure on a vane, the total force acts effectively at the halfway point. τ is related to power P by: $P = \tau \omega$. In terms of angular speed, the force F_c needed to hold the object of mass m in circular motion is: $F_c = r \omega^2 m$.

Relativistic Travel

Some mathematical details are given here for basic results in relativistic travel (v $\gtrsim$ 0.2c), keyed to chapter and section. Again, math expressions are to be read the way a computer would read them (with * omitted in lieu of a space for multiplication). Again, there are not enough roman and greek alphabet characters for all needed symbols. The same character is used for different concepts where unavoidable and in conformity with wide usage in physics books and journals.

Subscript 0 denotes a proper (rest frame) or other constant quantity. Being constant refers to in time only. Prime ' denotes the moving starship frame S' and its variables x', v', a', and t'. τ, another commonly used other symbol for proper time, is equivalent to t'. S' must start coincident with S at t = t' = 0 and x = x' = 0. S' moves at speed +v with respect to (and as measured in) S. Coordinates y and z are rarely needed here. Users should watch that units of measure are consistent. Referencing is minimal; see also Chapters 3 & 4.

3.2 GAMMA

The gamma factor γ which appears throughout relativity is a dimensionless number which shows how variables are changed between frames. It is defined: $\gamma = 1/\sqrt{[1 - (v^2/c^2)]}$. Using γ saves much writing of this cumbersome expression. Calculus operations with it are easier than appears, especially if a table of derivatives and integrals is prepared beforehand. When v is needed from γ, v is given by: $v = c\sqrt{[1 - (1/\gamma^2)]}$. When v is small γ is approximated by: $\gamma \cong 1 + (v^2/2\,c^2)$.

Length contraction: $L = L_0/\gamma$.

Time dilation: $t = \gamma\,t_0 = \gamma\,\tau$. Frequency change due to time dilation alone: $f = f_0/\gamma$.

Mass augmentation: $m = \gamma\,m_0$.

Regular doppler shift for light (no time dilation): $f = f_0 (1 \pm v/c)$. Use + for approach, − for recede.

Combined relativistic doppler shift: $f = f_0 \sqrt{[(1 \pm v/c)/(1 \mp v/c)]}$, where one of each sign is chosen as before. A somewhat more complex result is available for the shift as seen for any direction of travel with respect to the observer [e.g. b-French].

3.3 OPTICAL EFFECTS

General doppler shift for a source seen moving at angle θ with respect to starship velocity is given by: $f = f_0 \sqrt{[(1 - v^2/c^2)/(1 - v\cos\theta/c)]}$. Note $\theta = 0$ is approach, and $\theta = \pi$ is recession.

Relativistic aberration: For a star observed at angle θ in S from the direction a starship is headed at v, aboard a starship S' it is observed at θ' calculated thus: $\cos\theta' = (\cos\theta + v/c)/(1 + (v/c)\cos\theta)$. [b-French]

The intensity I of its light seen at angle θ' as compared to the intensity $I(\theta)$ in S is given by: $I(\theta')/I(\theta) = 1/\gamma^2 [1 + (v/c)\cos\theta]^2$. [a-Weisskopf] Intensity is compressed to an angular width approximately $1/\gamma$ in radians.

3.5 RELATIVISTIC MOTION

It is easy to lose track of reference frames in Newtonian analyses of rocket motion, where attention jumps back and forth between what happens on the starship and what is seen from outside. Solutions were obtained without regard to the frame in which a particular law was applied. Now greater care must be taken. The hyperbolic functions sinh, cosh, and tanh, often seen in tables and on calculators, are well used in relativity. Analytic solutions involving these appear in research papers but rarely in textbooks on relativity. Every solution must start with some physical law, often Newton's or relativistic equivalent. Problems without analytic solution can still be solved numerically by advanced readers. (In a possibly subtle surprise, the frequent γs in relativity often do not prevent analytic calculation; i.e. Einstein's math is doable.)

Speed and acceleration as seen in S (Earth frame) can be found in terms of proper time [a-McMillan; b-Belinfante]. Few sources of this derivation are known, and some detail is given here. Constant acceleration a_0 is assumed as measured aboard starship frame S' and is the physical input. This is an unnatural case, and fuel mass if any cannot be consumed linearly. It would be consumed more rapidly except that the remainder becomes easier to accelerate. Fortunately mass consumption need not be known to solve for the motion. The mass is found to decrease exponentially in time by the Rocket Equation later.

Analysis must start in S'. Frame S' moves at constant speed with respect to S and is momentarily alongside the starship. Acceleration a_0 adds speed $dv' = a_0 dt'$ to starship speed which was $v' = 0$ in S'. dv' can be transformed to S by $dv' = (v + dv - v)/[1 - (v + dv) v/c^2]$. In S we expect speed $v + dv$. The velocity transform is derived in textbooks from Lorentz transform and is given later. Another approach is simply to take differential of velocity transform. The needed result is $dv' \cong dv/[1 - (v^2/c^2)] = \gamma^2 dv$, the transform of an increment of velocity in S' to S. Another way to obtain this result is to note that in S the length units of v are contracted by γ and the time units of t dilated by γ, with net result γ^2. Thus $\int[a_0 dt'] = \int[\gamma^2 dv]$ results in $a_0 \tau = (c/2) \ln[(1 + v/c)/(1 - v/c)]$ from integral tables. Or $\tau = (c/2 a_0) \ln[(1 + v/c)/(1 - v/c)]$ gives proper time as a function of speed. Since this natural log function is the inverse tanh function, the final result for speed as a function of proper time is: $v(\tau) = c \tanh[a_0 \tau/c]$. Since tanh approaches 1 as its argument approaches infinity, the expected behavior of v is to approach c as τ approaches infinity.

Time in S and proper time are related at the differential level by time dilation: $dt = \gamma dt' = \gamma d\tau$. It is possible to get γ as a function of τ so that this can be integrated. v/c as a function of τ can be plugged into γ to get $\gamma = \cosh[a_0 \tau/c]$. Then $t = \int \cosh[a_0 \tau/c)] d\tau = (c/a_0)\sinh[a_0 \tau/c]$. Note that dt must be proportional to a cosh function. Using an inverse sinh, τ can be found in terms of t thus: $\tau = (c/a_0) \sinh^{-1}(a_0 t/c)$, or using another form $\tau = (c/a_0) \ln[(a_0 t/c) + \sqrt{[(a_0 t/c)^2 + 1)]}]$, which looks no simpler (unless $a_0 t/c$ is large).

Distance x can be found by $x = \int v \, dt = \int c \sinh[a_0\tau/c] \, d\tau$ after the identity "tanh = sinh/cosh" is used. The result is distance as a function of proper time: $x(\tau) = (c^2/a_0)[\cosh(a_0 \tau/c) - 1]$.

In the nonrelativistic limit, with $a_0 \tau/c << 1$, τ becomes t, and results for v and x reduce to their newtonian forms. (See power series for tanh and cosh.)

The term $a_0 \tau/c$ works out to be 1.032313 A_0 T when A_0 is expressed in units of gee and T is expressed in proper years (PY), to a limit of accuracy as determined by the value of 1 gee. c^2/a_0 works out to be 0.9687 LY when a_0 is 1 gee, so that x can be found easily in LY as X. Likewise $c/a_0 = 0.9687$ years so that time can be found in Earth years. v has been found in units of c, and all quantities here can be worked with in simple "interstellar" units. For large X, T is: $T \cong \ln(2 X)$. Appendix E gives one program to calculate motion numerically in terms of proper time as above and is the source of the table of results in Appendix E.

Speed and distance in terms of Earth time (S) can be found by similar methods [a-von Hoerner]. As before, the time parameter permitting simplest mathematics is t' or γ. Starting on the starship in S' again, $dv'/dt' = a'$. This time dt' is converted to dt by time dilation, and dv' is transformed as before to give: a_0 dt = γ^3 dv (correcting a probable misprint in the reference). Integrating gives a_0 t = v γ, and rearranging (never fun when γs are around) gives: v = a_0 t/$\sqrt{[(a_0 t/c)^2 + 1]}$. Note that v approaches c as t approaches infinity. Distance is found by integrating v dt to give: x = $(c^2/a_0)[\sqrt{(1 + (a_0 t/c)^2)} - 1]$. In the nonrelativistic limit, the results for v and x reduce to their newtonian forms. For x, the squareroot of 1 + o, where o $<<$ 1, must be approximated by 1 + o/2.

When v is very near c, approximation can be used to find the difference from c rather than going to double digit computation (15 places). For v/c = tanh(a_0 τ/c), τ can be eliminated by using the solution for distance in terms of the cosh of τ. An identity tells us that $1/\cosh^2 + \tanh^2 = 1$. Combining all this, dropping small terms, and approximating a squareroot gives: v $\cong$ c (1 - $c^4/2 a_0^2 x^2$). Along the way an always handy expression relating a_0, x, and v was obtained: v/c = $\sqrt{[1 - (1/(a_0 x/c^2 + 1)^2)]}$. Other forms of this are possible but not simpler. This same expression can be used to obtain an exact way to calculate gamma: γ = 1 + a_0 x/c^2. When γ is near 1 the small difference from 1 is easily found from the second term. When γ is large, the 1 can be ignored. For times longer than several years, γ can be approximated as: γ $\cong$ a_0 t/c, for a_0 $\cong$ 10 m/s^2. When A $\gtrsim$ 1 gee and X $\gtrsim$ 10 LY, the sinh and cosh functions can be approximated with the exponential 1/2 exp(a_0 τ/c) to give simpler forms for t and x as functions of τ. v has been approximated above.

The preceding assumption of constant acceleration is not usually realistic, since fuel mass is used nonlinearly. Constant thrust is realistic if the rate of mass ejection and the exhaust velocity are both constant or if one compensates the other. The motion for a rocket with constant thrust F_0 which consumes fuel mass at a linear rate μ has been solved in two references [a-Anderson 1971; a-Kooy; note also a-Pomeranz]. These sources differ on the answer, and [Anderson] appears to have omitted one γ, causing major changes in the solution presented (suppression of all constants also prevents general usefulness). The other [Kooy] appears correct and is done for very general variation in mass.

Here mass is assumed to decrease linearly according to M(τ) = M_0 - μ τ, where μ is a positive constant, the mass ejection rate. Newton's law becomes dv'/dτ = F_0/M(τ) in S'. When this equation is looked at from S, dv' is transformed as before (mass and force are not) to give: dv/dτ = F_0/γ^2M(τ). After integrating the partial solution is c tanh^{-1}(v/c) = - (F_0/μ) ln [M(τ)/M_0]. The tanh^{-1} side is equivalent to (c/2) ln [(1 + v/c)/(1 - v/c)]. F_0/μ can be replaced with v_e. The equation can be solved for v as a function of τ to get: v(τ) = $c[((M(\tau)/M_0)^{2v_e/c} - 1)/((M(\tau)/M_0)^{2v_e/c} + 1)]$. This resembles the rocket equation except M is now a function of τ and the exponents of 2 now carry information about thrust. This result holds until burnout time t_b = (M_0 - M')/μ, where the prime on M' is not to be confused with a frame prime. Like all relativistic results this one must give zero speed at zero τ and speed c as τ in this case goes to burnout time giving ln 0. The solution for x would be arduous to obtain. To see the nonrelativistic limit, note that ln[(1 + v/c)/(1 - v/c)] $\cong$ ln[1 + 2v/c] $\cong$ 2v/c for small v.

For constant thrust and linearly decreasing mass, v(t) in S frame time can be found from Newton's: γ^3 dv/dt = F_0/γM(t). Integrating: $\int\gamma^4$ dv = F_0 $\int$dt/M(t). Using F_0/μ = v_e, this equation becomes: v γ^2/2 + (c/4) ln[(1 + v/c)/(1 - v/c)] = v_e ln[M_0/(M_0 - μ t)]. Making everything into a logarithm with exponents and then exponentiating to eliminate logarithms, the equation becomes: $[M_0/(M_0 - \mu t)]^{v_e}$ = exp(v γ^2/2) $[(1 + v/c)/(1 - v/c)]^{c/4}$. Note all powers are speeds and therefore compatible. The powers can be rearranged to put v_e on the right as a power, but no simplification occurs. v(t) is unavailable explicitly. Time is limited to the burnout time as before. This result for v(t) reduces to the nonrelativistic result by noting that $[(1 + v/c)/(1 - v/c)]^{c/4v_e}$ in the limit as v/c goes small and with rearrangement of exponents becomes $[[1 + 2 v/c]^{c/2v}]^{v/2v_e}$, which is the definition of the exp function to the power v_e/2v. γ^2 becomes 1 and the mass ratio becomes equal to exp(v/2 v_e) exp(v/2 v_e) = exp(v/v_e). Obtaining the Rocket Equation here with variable mass confirms the relativistic result.

Solving for relativistic motion when power and therefore thrust decreases as $1/r^2$ is useless because no drive would stay near the Sun long enough to reach relativistic speed. More generally, input energy and power divide up as before to exhaust, starship, and losses. Transforming frames provides a new feature. Aboard (at any speed) one does not know one is gaining speed or energy despite feeling acceleration. The KE of the starship is "seen" only from frame S.

3.6 RELATIVISTIC ROCKET EQUATIONS AND EFFICIENCY

The rocket has mass ratio M_0/M' to start, and fuel mass is assumed totally converted to energy. Exhaust velocity is fraction f of c, or $v_e = f c$. The simplest form comes from integration of Newton's law [b-Goldstein; a-Oliver; a-Kooy]: γ^2 m dv = - v_e dm to give $M_0/M' = \exp[a_0 \tau/v_e]$, the same as the Newtonian form where v = $a_0 \tau$. If $v_e = c$, the ratio becomes approximately: $M_0/M' \cong 2 \gamma$. With T in proper years and A_0 in gees, the ratio is exp[1.0323 A_0 T]. The ratio rises rapidly after speed exceeds about 0.8c. The mass remaining at time τ is: $M(\tau) = M_0$ exp (-a τ/v_e), until burnout.

Velocity can be found from the ratio thus: $v = c[(1-(M_0/M')^{-2f})/(1 + (M_0/M')^{-2f})] = c[((M_0/M')^{2f} - 1)/((M_0/M')^{2f} + 1)]$. For large ratio this approximates to: $v \cong c[1 - 2(M'/M_0)^{2f})]$. The reference [a-von Hoerner] gives several relations and approximations for τ, t, x, and v in terms of mass ratio for photon drive using f = 1. Given desired final speed v, the required mass ratio can be found by: $M_0/M' = [(1 + v/c)/(1 - v/c)]^{1/2f}$.

A generally useful and complex formula for relativistic rocket efficiency η for constant exhaust velocity is found in two references [a-Marx; a-Oliver with nice graphs]. From comparing starship KE to fuel mass-energy (RE), efficiency is: $\eta = (\gamma - 1)/[(M_0/M' - 1)(1 - \sqrt{(1 - f^2)})]$. As with non-relativistic rockets, efficiency η starts at 0.65 but becomes nearly independent of speed, falling to 0.5 at c. For photon drive, rocket efficiency becomes: $\eta = 1/2 (1 - M'/M_0)$. Note that photon drive η is less than 1/2 and falls to zero for low speed.

3.7 TRANSFORMATIONS

The Lorentz transform is given here for change from S' to S: $x = \gamma (x' + v t')$ and $t = \gamma (t' + v x'/c^2)$. S' moves at speed v in x and x' direction as viewed from S. y and z coordinates are ignored here, and they do not change between frames. To change from S to S' use -v. It should be apparent that length contraction and time dilation are derived from the Lorentz transform with further careful analysis. In effect, the Lorentz transform applies γ to any coordinate going either way between frames in a symmetry that pervades relativity and shows that either S or S' sees the other frame with length contracted and time dilated.

When velocities are "added" or transformed, the velocity of interest in a frame is called u and distinguished from the relative velocity v between frames. u is found in S by: $u = (u' + v)/(1 + v u'/c^2)$. [See references for the different velocity addition in other directions.] To go from S to S', use -v.

Acceleration is transformed from S' to S by: $a = a'/[\gamma^3 (1 + v u/c^2)^3]$. The transform in other directions is more complex. If u' is zero, a is simply given by: $a = a'/\gamma^3$.

Force in the x direction is not changed by transformation, provided that there is no speed u in S'. Otherwise a power term appears.

At a simple level, electric force and field are transformed (reduced) by γ^2. More precisely, a moving electric field is found to be a magnetic field. Details are not needed in this book and are found in any advanced electromagnetic text.

It is often handy to differentiate a transform to accomplish a calculation. γ must not be ignored during differentiation; it contains v which is often not constant.

3.8 ENERGY AND MOMENTUM

Total energy TE = KE + RE = $(\gamma-1) m_0 c^2 + m_0 c^2 = \gamma m_0 c^2$. If KE and RE are known, then KE = $(\gamma - 1)$ RE, and $\gamma = 1 + (KE/RE)$.

Momentum is defined as $p = \gamma \, m_0 \, v$. Henceforth TE is called energy E here.

Momentum and energy are mixed together in relativity and transform thus from S' to S: $p = \gamma \, (p' + v \, E'/c^2)$ and $E = \gamma \, (E' + v \, p')$. These are for the x- component of momentum. The other components do not change. To go from S to S', use -v. Note that energy in S' acquires an additional part when viewed from S. If an increment of speed dv' is added in S', then in S an extra energy $v \, m_0 \, dv'$ appears. γ also applies.

Total energy is related to momentum p and RE like a vector: $E^2 = p^2 \, c^2 + m_0{}^2 \, c^4$. For photons there is no rest mass and $E = p \, c$.

4.1 DIFFRACTION OF LIGHT

For a circular source, full angular width α of Airy disk (to first minimum) is: $\alpha \cong 2.44 \, \lambda/D$, where D is diameter of source and λ is wavelength [b-Hecht]. This same angular width (in radians) is $\alpha = d/x$, where d is diameter of spot seen at distance x from source. Most of the light intensity comes within half this size. A fresnel diffraction lens with every other section opaque or filled with half-wave retarding material is discussed in [b-Hecht; a-Kirz].

4.3 PHOTON DRIVE

Stefan-Boltzmann radiation law: $I = e \, \sigma \, T^4$, where I is the total power radiated per unit area of source, e is emissivity between 0 and 1, σ is Stefan-Boltzmann constant, and T is temperature in °K. Wavelength at which most radiation occurs is $\lambda = a/T$, where a is Wien's constant [b-Hecht]. Pressure produced by radiation is $P = \sigma \, T^4/c$, and thrust is $F = P \, A$. Perfect emissivity is assumed here. A is area of radiator.

4.5 ANTI-MATTER

Energy available in center of mass frame S' in collision is found by $E' \cong \sqrt{(2 \, E \, m)}$, where E is energy in lab frame S and m is rest mass of particle. All are measured in energy units here such as GeV. The approximation is that E is significantly greater than m.

4.6 ANTI-MATTER PROPULSION

For a complex reaction such as p $\bar{\text{p}}$ annihilation used in a momentum drive, a more realistic equation for effective exhaust velocity has been found that can be applied to almost any rocket drive [a-Huth]. Although the result appears complex, relatively simple derivation incorporates these effects (explicit p $\bar{\text{p}}$ application given in parenthesis): the fraction ε of RE that goes to KE of exhaust (1/3); the fraction η of fuel mass exhausted from which energy was first extracted (all of it, or 1); the dilution factor α of extra reaction mass added to active fuel mass, a number from 0 to very large (how much extra hydrogen is added to the p $\bar{\text{p}}$ mixture); and the fraction ω of energy released which does not go to thrust (about 1/3 representing neutral pion loss). The result is: $v_e/c = \sqrt{[1 - [(\alpha + (1 - \varepsilon) \, \eta)/(1 + \alpha - (1 - \varepsilon)(1 - \eta) - \omega \, \varepsilon)]]}$. For a simpler rocket drive with no dilution, no waste, and all fuel mass used, but with only a fraction ε of fuel mass converted to energy, this result reduces to a simple alternative way to estimate effective exhaust velocity: $v_e = c \sqrt{[2 \, \varepsilon - \varepsilon^2]}$. Since $\varepsilon = $ KE/RE, this result picks up a small correction term over a previous such result.

4.7 GAS, DUST, AND DRAG (BRAKING)

Dust/atom collision rate (hits/second) with area A is $\gamma \, A \, v \, n$, where v is relative speed of dust/atom, and n is number density in space ($\#/m^3$).

Drag or braking force comes from Newton's law with force opposite to direction of (changed) momentum: $F = dp/dt \cong \gamma \, \rho \, v^2 \, A$, where ρ is mass density in space (kg/m^3). This is for particles which stop. For rebounding particles, double F. For particles which strike glancingly,

F is less. A particle initially at rest struck by massive starship M at speed V (interacting with its structure by any non-dissipative force) acquires speed ~2 V in direction of starship, and starship loses speed $dV \cong 2 \, m \, V/M$.

4.8 ELECTRIC AND MAGNETIC FIELDS

ε_0 is the permittivity for empty space, telling how much electric field there can be from charge q.

Capacitance of any object is $C = Q/V$ farads, where Q is total charge and V is potential in volts. Also found from geometry thus: For a long wire $C \cong 2 \, \pi \, \varepsilon_0 \, L$ where L is length. For two plates $C = \varepsilon_0 \, A/d$, where A is area and d is separation.

Electric field at distance r from point or spherical charge Q is: $E = Q/4 \, \pi \, \varepsilon_0 \, r^2$, pointing radially outward from + charge.

Electric field from sheet of charge is: $E = Q/2 \, \varepsilon_0 \, A$ pointing both ways from positive sheet. E reaches about as far as linear dimension of sheet.

Electric field at distance r from wire is: $E \cong Q \, /2 \, \pi \, \varepsilon_0 \, L \, r$, where L is length of wire. Accurate if wire is long compared to distances r of interest. E points away from positive wire.

Magnetic gradient force $F \cong A \, I \, dB/dz$, where A is area of coil carrying current I, and dB is change in field over distance dz. See textbooks for direction.

4.9 RAMJET

Synchrotron radiation frequency of gyration of charged particle in magnetic field B is $f_0 = q \, B/2 \, \pi \, m$, where q is charge (also called e if electron or proton unit of charge), and m is mass of particle. Total energy emitted per revolution by charge forced to follow circular path is $E = q^2 \, \gamma^4 \, v^2/3 \, \varepsilon_0 \, R \, c^2$, where R is radius of gyration [see Appendix B], v is its speed. Total power emitted is $P = E \, f_0$. Band of most intense radiation is from about $0.1 \, f_0$ to f_0.

Magnetic pressure $P_m = B^2/2 \, \mu_0$.

Non-relativistic rate of mass collection by ramjet is: $dm/dt = \rho \, A \, v$, where v is starship speed, ρ is mass density of space (kg/m^3), and A is effective scoop area. Power collected is $P = e \, dm/dt$, where e is available fusion energy in J/kg of mass collected. At high speed where $v > v_e$ as desired, most of power goes to starship as: $P = M \, a \, v$, where M is mass of starship. m can be eliminated and the required area of collection for a given acceleration found: $A = M \, a/e \, \rho$, independent of speed. Or the possible maximum acceleration is: $a = \rho \, A \, e/M$. Exhaust velocity can be found using Newton's law for rocket: $dm/dt \, v_e = a \, M$, so that: $v_e = A \, e \, \rho/\rho \, A \, v = e/v$. At high speed it decreases with starship speed.

To analyze ramjet operation further, use starship frame S' (do not mix frames) and assume mass is collected at rate dm/dt with power input P as before. Initial KE collected is assumed (somehow) to be restored to exhaust, essential for efficient operation. Initial total power collected is: $(1/2) \, dm/dt \, v^2 + e \, dm/dt$ and equals final power distributed to starship and exhaust: $M \, a \, dv + (1/2) \, dm/dt \, v_e^2$. M a dv can be shown to be negligible. An equation for v_e is obtained: $v^2 + 2 \, e = v_e^2$, or $v_e = \sqrt{(v^2 + 2 \, e)}$. As v increases v_e increases according to added energy per unit mass e. Since fusion e is limited to less than 0.01 in terms of c^2, little is added to v_e after v exceeds 0.1c. Note if $e = 0$, incoming KE is fed directly to exhaust so that $v_e = v$. The result for v_e differs from some in the literature. All this assumes perfect efficiency and no losses.

Other Selected Topics

The math and physics background for selected topics of Chapters 5 through 12 is provided here, keyed by section number and in brief form, with conventions as before. Again, the same symbols are commonly used for quite different variables.

5.4 MAGNETIC INDUCTION

The induced voltage in a loop of wire of area A moving through magnetic field B (perpendicular to field lines) is given by: $V = B\, dA/dt$. dA/dt tells how much area per second encounters new field lines. A wire of length L drawn similarly through B field gets a voltage across its ends $V = B\, L\, v$, where v is its speed across the field. Current that flows in the circuit due to this voltage depends on impedance in circuit. Any current flow also produces a counter-acting B field that partially cancels the effect, worse at lower impedance (resistance).

5.5 (ALSO 9.2) ROTATING WHEELS

Period of rotation T of wheel of radius R is circumference divided by speed at rim: $T = 2\pi R/v$. From the centripetal acceleration to hold rim on, $a_c = v^2/R$, period can be found as: $T = 2\pi \sqrt{(R/a_c)}$. Period increases slowly with R. For a given period, R can also be found by: $R = a_c T^2/4\pi^2$.

5.8 OPTIMIZATION OF STARSHIP MASS

Let T be duration of trip. Mass M is proportional to trip time: $M \sim T$, or $M = b_1 T + b_0$. Also duration $T \sim 1/v$, where v is speed. Since $M \sim v^2$, $M \sim 1/T^2$, or $M = b_2/T^2$. One function increases with T, one decreases. Add these functions of T, and result has a minimum at $dM/dT = 0$ at M_m, the optimum mass, at duration $T_m = (2\, b_2/b_1)^{1/3}$. Use this T_m to find M_m. Before combining equations, use best guesses for M and T in each case to determine multiplicative constants. Constant b_0 is found to be negligible for practical cases. Accuracy is not much better than a power of 10. $b_2 \cong (10)^{10}$; $b_1 \cong 10$.

6.7 SPREAD OF SELF-REPRODUCING PROBES AND COLONIES

Let d be density of stars worth visiting (about 0.003 to 0.004 per CLY). Assume stars are distributed uniformly and quasi-randomly in shells of radius r_i around origin, with number at each shell determined by volume of shell. Let r_i be measured in LY and increase in units of 10 LY to represent distance a probe could travel. Number of stars and therefore probes needed for visiting shell i is $n_i = 4\pi d\, r_i^3/3$ minus number already visited in shell i - 1. That number is $n_{i-1} = 4\pi d\, (r_{i-1})^3/3$ and is the number of probes available to reproduce. Earlier generation i-1 of probes must multiply its number by $m_i = n_i/n_{i-1}$ to produce necessary number of new probes. Result: $m_i = (r_i/r_{i-1})^3$. As i approaches infinity, m_i approaches 1, and probes need reproduce less and less, approaching 1:1. Alternatively, if each probe produces k probes at shell i, number of probes increases approx. as k^i. Beyond some r, this number grows faster than the need for probes n_i, which grows at a fixed power of r.

7.2 STELLAR BRIGHTNESS

Magnitude (absolute, as seen from 32.6 LY) for any star of known luminosity L can be found with respect to solar luminosity L_0 by: $M = -2.512 \log(L/L_0) + M_0$, where magnitude of Sun is taken visually as $M_0 = 4.8$ from a reference distance of 32.6 LY (10 parsecs). Since L ~ $1/r^2$, where r is distance, M at any distance can be found relative to 32.6 LY. Usually this relation is used the other way to determine L, given absolute M. For an observed star at distance r, absolute M and visual M' magnitudes are related according to distance thus: $M = M' + 5 - 5 \log(r/3.26) = M' + 5 - 5 \log(r) + 2.57$, where r is measured in LY. [b-Baker]

7.3 CHANGING COURSE

For a starship proceeding at speed v_x and wanting to swing its course by angle θ (in radians), the transverse speed in y-direction required is $v_y = v_x \tan\theta$. For small angles $\tan\theta \sim \theta$. Transverse speed is built up by acceleration in y-direction. New non-relativistic total speed is: $v = \sqrt{(v_x^2 + v_y^2)}$.

7.5 & 7.6 RADIO COMMUNICATION

The noise level of a system, and therefore the minimum power Po needed for detection of a signal, is given by: $P_0 = k T w$, where k is Boltzmann's constant, T is absolute effective temperature of system (not of space), and w is bandwidth considered, in Hz [b-Pierce]. Minimum signal energy per photon is $E = h f$, where h is Planck's constant, and f is frequency.

Transmitter power P_t required to provide received signal power P_r is given by: $P_t = 4 \pi r^2/G A_r$, where A_r is area used by receiver, G is gain of transmitter antenna, and $4 \pi r^2$ is area of sphere in space that P_t is spread over when it reaches receiver. Therefore r is distance for transmission. Gain G concentrates P_t and is given by ratio of total solid angle to solid angle covered by "beam" of transmitter: $G \cong 4 \pi A_t/\lambda^2$, where λ is wavelength [b-Sagan]. Beam size is given by diffraction limit as before. For minimum P_t use $P_r = P_0$ from above.

Minimum energy to send 1 bit is $E = k T \ln 2 = 0.693 k T$, where T is the effective system temperature, and ideal encoding is used.

Information channel capacity in bits per second is: $i = w \log_2(1 + P/P_n)$, where w is bandwidth, P is power used, and P_n is noise power. Log_2 is logarithm base 2. Note for large ratio $P/P_n = 1000/1$, $\log_2 \sim 10$. Typical ratio 10/1 gives $\log_2 \sim 3$. [b-Pierce]

8.3 ENTROPY AND ATOMS

The entropy or amount of disorder of a system is defined as: $S = k \ln w$, where k is Boltzmann's constant and w is the number of possible states. The energy represented is $E = S T$, where T is absolute temperature. The number of states is often very very large, but the logarithm function (base e) cuts their effect down to ordinary numbers. If N particles of M different kinds are mixed randomly, the number of permutations (or states) is: $w \cong (M + N)!/M! N!$, where ! denotes the factorial function. An approximation called Stirling's allows factorials to be simplified if $N \gtrsim 10$: $N! \cong (\sqrt{N}) N^N e^{-N}$. When logs are used, the N^N term is usually larger than the others. When $N >> M$, then the final approximation is: $S \cong k M \ln(N/M)$ [b-Morse].

To find the number of atoms or molecules in a kilogram of a given element or compound, use the fact that one mole or Avogadro number $6(10)^{23}$ atoms (or molecules) are in one gram-mole of material, with the latter defined with the number of grams set numerically equal to the atomic weight of the material. Thus water has atomic weight 18, and 18 grams of water contain one mole of molecules. A kilogram of a light element like lithium (atomic weight 6) has ~$(10)26$ atoms.

8.5 & 9.6 RADIATION MEASUREMENT AND ELECTRIC POTENTIAL

Radiation (particles, photons) is absorbed in an approximate depth of material, depending on incident KE, which is best obtained from tables of measurements. The depth might be given as a distance, which represents the depth at which 63% (1/e) is absorbed, or as the mass per unit area required to absorb it. The intensity decreases as an exponential function with distance, a law not needed explicitly in this book. When the depth is x, the mass required per area is given by: $m = \rho\, x$, where ρ is the density of the material (mass per volume). Water has density 1000 kg/m^3, rock typically 2 times greater, and iron 8 times greater.

When the flux of particles is n per m^{2}s, each carrying KE, the radiation dose ϕ absorbed in depth x per second is given by: $\phi = n\, KE/m$, where m is the mass in area 1 m^2. KE is often in MeV and must be converted to joules. The SI unit of ϕ is grays (Gy), and 1 gray is 1 J/kg absorbed. But rads is more commonly used, where 1 rad is 0.01 gray. The dose is cumulative over time, and attention must be given to the duration of radiation. Rads could be per second or per year with the time unspecified.

In biological material, different kinds and energies of radiation have different Relative Biological Efficiency (RBE), meaning different devasting effects enhanced over simple energy absorption. RBE is also called quality factor QF. Photons and electrons have RBE ~ 1. Neutrons and protons to 10 MeV have RBE ~ 10. Heavy nuclei have the largest RBE ~ 20. The roentgen equivalent man dose (rem) is found by multiplying the absorbed radiation in rads by RBE. The SI unit for biological dose is sievert (Sv), where 1 rem is 0.01 sievert. [b-Rees]

Given electric potential (voltage) V varying with distance as $V(r) = q/(4\pi\varepsilon_0 r)$ for a spherically symmetric situation, electric field E, or charge q, or radius R of sphere on which q is placed can be found when the remainder of these quantities are known. In this simple case $E = V\, r$ (from $E = -dV/dr$). Value of $1/4\pi\varepsilon_0$ is approx. (10)10 in SI units. For potential on, or field at, sphere of radius R with charge q, use r = R.

11.1 DRAKE EQUATION

More accurately called the Drake formula, it estimates the number N of (communicative) civilizations from a product of probability factors thus: $N = R_s\, f_p\, n_e\, f_l\, f_i\, f_c\, L$. R_s is the rate of star formation in our galaxy (about 1 to 10 per year for many billion years), f_p is the fraction of those stars having planets (perhaps 0.1), n_e is the number of those planets being habitable (perhaps 0.1 to 1), f_l is the probability of life arising on a habitable planet (thought to be 0.01 to 1), f_i is the probability of intelligence arising from the life (very uncertain, perhaps 0.01 to 1), f_c is the probability of intelligent beings developing technology for communication (thought high), and L is the lifetime of a technical civilization in (Earth) years. $f_p\, n_e$ together is the probability of a habitable planet at a star (0.01 to 1). The probability factors must be independent, and each must fully represent all possibilities. The units of R_s and L together cancel so that N comes out a pure number. Unless the probabilities in a product are each near 1, the product comes out much smaller than 1, producing a small result for N. E.g. (0.01)(0.01)(0.01) = 0.000001.

If L is taken as approx. 5 billion years presence of 2nd and 3rd generation stars, then N estimates the total number of civilizations ever in our galaxy.

Separation of civilizations: Given N civilizations in our galactic disk, excluding core, with volume $V_g \cong (10)^{13}$ CLY, volume per civilization is V_g/N in CLY, and the average separation s is approximately: $s \cong (V_g/N)^{1/3}$ in LY.

11.5 BIOZONE SIZE

The average radius of the spherical biozone around a star (used in the plane containing planets) is given by $R = R_o\,\sqrt{(L'/L_o)}$, where R_o is 150 million km for Earth, L' is luminosity of star, and L_o is luminosity of our Sun. The width of biozone could be based on our range of 100 to 200 million km.

11.7 FILLING GALAXY

Probability P of our contact with another one of N civilizations can be estimated from the volume of galaxy filled thus: $P \cong 4 N R^3/V_g$, where R is typical radius of an expanding civilization and 4 is from spherical approximation. R (in LY) can be estimated from: $R = V T$, where V is typical speed of expansion (in LY/year) and T is typical age of civilization (years).

12.2 ENERGY FROM UNCERTAINTY

The uncertainty, fluctuation, variation, or deviance that nature permits is described by the Heisenberg uncertainty relations. For energy and time the relation is: $\Delta E \, \Delta t > h/2$, where Δ denotes the variation in energy E or time t. h is Planck's constant. If a fluctuation ΔE occurs, then it can occur only for a time that is interpreted as less than $\Delta t = h/(2 \, \Delta E)$. (No, there is no error here with the > sign.)

12.3 BLACK HOLES AND GRAVITY MACHINES

Radius R of "event horizon" (spherical boundary inside of which there is no return) for simple black hole (no charge, no spin) is given by: $R = 2 G M/c^2$, where G is gravitational constant and M is mass of black hole.

Artificially generated gravity field g (in m/s^2) is smaller by factor $G/c^2 \sim (10)^{-27}$ in ordinary mechanical motions of masses.

Changing mass current in one second in a mass-carrying tube wound around a toroid gives field strength g at the center as: $g \sim (10)^{-26} \, \mu \, r^2/4 \, \pi \, R^2$, where μ is mass flowing per second, r is minor radius of toroid, R is major radius, and $(10)^{-26}$ is a special constant in SI units [a-Forward 1963].

Programs and Printout Tables of Results for Rocket Motion

These programs are for the BASIC language. Some forms of BASIC will run without the line numbers. The first program (newtonian motion, constant thrust) illustrates an incremental method of calculating motion when an analytic solution is unavailable and produces Figure 1.5.

NEWTONIAN MOTION FOR CONSTANT THRUST

```
10 PRINT "Newtonian motion, launched rocket, gravity optional, constant thrust. (c)JHM"
15 PRINT "Plots a,v,y versus t; MKS units. Picture only, for VGA screen."
20 PRINT "For no gravity enter g=0; for Earth gravity enter g=9.8 or g=10."
25 PRINT "Thrust ve times mu required to be greater than weight M0 times g."
30 SCREEN 12
70 LINE (120, 440)-(520, 440)   'horizontal t axis
75 LINE (120, 40)-(120, 440)    'vertical y,v,a axis
80 FOR i = 1 TO 40
85   LINE (120 + 10 * i, 440)-(120 + 10 * i, 438) 'mark t axis every 10 seconds
90 NEXT i
95 FOR i = 1 TO 4
100  LINE (120 + 100 * i, 440)-(120 + 100 * i, 436) 'mark every 100 sec
105 NEXT i
110 FOR j = 1 TO 40
115  LINE (120, 40 + 10 * j)-(122, 40 + 10 * j) 'mark every 10
120 NEXT j
125 FOR j = 0 TO 4
130  LINE (120, 40 + 100 * j)-(124, 40 + 100 * j) 'mark every 100 units
135 NEXT j
140 LOCATE 26, 60: PRINT "time (s)"
145 LOCATE 4, 19: PRINT "position y (x100 km)"
150 LOCATE 6, 19: PRINT "speed v (x10 km/s)"
155 LOCATE 8, 19: PRINT "acceleration a (x100 m/s2)"
160 LOCATE 1, 10: INPUT "ve,mu,M0,M',g"; ve, mu, M0, Mb, g
170 dt = 1 'time interval 1 second
180 tb = (M0 - Mb) / mu 'stops at burnout; need integer
190 v = 0: y = 0 'start from rest at t=0
200 FOR t = 1 TO tb
210  dv = ((ve * mu * dt) / (M0 - mu * t)) - g * dt 'dv from Newton's law
220  v = v + dv
230  a = dv / dt
240  dy = v * dt
250  y = y + dy
260  PSET (120 + t, 440 - (y / 1000)), 15 'height
270  PSET (120 + t, 440 - (v / 100)), 13 'speed; different shade/color
280  PSET (120 + t, 440 - a), 11 'accel; ditto
300 NEXT t
305 LOCATE 2, 10
310 PRINT USING "\     \###.#\    \##.##\     \###.###\    \###\ \";
"burnout: a="; a; " m/s2, v="; v / 1000; " km/s, y="; y / 1000; " km, t="; tb; " s"
340 END
```

NON-RELATIVISTIC TRAVEL FOR CONSTANT ACCELERATION

```
10 PRINT "Non-relativistic travel, starting from zero position & speed. (c)JHM"
20 PRINT "In MKS units and Earth Year units, for selected gee values & times."
25 PRINT "Constant mass and acceleration. Useful for braking too."
30 PRINT "Calculation cutoff at (v/c) = beta = .15 (1% error in gamma)."
35 PRINT "Computer use of E equivalent to power of (10)."
40 PRINT
50 FOR i = 0 TO 3  'from .001 gee to 1 gee
60   g = (.001) * (10 ^ i)  'log scaled
70   a = g * 9.80665 'meters/sec2
80   PRINT "gee ="; g; "     acceleration ="; a; "in mks units"
90   PRINT "time(Y)   "; "time(sec)   "; "speed(m/s) "; "speed(v/c) "; "dist.(m)   "; "dist.(LY)  "
100   PRINT "  0   "; "   0   "; "    0   "; "     0   "; "    0    "; " 0  "
110   FOR k = 0 TO 4  'from .01 year (~4 days) to 100 years
120     FOR j = 1 TO 9  'decimal years
130     y = .01 * (10 ^ k) * j
140     t = y * 3.155815E + 07
150     v = a * t
160     vc = v / 2.997925E + 08
170     IF vc = .15 THEN 210
180     x = .5 * a * t * t
190     xly = x / 9.460895E + 15
200     PRINT USING "####.### ##.###^^^^  ##.###^^^^  .######
##.###^^^^  ###.###"; y; t; v; vc; x; xly
210     NEXT j
220   NEXT k
230 PRINT
240 NEXT i
250 END
```

```
Non-relativistic travel, starting from zero position & speed.
In MKS units and Earth Year units, for selected gee values & times.
Constant mass and acceleration. Useful for braking too.
Calculation cutoff at (v/c)=beta=.15 (1% error in gamma)
Computer use of E equivalent to power of (10).
```

gee = .001   acceleration= 9.806651E-03 in mks units

time(Y)	time(sec)	speed(m/s)	speed(v/c)	dist.(m)	dist.(LY)
0	0	0	0	0	0
0.010	3.156E+05	3.095E+03	.0000103	4.883E+08	0.000
0.020	6.312E+05	6.190E+03	.0000206	1.953E+09	0.000
0.030	9.467E+05	9.284E+03	.0000310	4.395E+09	0.000
0.040	1.262E+06	1.238E+04	.0000413	7.813E+09	0.000
0.050	1.578E+06	1.547E+04	.0000516	1.221E+10	0.000
0.060	1.893E+06	1.857E+04	.0000619	1.758E+10	0.000
0.070	2.209E+06	2.166E+04	.0000723	2.393E+10	0.000
0.080	2.525E+06	2.476E+04	.0000826	3.125E+10	0.000
0.090	2.840E+06	2.785E+04	.0000929	3.955E+10	0.000
0.100	3.156E+06	3.095E+04	.0001032	4.883E+10	0.000
0.200	6.312E+06	6.190E+04	.0002065	1.953E+11	0.000
0.300	9.467E+06	9.284E+04	.0003097	4.395E+11	0.000
0.400	1.262E+07	1.238E+05	.0004129	7.813E+11	0.000
0.500	1.578E+07	1.547E+05	.0005162	1.221E+12	0.000
0.600	1.893E+07	1.857E+05	.0006194	1.758E+12	0.000
0.700	2.209E+07	2.166E+05	.0007226	2.393E+12	0.000
0.800	2.525E+07	2.476E+05	.0008259	3.125E+12	0.000
0.900	2.840E+07	2.785E+05	.0009291	3.955E+12	0.000
1.000	3.156E+07	3.095E+05	.0010323	4.883E+12	0.001
2.000	6.312E+07	6.190E+05	.0020646	1.953E+13	0.002
3.000	9.467E+07	9.284E+05	.0030969	4.395E+13	0.005
4.000	1.262E+08	1.238E+06	.0041293	7.813E+13	0.008
5.000	1.578E+08	1.547E+06	.0051616	1.221E+14	0.013
6.000	1.893E+08	1.857E+06	.0061939	1.758E+14	0.019
7.000	2.209E+08	2.166E+06	.0072262	2.393E+14	0.025
8.000	2.525E+08	2.476E+06	.0082585	3.125E+14	0.033
9.000	2.840E+08	2.785E+06	.0092908	3.955E+14	0.042
10.000	3.156E+08	3.095E+06	.0103231	4.883E+14	0.052
20.000	6.312E+08	6.190E+06	.0206463	1.953E+15	0.206
30.000	9.467E+08	9.284E+06	.0309694	4.395E+15	0.465
40.000	1.262E+09	1.238E+07	.0412925	7.813E+15	0.826
50.000	1.578E+09	1.547E+07	.0516157	1.221E+16	1.290
60.000	1.893E+09	1.857E+07	.0619388	1.758E+16	1.858
70.000	2.209E+09	2.166E+07	.0722619	2.393E+16	2.529
80.000	2.525E+09	2.476E+07	.0825851	3.125E+16	3.303
90.000	2.840E+09	2.785E+07	.0929082	3.955E+16	4.181
100.000	3.156E+09	3.095E+07	.1032313	4.883E+16	5.162

gee = .01   acceleration= 9.806651E-02 in mks units

time(Y)	time(sec)	speed(m/s)	speed(v/c)	dist.(m)	dist.(LY)
0	0	0	0	0	0
0.010	3.156E+05	3.095E+04	.0001032	4.883E+09	0.000
0.020	6.312E+05	6.190E+04	.0002065	1.953E+10	0.000
0.030	9.467E+05	9.284E+04	.0003097	4.395E+10	0.000
0.040	1.262E+06	1.238E+05	.0004129	7.813E+10	0.000
0.050	1.578E+06	1.547E+05	.0005162	1.221E+11	0.000
0.060	1.893E+06	1.857E+05	.0006194	1.758E+11	0.000
0.070	2.209E+06	2.166E+05	.0007226	2.393E+11	0.000
0.080	2.525E+06	2.476E+05	.0008259	3.125E+11	0.000
0.090	2.840E+06	2.785E+05	.0009291	3.955E+11	0.000
0.100	3.156E+06	3.095E+05	.0010323	4.883E+11	0.000
0.200	6.312E+06	6.190E+05	.0020646	1.953E+12	0.000
0.300	9.467E+06	9.284E+05	.0030969	4.395E+12	0.000
0.400	1.262E+07	1.238E+06	.0041293	7.813E+12	0.001
0.500	1.578E+07	1.547E+06	.0051616	1.221E+13	0.001
0.600	1.893E+07	1.857E+06	.0061939	1.758E+13	0.002
0.700	2.209E+07	2.166E+06	.0072262	2.393E+13	0.003
0.800	2.525E+07	2.476E+06	.0082585	3.125E+13	0.003
0.900	2.840E+07	2.785E+06	.0092908	3.955E+13	0.004
1.000	3.156E+07	3.095E+06	.0103231	4.883E+13	0.005
2.000	6.312E+07	6.190E+06	.0206463	1.953E+14	0.021
3.000	9.467E+07	9.284E+06	.0309694	4.395E+14	0.046
4.000	1.262E+08	1.238E+07	.0412925	7.813E+14	0.083
5.000	1.578E+08	1.547E+07	.0516157	1.221E+15	0.129
6.000	1.893E+08	1.857E+07	.0619388	1.758E+15	0.186
7.000	2.209E+08	2.166E+07	.0722619	2.393E+15	0.253
8.000	2.525E+08	2.476E+07	.0825851	3.125E+15	0.330
9.000	2.840E+08	2.785E+07	.0929082	3.955E+15	0.418
10.000	3.156E+08	3.095E+07	.1032313	4.883E+15	0.516

```
gee = .1          acceleration= .980665 in mks units
time(Y)    time(sec)    speed(m/s)   speed(v/c)   dist.(m)    dist.(LY)
   0           0            0            0           0           0
  0.010     3.156E+05    3.095E+05    .0010323     4.883E+10    0.000
  0.020     6.312E+05    6.190E+05    .0020646     1.953E+11    0.000
  0.030     9.467E+05    9.284E+05    .0030969     4.395E+11    0.000
  0.040     1.262E+06    1.238E+06    .0041293     7.813E+11    0.000
  0.050     1.578E+06    1.547E+06    .0051616     1.221E+12    0.000
  0.060     1.893E+06    1.857E+06    .0061939     1.758E+12    0.000
  0.070     2.209E+06    2.166E+06    .0072262     2.393E+12    0.000
  0.080     2.525E+06    2.476E+06    .0082585     3.125E+12    0.000
  0.090     2.840E+06    2.785E+06    .0092908     3.955E+12    0.000
  0.100     3.156E+06    3.095E+06    .0103231     4.883E+12    0.001
  0.200     6.312E+06    6.190E+06    .0206463     1.953E+13    0.002
  0.300     9.467E+06    9.284E+06    .0309694     4.395E+13    0.005
  0.400     1.262E+07    1.238E+07    .0412925     7.813E+13    0.008
  0.500     1.578E+07    1.547E+07    .0516157     1.221E+14    0.013
  0.600     1.893E+07    1.857E+07    .0619388     1.758E+14    0.019
  0.700     2.209E+07    2.166E+07    .0722619     2.393E+14    0.025
  0.800     2.525E+07    2.476E+07    .0825850     3.125E+14    0.033
  0.900     2.840E+07    2.785E+07    .0929082     3.955E+14    0.042
  1.000     3.156E+07    3.095E+07    .1032313     4.883E+14    0.052


gee = 1           acceleration= 9.80665 in mks units
time(Y)    time(sec)    speed(m/s)   speed(v/c)   dist.(m)    dist.(LY)
   0           0            0            0           0           0
  0.010     3.156E+05    3.095E+06    .0103231     4.883E+11    0.000
  0.020     6.312E+05    6.190E+06    .0206463     1.953E+12    0.000
  0.030     9.467E+05    9.284E+06    .0309694     4.395E+12    0.000
  0.040     1.262E+06    1.238E+07    .0412925     7.813E+12    0.001
  0.050     1.578E+06    1.547E+07    .0516157     1.221E+13    0.001
  0.060     1.893E+06    1.857E+07    .0619388     1.758E+13    0.002
  0.070     2.209E+06    2.166E+07    .0722619     2.393E+13    0.003
  0.080     2.525E+06    2.476E+07    .0825850     3.125E+13    0.003
  0.090     2.840E+06    2.785E+07    .0929082     3.955E+13    0.004
  0.100     3.156E+06    3.095E+07    .1032313     4.883E+13    0.005
```

RELATIVISTIC TRAVEL FOR CONSTANT ACCELERATION

```
10 PRINT "Relativistic travel, starting from zero position & speed. (c)JHM"
15 PRINT "In MKS units and Earth Years & Light Years."
20 PRINT "For selected gee values and Proper Years aboard ship."
25 PRINT "Avoids v/c=beta beyond .9999999; uses single precision."
30 PRINT "Gamma shown is gamma reached after that proper time."
35 PRINT "Computer use of E is equivalent to power of (10)."
40 PRINT
45 FOR i = 0 TO 3  'from .001 gee to 1 gee
50   g = (.001) * (10 ^ i)
60   pa = g * 9.80665 'meters/sec2
70   PRINT USING "\         \##.###\        \###.####\ \"; "proper gee="; g; "   proper
a="; pa; " mks"
80   PRINT "time(PY)    "; "time(EY)    "; "speed(m/s) "; "speed(v/c)   "; "gamma   "; "dist.(m)     ";
"dist.(LY)"
90   PRINT " 0     "; " 0     "; " 0     "; " 0      "; " 1    "; " 0      "; " 0    "
100  FOR k = 1 TO 4 'for .01 to 10 years, times powers of ten
110    FOR j = 1 TO 9'decimal year
120      py = .001 * (10 ^ k) * j / g 'proper years
130      pt = py * 3.155815E+07 'proper time(sec)
140      atc = pa * pt / 2.997925E+08: ca = 2.997925E+08 / pa 'spec. var.
150      IF atc 9 THEN 270 'after this tanh is .9999999; limits v/c
160      COSH = (EXP(atc) + EXP(-atc)): SINH = (EXP(atc) - EXP(-atc)) 'def.
170      et = ca * SINH / 2: ey = et / 3.155815E+07' earth time seconds, years
180      x = (8.987552E+16 / pa) * ((COSH / 2) - 1) 'distance meters
190      xly = x / 9.460895E+15' distance LY
200      beta = (SINH) / (COSH) 'def. tanh
210      v = beta * 2.997925E+08 'speed(m/s)
230      b2 = beta * beta
240      gamma = 1 / (SQR(1 - b2))
260      PRINT USING "####.### #######.### ##.###^^^^  .#######
#####.### ##.###^^^^  #######.###"; py; ey; v; beta; gamma; x; xly
270    NEXT j
280  NEXT k
290 PRINT
300 NEXT i
310 END
```

Relativistic travel, starting from zero position & speed.
In MKS units and Earth Years & Light Years.
For selected gee values and Proper Years aboard ship.
Avoids v/c=beta beyond .9999999; uses single precision.
Gamma shown is gamma reached after that proper time.
Computer use of E is equivalent to power of (10).

proper gee= 0.001   proper a=   0.0098 mks

time(PY)	time(EY)	speed(m/s)	speed(v/c)	gamma	dist.(m)	dist.(LY)
0	0	0	0	1	0	0
10.000	10.000	3.095E+06	.0103228	1.000	4.884E+14	0.052
20.000	20.001	6.189E+06	.0206434	1.000	1.953E+15	0.206
30.000	30.005	9.281E+06	.0309595	1.000	4.395E+15	0.465
40.000	40.011	1.237E+07	.0412691	1.001	7.814E+15	0.826
50.000	50.022	1.546E+07	.0515699	1.001	1.221E+16	1.291
60.000	60.038	1.855E+07	.0618597	1.002	1.759E+16	1.859
70.000	70.061	2.163E+07	.0721364	1.003	2.394E+16	2.530
80.000	80.091	2.470E+07	.0823978	1.003	3.127E+16	3.305
90.000	90.130	2.777E+07	.0926418	1.004	3.958E+16	4.184
100.000	100.178	3.084E+07	.1028662	1.005	4.888E+16	5.166
200.000	201.424	6.103E+07	.2035782	1.021	1.960E+17	20.720
300.000	304.819	8.999E+07	.3001586	1.048	4.430E+17	46.827
400.000	411.464	1.172E+08	.3909536	1.086	7.925E+17	83.765
500.000	522.499	1.423E+08	.4747283	1.136	1.248E+18	131.929
600.000	639.107	1.651E+08	.5507018	1.198	1.815E+18	191.833
700.000	762.531	1.854E+08	.6185290	1.273	2.499E+18	264.116
800.000	894.089	2.033E+08	.6782417	1.361	3.307E+18	349.547
900.000	1035.184	2.189E+08	.7301656	1.464	4.248E+18	449.040
1000.000	1187.319	2.323E+08	.7748344	1.582	5.333E+18	563.654
2000.000	3756.363	2.903E+08	.9683201	4.005	2.754E+19	2910.559
3000.000	10696.810	2.986E+08	.9959246	11.088	9.245E+19	9771.886
4000.000	30085.530	2.996E+08	.9994821	31.073	2.756E+20	29132.420
5000.000	84485.800	2.998E+08	.9999342	87.169	7.902E+20	83522.660
6000.000	237204.900	2.998E+08	.9999916	244.783	2.235E+21	236238.200
7000.000	665967.800	2.998E+08	.9999989	682.667	6.291E+21	664999.800
8000.000	1869742.000	2.998E+08	.9999999	2048.000	1.768E+22	1868773.000

proper gee= 0.010   proper a= 0.0981 mks

time(PY)	time(EY)	speed(m/s)	speed(v/c)	gamma	dist.(m)	dist.(LY)
0	0	0	0	1	0	0
1.000	1.000	3.095E+06	.0103228	1.000	4.884E+13	0.005
2.000	2.000	6.189E+06	.0206434	1.000	1.953E+14	0.021
3.000	3.000	9.281E+06	.0309595	1.000	4.395E+14	0.046
4.000	4.001	1.237E+07	.0412691	1.001	7.814E+14	0.083
5.000	5.002	1.546E+07	.0515699	1.001	1.221E+15	0.129
6.000	6.004	1.855E+07	.0618597	1.002	1.759E+15	0.186
7.000	7.006	2.163E+07	.0721364	1.003	2.394E+15	0.253
8.000	8.009	2.470E+07	.0823978	1.003	3.127E+15	0.331
9.000	9.013	2.777E+07	.0926418	1.004	3.958E+15	0.418
10.000	10.018	3.084E+07	.1028662	1.005	4.888E+15	0.517
20.000	20.142	6.103E+07	.2035782	1.021	1.960E+16	2.072
30.000	30.482	8.999E+07	.3001586	1.048	4.430E+16	4.683
40.000	41.146	1.172E+08	.3909536	1.086	7.925E+16	8.377
50.000	52.250	1.423E+08	.4747282	1.136	1.248E+17	13.193
60.000	63.911	1.651E+08	.5507018	1.198	1.815E+17	19.183
70.000	76.253	1.854E+08	.6185290	1.273	2.499E+17	26.412
80.000	89.409	2.033E+08	.6782417	1.361	3.307E+17	34.955
90.000	103.518	2.189E+08	.7301656	1.464	4.248E+17	44.904
100.000	118.732	2.323E+08	.7748344	1.582	5.333E+17	56.365
200.000	375.636	2.903E+08	.9683200	4.005	2.754E+18	291.056
300.000	1069.681	2.986E+08	.9959246	11.088	9.245E+18	977.189
400.000	3008.550	2.996E+08	.9994821	31.073	2.756E+19	2913.239
500.000	8448.577	2.998E+08	.9999342	87.169	7.902E+19	8352.263
600.000	23720.490	2.998E+08	.9999916	244.783	2.235E+20	23623.820
700.000	66596.780	2.998E+08	.9999989	682.667	6.291E+20	66499.970
800.000	186973.900	2.998E+08	.9999999	2048.000	1.768E+21	186877.100

proper gee= 0.100   proper a= 0.9807 mks

time(PY)	time(EY)	speed(m/s)	speed(v/c)	gamma	dist.(m)	dist.(LY)
0	0	0	0	1	0	0
0.100	0.100	3.095E+06	.0103228	1.000	4.884E+12	0.001
0.200	0.200	6.189E+06	.0206434	1.000	1.953E+13	0.002
0.300	0.300	9.281E+06	.0309595	1.000	4.395E+13	0.005
0.400	0.400	1.237E+07	.0412691	1.001	7.814E+13	0.008
0.500	0.500	1.546E+07	.0515699	1.001	1.221E+14	0.013
0.600	0.600	1.855E+07	.0618597	1.002	1.759E+14	0.019
0.700	0.701	2.163E+07	.0721364	1.003	2.394E+14	0.025
0.800	0.801	2.470E+07	.0823978	1.003	3.127E+14	0.033
0.900	0.901	2.777E+07	.0926418	1.004	3.958E+14	0.042
1.000	1.002	3.084E+07	.1028662	1.005	4.888E+14	0.052
2.000	2.014	6.103E+07	.2035782	1.021	1.960E+15	0.207
3.000	3.048	8.999E+07	.3001586	1.048	4.430E+15	0.468
4.000	4.115	1.172E+08	.3909536	1.086	7.925E+15	0.838
5.000	5.225	1.423E+08	.4747283	1.136	1.248E+16	1.319
6.000	6.391	1.651E+08	.5507017	1.198	1.815E+16	1.918
7.000	7.625	1.854E+08	.6185290	1.273	2.499E+16	2.641
8.000	8.941	2.033E+08	.6782417	1.361	3.307E+16	3.495
9.000	10.352	2.189E+08	.7301656	1.464	4.248E+16	4.490
10.000	11.873	2.323E+08	.7748344	1.582	5.333E+16	5.637
20.000	37.564	2.903E+08	.9683201	4.005	2.754E+17	29.106
30.000	106.968	2.986E+08	.9959246	11.088	9.245E+17	97.719
40.000	300.855	2.996E+08	.9994821	31.073	2.756E+18	291.324
50.000	844.858	2.998E+08	.9999342	87.169	7.902E+18	835.226
60.000	2372.051	2.998E+08	.9999916	244.783	2.235E+19	2362.383
70.000	6659.682	2.998E+08	.9999989	682.667	6.291E+19	6650.002
80.000	18697.420	2.998E+08	.9999999	2048.000	1.768E+20	18687.730

```
proper gee=  1.000    proper a=  9.8067 mks
```

time(PY)	time(EY)	speed(m/s)	speed(v/c)	gamma	dist.(m)	dist.(LY)
0	0	0	0	1	0	0
0.010	0.010	3.095E+06	.0103228	1.000	4.884E+11	0.000
0.020	0.020	6.189E+06	.0206434	1.000	1.953E+12	0.000
0.030	0.030	9.281E+06	.0309595	1.000	4.395E+12	0.000
0.040	0.040	1.237E+07	.0412691	1.001	7.815E+12	0.001
0.050	0.050	1.546E+07	.0515699	1.001	1.221E+13	0.001
0.060	0.060	1.855E+07	.0618597	1.002	1.759E+13	0.002
0.070	0.070	2.163E+07	.0721364	1.003	2.394E+13	0.003
0.080	0.080	2.470E+07	.0823978	1.003	3.127E+13	0.003
0.090	0.090	2.777E+07	.0926418	1.004	3.958E+13	0.004
0.100	0.100	3.084E+07	.1028662	1.005	4.888E+13	0.005
0.200	0.201	6.103E+07	.2035782	1.021	1.960E+14	0.021
0.300	0.305	8.999E+07	.3001586	1.048	4.430E+14	0.047
0.400	0.411	1.172E+08	.3909536	1.086	7.925E+14	0.084
0.500	0.522	1.423E+08	.4747283	1.136	1.248E+15	0.132
0.600	0.639	1.651E+08	.5507017	1.198	1.815E+15	0.192
0.700	0.763	1.854E+08	.6185290	1.273	2.499E+15	0.264
0.800	0.894	2.033E+08	.6782417	1.361	3.307E+15	0.350
0.900	1.035	2.189E+08	.7301656	1.464	4.248E+15	0.449
1.000	1.187	2.323E+08	.7748344	1.582	5.333E+15	0.564
2.000	3.756	2.903E+08	.9683201	4.005	2.754E+16	2.911
3.000	10.697	2.986E+08	.9959246	11.088	9.245E+16	9.772
4.000	30.086	2.996E+08	.9994821	31.073	2.756E+17	29.132
5.000	84.486	2.998E+08	.9999342	87.169	7.902E+17	83.523
6.000	237.205	2.998E+08	.9999916	244.783	2.235E+18	236.238
7.000	665.968	2.998E+08	.9999989	682.667	6.291E+18	665.000
8.000	1869.742	2.998E+08	.9999999	2048.000	1.768E+19	1868.773

Bibliography

NOTES:

Articles (first) and books (second) are listed by chapter and thus appear under the subject categories that chapters represent. When an article or book covers several subject areas, it is listed with the relevant chapters. General books and articles are listed for the Introduction. General articles and books on propulsion are listed for Chapter 4. Some articles come from collections in books, and such books are identified with code names given below. A few uncited articles and books are included for more breadth. References are listed alphabetically by first author, with unauthored listed with an identifier within the list. Under an author, single authors are listed first, and articles are alphabetized by title. When there are more than two authors, the first one is listed with et al. Some journal title abbreviations are also listed below. # denotes a publication, report, or paper number.

Minimal reference data is given as needed to specify a reference to any librarian. Long titles may be shortened by omitting subtitles and occasionally truncated (for articles) by omitting uninformative phrases. Sometimes the article title does not seem to match the subject for which it was cited, but the article does contain material on the subject at hand. Occasionally extra notes are provided to clarify the subject. Some articles can be found in several sources, not all of which are listed here. Some hard-to-find sources for articles (e.g. the original foreign publication) are supplanted with more easily-found sources. See also remarks on the literature in the Introduction.

Unaccountably, the term "manned" is used often even in current space travel literature to refer to missions carrying humans (presumably male and female) in the past and future. Bibliographies are very difficult to produce error-free, and apologies are offered in advance for any inconvenience when an error in referencing prevents finding the source. The reversal of the title of the journal *Astronautica Acta* to *Acta Astronautica* is not accidental; it changed twice.

The *Journal of the British Interplanetary Society* (JBIS) can be found in major US research libraries, and the Library of Congress is authorized to provide article copies on request. JBIS has published a major bibliography of about 2700 items in 13 categories and many sub-categories of interstellar studies ("Interstellar Travel and Communication: a bibliography" by Mallove et al., June 1980 whole issue, vol 33 p 201-248). An update of about 700 items through 1986 appeared in 1987, vol 40 p 353-364. More may appear.

Source abbreviations (organizations, journal titles, books):

AIAA = American Institute of Aeronautics and Astronautics

AAS = American Astronautical Society

AASA = American Astronautical Society Advances Series

AASH = American Astronautical Society History Series

AASST = American Astronautical Society Science & Technology Series

B&P = Bova, Ben, & Preiss, Byron, eds. *First Contact* 1990 New American Lib.

Billingham = Billingham, John, ed. *Life in the Universe* 1981 MIT Pr.

Cameron = Cameron, A. G. W., ed. *Interstellar Communication* 1963 Benjamin.

F&J = Finney, Ben, & Jones, Eric, eds. *Interstellar Migration and the Human Experience* 1985 Univ. of California Pr.

Goldsmith = Goldsmith, Donald, ed. *The Quest for Extraterrestrial Life* 1980 University Science Bks.

H&Z = Hart, Michael, & Zuckerman, Ben, eds. *Extraterrestrials* 1982 Pergamon.

JBIS = *Journal of the British Interplanetary Society*

JPL = Jet Propulsion Laboratory

NYT = New York Times

PD = Project Daedalus supplement to JBIS 1978.

P80 = Papagiannis, M. D., ed. *Strategies for the Search for Life in the Universe* 1980 Reidel.

P85 = Papagiannis, Michael, ed. *The Search for Extraterrestrial Life* 1985 Reidel.

P&C = Ponnamperuma, Cyril, & Cameron, A. G. W. *Interstellar Communication: Scientific Perspectives* 1974 Houghton Mifflin.

QJRAS = *Quarterly Journal of Royal Astronomical Society*

Regis = Regis, Edward, ed. *Extraterrestrials* 1985 Cambridge UP.

Rothman = Rothman, Tony, ed. *Frontiers of Modern Physics* 1985 Dover.

ScAm = *Scientific American*

INTRODUCTION: GENERAL ARTICLES ON SPACE AND INTERSTELLAR TRAVEL:

Bainbridge, William. "Impact of science fiction on attitudes toward technology" in AASH 1982 vol 5 p 121-135.

Cesarone, Robert, et al. "Prospects for Voyager" JBIS 1984 vol 37 p 99-116.

Drake, Frank. "N is neither very small nor very large" in P80 p 27-34. SETI.

Ehricke, K. A. "Evolution of interstellar operations" in AASA 1969 vol 14 microfiche #69-387.

Finney, B. R., & Jones, E. M. "From Africa to the stars" in AASA 1983 vol 53 #83-205.

Forward, R. L. "Program for interstellar exploration" JBIS 1976 vol 29 p 611-632.

Markowitz, William. "Physics and metaphysics of unidentified flying objects" *Science* 15 Sep 1967 vol 157 p 1274-1279.

Mattinson, H. R. "Astronomical data on nearby stellar systems" JBIS PD 1978 p S8-S18.

Pierce, J. R. "Relativity and space travel" *IRE Proceedings* Jun 1959 p 1053-1061.

Purcell, Edward. "Radioastronomy and communication through space" USAEC 1961 #BNL-658. Also in Goldsmith p 188-196.

Sagan, Carl. "A Pale blue dot" *Parade* 9 Sep 1990 (Earth seen from Voyager).

Shepherd, L. R. "Interstellar flight" JBIS 1952 vol 11 p 149-167.

Singer, Cliff. "Settlements in space and interstellar travel" in H&Z p 46-61.

von Hoerner, Sebastian. "General limits of space travel" *Science* 6 Jul 1962 vol 137 p 18-23.

INTRODUCTION: GENERAL BOOKS ON SPACE AND INTERSTELLAR TRAVEL:

[some are not referenced in the text]

Abell, George. *Realm of the Universe* 1976 Holt Rinehart Winston.

Berman, Arthur. *Space Flight* 1979 Doubleday (Science Study Series).

Bernal, J. D. *The World, the Flesh and the Devil* 1929 Methuen, 1969 Indiana UP.

Blaine, J. C. D. *The End of an Era in Space Exploration* 1976 AAS/Univelt.

Calder, Nigel. *Spaceships of the Mind* 1978 Viking.

Cameron, A. G. W., ed. *Interstellar Communication* 1963 Benjamin.

Clarke, Arthur. *Profiles of the Future* 1977 Fawcett.

Clarke, Arthur. *The Promise of Space* 1968 Harper & Row.

Dyson, Freeman. *Disturbing the Universe* 1979 Harper & Row.

Finney, Ben, & Jones, Eric, eds. *Interstellar Migration and the Human Experience* 1985 Univ. of California Pr.

Goldsmith, Donald, ed. *The Quest for Extraterrestrial Life* 1980 University Science Bks.

Goswami, Amit. *Cosmic Dancers* 1983 Harper & Row.

Hart, Michael, & Zuckerman, Ben, eds. *Extraterrestrials* 1982 Pergamon.

Heppenheimer, T. A. *Toward Distant Suns* 1979 Stackpole.

Macvey, John. *Journey to Alpha Centauri* 1965 Macmillan.

Mallove, Eugene, & Matloff, Gregory. *Starflight Handbook* 1989 Wiley.

Moore, Patrick. *New Concise Atlas of the Universe* 1978 Rand McNally.

Moore, Patrick. *Travellers in Space and Time* 1983 Doubleday.

Nicholls, Peter. *The Science in Science Fiction* 1982 Knopf.

Nicolson, Ian. *The Road to the Stars* 1978 Wm Morrow.

O'Neill, Gerard. *2081* 1981 Simon & Schuster (life projected in 2081).

Papagiannis, Michael, ed. *The Search for Extraterrestrial Life* 1985 D. Reidel.

Papagiannis, M. D., ed. *Strategies for the Search for Life in the Universe* 1980 Reidel.

Ridpath, Ian. *Messages from the Stars* 1978 Harper & Row.

Sagan, Carl. *Communication with Extraterrestrial Intelligence* 1973 MIT Pr.

Sagan, Carl. *Cosmos* 1980 Random House.

Shklovskii, I. S., & Sagan, Carl. *Intelligent Life in the Universe* 1966 Dell.

Strong, James. *Flight to the Stars* 1965 Hart.

CHAPTER 1: ARTICLES ON BASICS OF SPACE TRAVEL:

Dyson, Freeman. "Gravitational machines" in Cameron p 115-120 (boosts).

Ehricke, Krafft. "Saturn Jupiter rebound" JBIS 1972 vol 25 p 561-571.

Heacock, Raymond. "The Voyager Spacecraft" IME Proceedings 1980 vol 194 p 211-224.

Jaffe, L. D., et al. "Interstellar precursor mission" JBIS 1980 vol 33 p 3-26.

Marx, G. "Mechanical efficiency of interstellar vehicles" *Astron. Acta* 1963 vol 9 p 131-139.

Matloff, G., & Mallove, E. "Non-nuclear interstellar flight: application of planetary gravitational assists" JBIS 1985 vol 38 p 133-136.

Matloff, G., & Parks, K. "Interstellar gravity assist propulsion: correction and new application" JBIS 1988 vol 41 p 519-526.

Matloff, G., & Ubell, C. "Worldships: prospects for non-nuclear propulsion and power sources" JBIS 1985 vol 38 p 253-261.

Oliver, B. M. "Review of interstellar rocketry fundamentals" JBIS 1990 vol 43 p 259-264.

von Hoerner, Sebastian. "General limits of space travel" *Science* 6 Jul 1962 vol 137 p 18-23.

CHAPTER 1: BOOKS INCLUDING BASICS OF SPACE TRAVEL:

Berman, Arthur. *Space Flight* 1979 Doubleday (Science Study Series).

Blaine, J. C. D. *The End of an Era in Space Exploration* 1976 AAS/Univelt.

Glasstone, Samuel. *Sourcebook on the Space Sciences* 1965 Van Nostrand.

Goldstein, Herbert. *Classical Mechanics* 1950 Addison-Wesley.

Halliday, David, & Resnick, Robert. *Physics* 1978 Wiley.

McAleer, Neil. *The Omni Space Almanac* 1987 Omni/World Almanac.

Romer, Robert. *Energy an Introduction to Physics* 1976 Freeman.

CHAPTER 2: ARTICLES ON ADVANCED PROPULSION (NON-RELATIVISTIC):

Aftergood, Steven, et al. "Nuclear power in space" ScAm Jun 1991.

Alfven, Hannes. "Spacecraft propulsion: new methods" *Science* 14 Apr 1972 (plasma methods).

Aston, Graeme. "Electric propulsion" JBIS 1986 vol 39 p 503-507.

Aston, Graeme. "Transportation applications of electric propulsion" AASST 1987 vol 69 #87-128.

Bond, Alan. "Analysis of potential performance of ram-augmented interstellar rocket" JBIS 1974 vol 27 p 674-685.

Bond, Alan. "Problems of interstellar propulsion" *Spaceflight* 1971 vol 13 p 245-251.

Bond, Alan, & Martin, Anthony. "Project Daedalus reviewed" JBIS 1986 vol 39 p 385-390.

Bond, Alan, & Martin, Anthony. "Propulsion: Part II engineering design..." JBIS PD 1978 p S63-S82.

Bond, Alan, & Martin, Anthony. "Worldships: assessment of engineering feasibility" JBIS 1984 vol 37 p 254-266.

Boyer, K., & Balcomb, J. D. "System studies of fusion powered propulsion systems" AIAA 1971 #71-636.

Brown, William. "LEO to GEO transportation system combining electric propulsion with beamed microwave power from Earth" AASST 1987 vol 69 #87-126.

Cassenti, B. N. "Comparison of interstellar propulsion methods" JBIS 1982 vol 35 p 116-124; also AIAA 1980 #80-1229.

Cheng, Dah Yu. "Application of deflagration plasma gun as space propulsion thruster" *AIAA J.* 1971 vol 9 p 1681-1685.

Cherfas, Jeremy. "European superlaser" *Science* 1 Jun 1990 news.

Chou, Tsu-Wei, et al. "Composites" ScAm Oct 1986 (materials).

Clarke, Arthur. "Electromagnetic launching as a major contributor to spaceflight" JBIS 1950 vol 9 p 261-267.

Clauser, H. R. "Advanced composite materials" ScAm Jul 1973.

Conn, Robert. "Engineering of magnetic fusion reactors" ScAm Oct 1983.

Dawson, John. "On the production of plasma by giant pulse lasers" *Phys. Fluids* 1964 vol 7 p 981-987.

de Bergh, Catherine, et al. "Deuterium on Venus" *Science* 1 Feb 1991.

Drexler, K. E. "High performance solar sails and related reflecting devices" AIAA 1979 #79-1418.

Dushman, Saul. "Mass-energy relation" *GE Rev.* Oct 1944 p 6-13 (fusion).

Drexler, Eric. "Sailing through space on sunlight" *Smithsonian* Feb 1982.

Dyson, Freeman. "Death of a project" *Science* 9 Jul 1965.

Dyson, Freeman. "Interstellar transport" *Phys. Today* Oct 1968.

Englert, Gerald. "Towards thermonuclear rocket propulsion" *New Sc.* Oct 1962.

Flam, Faye. "Plastics get oriented and get new properties" *Science* 22 Feb 1991 news (strength).

Forward, R. L. "Program for interstellar exploration" JBIS 1976 vol 29 p 611-632.

Furth, H. P. "Magnetic confinement fusion" *Science* 28 Sep 1990.

Gough, William, & Eastlund, Bernard. "Prospects of fusion power" ScAm Feb 1971.

Hamakawa, Yoshihiro. "Photovoltaic power" ScAm Apr 1987.

Hamilton, David. "Energy science takes a heavy budget hit" *Science* 26 Oct 1990 news (fusion machines).

Heacock, Raymond. "The Voyager Spacecraft" IME Proceedings 1980 vol 194 p 211-224.

Heppenheimer, T. A. "Some advanced applications of a 1-million second specific impulse rocket engine" JBIS 1975 vol 28 p 175-181.

Hyde, Roderick, et al. "Prospects for rocket propulsion with laser-induced fusion microexplosions" AIAA 1972 #72-1063.

Jaffe, L. D., et al. "Interstellar precursor mission" JBIS 1980 vol 33 p 3-26.

Kantrowitz, Arthur. "Laser propulsion to Earth orbit..." AASST 1987 vol 69 #87-129.

Kantrowitz, Arthur. "Propulsion to orbit by ground based lasers" *Astron. & Aeron.* 1972 vol 10 p 74-76.

Kelly, Anthony. "Fiber reinforced metals" ScAm Feb 1965.

Kerr, Richard. "Asteroid and comet dust in space" *Science* 23 Sep 1988 news.

Kiernan, Vincent. "Reactor project hitches onto Moon-Mars effort" *Science* 22 Jun 1990 news.

Kim, Kwang-je, & Sessler, Andrew. "Free electron lasers" *Science* 5 Oct 1990.

Lederman, Leon. "Value of fundamental science" ScAm Nov 1984 (muon catalyst).

Lemke, E. H. "Magnetic acceleration of interstellar probes" JBIS 1982 vol 35 p 498-503.

Martin, Anthony, & Bond, Alan. "Propulsion system: Part I theoretical..." JBIS PD 1978 p S44-S62.

Matloff, G. L. "Beyond the 1000 year ark: further studies of non-nuclear interstellar flight" JBIS 1983 vol 36 p 483-489.

Matloff, G. L. "Electric propulsion and interstellar flight" AIAA 1987 #87-1052.

Matloff, Gregory. "Faster non-nuclear world ships" JBIS 1986 vol 39 p 475-485.

Matloff, Gregory. "Interstellar ramjet acceleration runway" JBIS 1979 vol 32 p 219-220.

Matloff, Gregory. "Interstellar solar sailing: consideration of real and projected sail materials" JBIS 1984 vol 37 p 135-141.

Matloff, G. L., & Fennelly, A. J. "Interstellar applications and limitations of several electrostatic/electromagnetic ion collection techniques" JBIS 1977 vol 30 p 213-222.

Matloff, G. L., & Mallove, E. S. "Solar sail starships: clipper ships of the galaxy" JBIS 1981 vol 34 p 371-380.

Matloff, G., & Parks, K. "Interstellar gravity assist propulsion: correction and new application" JBIS 1988 vol 41 p 519-526.

Matloff, G., & Ubell, C. "Worldships: prospects for non-nuclear propulsion and power sources" JBIS 1985 vol 38 p 253-261.

McAleer, Neil. "The Light stuff: laser propulsion" *Spaceworld* Jul 1987.

Nordley, Gerald. "Application of antimatter-electric power to interstellar travel" JBIS 1990 vol 43 p 241-258.

Nyberg, G. A., et al. "Performance analysis of laser-powered SSTO shuttle" AASST 1987 vol 69 #87-130.

O'Neill, Gerard. "Colonization of space" *Physics Today* Sep 1974.

Parkinson, R. C. "Propellent acquisition techniques" JBIS PD 1978 S83-S89.

Pirri, A. N., et al. "Propulsion by absorption of laser radiation" AIAA 1973 #73-624.

Pool, Robert. "Brookhaven chemists find new fusion method" *Science* 29 Sep 1989.

Powell, C., et al. "Energy balance in fusion rockets" *Astron. Acta* 1973 vol 18 p 59-69.

Rafelski, J., & Jones, S. "Cold nuclear fusion" ScAm Jul 1987.

Rasleigh, S. E., & Marshall, R. A. "Electromagnetic acceleration of macro-particles to high velocities" *J. Appl. Phys.* 1978 vol 49 p 2540-.

Reupke, W. A. "Implications of inertial fusion system studies for rocket propulsion" AIAA 1982 #82-1213.

Reupke, W. A. "Inertial fusion system studies and nuclear pulse propulsion" JBIS 1985 vol 38 p 483-493 (based on his AIAA paper).

Roth, J. R., et al. "Fusion power for space propulsion" *New Scientist* 20 Apr 1972.

Ruppe, H. O. "Comments on...Singer JBIS 1980..." JBIS 1981 vol 34 p 115-116.

Shoji, J. M. & Larson, V. R. "Performance and heat transfer characteristics of laser heated rocket" AIAA 1976 # 76-1044.

Singer, Clifford. "Interstellar propulsion using a pellet stream for momentum transfer" JBIS 1980 vol 33 p 107-115.

Singer, C. E. "Questions concerning pellet-stream propulsion" JBIS 1981 vol 34 p 117-119.

Slocum, R. W., & Zapola, A. G. "Microwave rocket propulsion" *Am. Rocket Soc. J.* 1961 vol 31 p 657-658.

Spencer, D., & Jaffe, L. "Feasibility of interstellar travel" JPL 1962 #32-233; also *Astron. Acta* 1963 vol 9 p 49-58 (fission & fusion only).

Subotowicz, M. "Energy for interstellar flights delivered through nuclear matter compression..." JBIS 1986 vol 39 p 312-316.

Vulpetti, G. "Starting model for fission engine" JBIS 1975 vol 28 p 573-578.

Weiss, R. F., et al. "Laser propulsion" *Astron. & Aeron.* 1979 vol 17 p 50-58.

Wilson, Howard, et al. "Lessons of Sunraycer" ScAm Mar 1989.

Winterberg, F. "Electromagnetic rocket gun" *Acta Astron.* 1985 vol 12 p 155-161.

Winterberg, F. "Rocket propulsion by nuclear microexplosions..." JBIS 1979 vol 32 p 403-409.

Winterberg, F. "Rocket propulsion by staged thermonuclear micro-explosions" JBIS 1977 vol 30 p 333-340.

CHAPTER 2: BOOKS ON ADVANCED PROPULSION (NON-RELATIVISTIC):

CRC. *Handbook of Chemistry and Physics* 49th ed. Chemical Rubber Co.

Dyson, Freeman. *Infinite in All Directions* 1988 Harper & Row.

Friedman, Louis. *Starsailing* 1988 Wiley.

Jackson, J. D. *Classical Electrodynamics* 1962 Wiley.

Jahn, Robert. *Physics of Electric Propulsion* 1968 McGraw-Hill.

Johnson, Richard, & Holbrow, Charles, eds. *Space Settlements* 1977 NASA #SP-413.

Mauldin, John. *Particles in Nature* 1986 TAB.

Nicolson, Ian. *The Road to the Stars* 1978 Wm Morrow.

O'Neill, Gerard. *The High Frontier: Human Colonies in Space* 1976 Wm Morrow.

O'Neill, Gerard. *2081* 1981 Simon & Schuster.

Popov, E. P. *Mechanics of Materials* 1952 Prentice-Hall.

Reitz, John, & Milford, Frederick. *Foundations of Electromagnetic Theory* 1960 Addison-Wesley.

Sänger, Eugen. *Space Flight* 1965 McGraw-Hill.

Segre, Emilio. *Nuclei and Particles* 1964 Benjamin.

CHAPTER 3: ARTICLES ON RELATIVITY:

Ackeret, J. "On the theory of rockets" JBIS 1947 vol 6 p 116-123.

Anderson, Gerald. "Optimal interstellar relativistic rocket trajectories with both thrust and acceleration constraints" JBIS 1974 vol 27 p 273-285.

Anderson, Gerald, & Greenwood, Donald. "Relativistic flight with constant thrust rocket" *Astron. Acta* 1971 vol 16 p 153-158.

Bondi, H. "Space traveller's youth" *Discovery* Dec 1957.

Brigman, G. H. "Four vector kinematics of interstellar travel" AASA 1963 vol 8 p 505-518.

Hafele, J., & Keating Richard. "Around the world atomic clocks: observed relativistic time gains" *Science* 14 Jul 1972.

Huth, John. "Relativistic theory of rocket flight with advanced propulsion systems" *Am. Rocket. Soc. J.* Mar 1960 p 250-253.

Kooy, J. M. "On relativistic rocket mechanics" *Astron. Acta* 1958 vol 4 p 31-58.

Marx, G. "Mechanical efficiency of interstellar vehicles" *Astron. Acta* 1963 vol 9 p 131-139.

McMillan, E. M. "Clock paradox and space travel" *Science* 30 Aug 1957.

Moskowitz, Saul, & Devereux, William. "Trans-stellar space navigation" *AIAA J.* 1968 vol 6 p 1021-1029.

Muller, Richard. "Cosmic background radiation and new aether" ScAm May 1978.

Oliver, B. M. "Review of interstellar rocketry fundamentals" JBIS 1990 vol 43 p 259-264.

Peschka, W. "Uber die überbrücknung interstellaren entfernungen" *Astron. Acta* 1956 vol 2 p 191-200 (German).

Pierce, J. R. "Relativity and space travel" *IRE Proceedings* Jun 1959 p 1053-1061.

Pomeranz, Kalman. "Equation of motion for relativistic particles and systems with variable rest mass" *Am. J. Phys.* 1964 vol 32 p 955-958.

Powell, Conley. "Optimum exhaust velocity programming for...relativistic rocket" JBIS 1974 vol 27 p 263-266.

Powell, Conley, & Mikkilineni, R. P. "Optimal exhaust velocity programming for a single stage constant power starship" JBIS 1977 vol 30 p 460-462.

Purcell, Edward. "Radioastronomy and communication through space" USAEC 1961 #BNL-658; in Goldsmith p 188-196; in Cameron p 121-143.

Sagan, Carl. "Direct contact among galactic civilizations by relativistic interstellar spaceflight" *Plan. & Space Sc.* 1963 vol 11 p 485-498; also in Goldsmith p 205-213.

Sänger, E. "Some optical and kinematic effects in interstellar astronautics" JBIS 1961 vol 18 p 273-277.

Sänger, Eugen. "Zur flugmechanik der photonenraketen" *Astron. Acta* 1957 vol 3 p 89-99 (German).

Scott, G. D., & van Driel, H. J. "Geometrical appearances at relativistic speeds" *Am. J. Phys.* Aug 1970 vol 38 p 971-977.

Scott, G. D., & Viner, M. R. "Geometrical appearance of large objects moving at relativistic speeds" *Am. J. Phys.* Jul 1965 vol 33 p 534-536.

Shepherd, L. R. "Interstellar flight" JBIS 1952 vol 11 p 149-167.

Stimets, R. W., & Sheldon, E. "Celestial view from a relativistic starship" JBIS 1981 vol 34 p 83-99.

Terrell, James. "Invisibility of Lorentz contraction" *Phys. Rev.* 1959 vol 116 p 1041-1045.

von Hoerner, Sebastian. "General limits of space travel" *Science* 6 Jul 1962 vol 137 p 18-23.

Weisskopf, Victor. "Visual appearance of rapidly moving objects" *Phys. Today* Sep 1960.

CHAPTER 3: BOOKS ON RELATIVITY:

Belinfante, F. J. *Introduction to Special Relativity Theory* circa 1967 unpub. notes.

Bondi, Hermann. *Relativity and Common Sense* 1964 Doubleday.

Born, Max. *Einstein's Theory of Relativity* 1924 Methuen.

Calder, Nigel. *Einstein's Universe* 1979 Viking.

Einstein, Albert. *Relativity* 1961 Crown.

French, A. P. *Special Relativity* 1968 Norton.

Goldstein, Herbert. *Classical Mechanics* 1950 Addison-Wesley.

Kaufmann, William. *The Cosmic Frontiers of General Relativity* 1977 Little, Brown.

Russell, Bertrand. *The ABC of Relativity* 1958 Allen & Unwin.

Sänger, Eugen. *Space Flight* 1965 McGraw-Hill.

Taylor, Edwin, & Wheeler, John. *Spacetime Physics* 1966 Freeman.

CHAPTER 4: ARTICLES ON RELATIVISTIC PROPULSION AND INTERSTELLAR SPACE:

Andrews, Dana, & Zubrin, Robert. "Magnetic sails and interstellar travel" JBIS 1990 vol 43 p 265-272.

Archer, J. L., & O'Donnell, A. J. "Scientific exploration of near stellar systems" in AASA 1971 vol 29 p 691-724.

Baker, W. F., et al. "Measurement of [pion, kaon, proton, and antiproton] production by 200 and 300 GeV protons" *Phys. Ltrs.* 5 Aug 1974 vol 51B p 303-305.

Benedikt, E. T. "Disintegration barriers to extremely high speed space travel" AASA 1961 vol 6 p 571-588.

Berge, Glenn, & Seielstad, George. "Magnetic field of the galaxy" ScAm Jun 1965.

Blitz, Leo. "Giant molecular cloud complexes in the galaxy" ScAm Apr 1982.

Bolie, V. W. "Cosmic background radiation drag effects at relativistic speeds" JBIS 1981 vol 34 p 499-500.

Bond, Alan. "Analysis of potential performance of ram-augmented interstellar rocket" JBIS 1974 vol 27 p 674-685.

Bond, Alan. "Problems of interstellar propulsion" *Spaceflight* 1971 vol 13 p 245-251.

Bond, Alan, & Martin, Anthony. "Project Daedalus reviewed" JBIS 1986 vol 39 p 385-390.

Bussard, R. W. "Galactic matter and interstellar flight" *Astron. Acta* 1960 vol 6 p 179-194.

Cassenti, B. N. "Comparison of interstellar propulsion methods" JBIS 1982 vol 35 p 116-124; also AIAA 1980 #80-1229.

Cassenti, B. N. "Design considerations for relativistic antimatter rockets" JBIS 1982 vol 35 p 396-404.

Cassenti, B. N. "Optimization of relativistic antimatter rockets" AIAA 1983 #83-1343.

Chapline, George. "Antimatter breeders?" JBIS 1982 vol 35 p 423-424.

Crowe, Devon. "Laser induced pair production as a matter-antimatter source" JBIS 1983 vol 36 p 507-508.

Dyson, Freeman. "Interstellar propulsion systems" in H&Z p 41-45.

Engelberger, J. F. "Space propulsion by magnetic field interaction" *J. Spacecraft* 1964 vol 1 p 347-349.

Ehricke, K. A. "Evolution of interstellar operations" in AASA 1969 vol 14 microfiche #69-387.

Eklund, Philip. "Rocket flight" *L-5 News* Feb 1982 (broad non-tech. review).

Fennelly, A. J. "Some solar system applications of an interstellar ion scoop" 1976 vol 29 p 489-493.

Fishback, John. "Relativistic interstellar spaceflight" *Astron. Acta* 1969 vol 15 p 25-35.

Forward, R. L. "Antimatter propulsion" JBIS 1982 vol 35 p 391-395.

Forward, Robert. "Feasibility of interstellar travel: a review" JBIS 1986 vol 39 p 379-384.

Forward, R. L. "Interstellar flight systems" AIAA 1980 #80-0823.

Forward, R. L. letter on light sailing JBIS 1986 vol 39 p 328.

Forward, R. L. "Program for interstellar exploration" JBIS 1976 vol 29 p 611-632.

Forward, R. L. "Roundtrip interstellar travel using laser-pushed lightsails" *J. Spacecraft* 1984 vol 21 p 187-195.

Forward, Robert. "Starwisp: an ultralight interstellar probe" *J. Spacecraft* 1985 vol 22 p 345-350.

Forward, R. L. "Zero thrust velocity vector control for interstellar probes" *AIAA J.* 1964 vol 2 p 885-889; also AIAA 1964 #64-53.

Forward, Robert, & Davis, Joel. "Ride a laser to the stars" *New Scientist* 2 Oct 1986.

Gordon, M. A., & Burton, W. B. "Carbon monoxide in the galaxy" ScAm May 1979.

Greenberg, J. M. "Interstellar grains" ScAm Oct 1967.

Greenberg, J. M. "Structure and evolution of interstellar grains" ScAm Jun 1984.

Hamilton, David. "Magnet lab decision repels MIT" *Science* 21 Sep 1990 news.

Heppenheimer, T. A. "On infeasibility of interstellar ramjets" JBIS 1978 vol 31 p 222-224.

Huth, John. "Relativistic theory of rocket flight with advanced propulsion systems" *Am. Rocket. Soc. J.* Mar 1960 p 250-253.

Jackson, A. A. "Some considerations of antimatter and fusion ram-augmented interstellar rocket" JBIS 1980 vol 33 p 117-120.

Jackson, A. A., & Whitmire, Daniel. "Laser powered interstellar rocket" JBIS 1978 vol 31 p 335-337.

Jones, Eric. "Manned interstellar vessel using microwave propulsion" JBIS 1985 vol 38 p 270-273.

Jones, Eric. "Where are they? Implications of ancient and future migrations" in P85 p 465-476 (microwaves).

Jones, Eric, & Finney, Ben. "Fast ships and nomads: two roads to the stars" in F&J p 88-103.

Kirz, Janos. "Phase zone plates for x-rays and extreme UV" *J Optical Soc Am* 1974 vol 64 p 301-309.

Kooy, J. M. "On relativistic rocket mechanics" *Astron. Acta* 1958 vol 4 p 31-58.

Lada, Charles, & Shu, Frank. "Formation of sunlike stars" *Science* 4 May 1990.

Langton, N. H. "Erosion of interstellar drag screens" JBIS 1973 vol 26 p 481-484.

Langton, N. H., & Oliver, W. R. "Materials in interstellar flight" JBIS 1977 vol 30 p 109-111.

Lederman, Leon. "The Tevatron" ScAm Mar 1991.

Mallove, Eugene. "Forward motion" *Air & Space* Jun 1987.

Martin, A. R. "Bombardment by interstellar material and its effects on the vehicle" JBIS PD 1978 p S116-S121.

Martin, A. R. "Effects of drag on relativistic spaceflight" JBIS 1972 vol 25 p 643-653.

Martin, A. R. "Magnetic intake limitations on interstellar ramjets" *Astron. Acta* 1973 vol 18 p 1-9.

Martin, A. R. "Some limitations of the interstellar ramjet" *Spaceflight* 1972 vol 14 p 21-25.

Martin, A. R. "Structural limitations on interstellar spaceflight" *Acta Astron.* 1971 vol 16 p 353-357.

Marx, G. "Interstellar vehicle propelled by terrestrial laser beam" *Nature* 2 Jul 1966.

Marx, G. "Mechanical efficiency of interstellar vehicles" *Astron. Acta* 1963 vol 9 p 131-139.

Marx, G. "Uber energieprobleme der interstellaren raumfahrt" *Astron. Acta* 1960 vol 6 p 366-372 (German).

Massier, P. F. "Need for expanded exploration of matter-antimatter annihilation for propulsion application" JBIS 1982 vol 35 p 387-390.

Matloff, Gregory. "Cosmic ray shielding for manned interstellar arks and mobile habitats" JBIS 1977 vol 30 p 96-98.

Matloff, Gregory. "Faster non-nuclear world ships" JBIS 1986 vol 39 p 475-485.

Matloff, G. L. "Superconducting ion scoop and its applications to interstellar flight" JBIS 1974 vol 27 p 663-673.

Matloff, Gregory. "Utilization of O'Neill's Model 1 Lagrange point colony as interstellar ark" JBIS 1976 vol 29 p 775-785.

Matloff, G. L., & Fennelly, A. J. "Interstellar applications and limitations of several electrostatic/ electromagnetic ion collection techniques" JBIS 1977 vol 30 p 213-222.

Matloff, G., & Ubell, C. "Worldships: prospects for non-nuclear propulsion and power sources" JBIS 1985 vol 38 p 253-261.

McKee, Christopher, & Draine, Bruce. "Interstellar shock waves" *Science* 19 Apr 1991.

Morgan, D. L. "Concept for design of antimatter annihilation rocket" JBIS 1982 vol 35 p 405-412.

Ney, Edward. "Star dust" *Science* 11 Feb 1977.

Nordley, Gerald. "Application of antimatter-electric power to interstellar travel" JBIS 1990 vol 43 p 241-258.

Norem, Philip. "Interstellar travel: a round trip propulsion system with relativistic velocity capabilities" AASST 1969 #69-388.

Oliver, B. M. "Review of interstellar rocketry fundamentals" JBIS 1990 vol 43 p 259-264.

Parker, E. N. "Magnetic fields in the cosmos" ScAm Aug 1983.

Paresce, Francesco, & Bowyer, Stuart. "Sun and interstellar medium" ScAm Sep 1986.

Pasachoff, Jay, & Fowler, William. "Deuterium in the universe" ScAm May 1974.

Pierce, J. R. "Relativity and space travel" *IRE Proceedings* Jun 1959 p 1053-1061.

Powell, Conley "Effect of subsystem inefficiencies upon performance of ram-augmented interstellar rocket" JBIS 1976 vol 29 p 786-794.

Powell, Conley. "Flight dynamics of ram-augmented interstellar rocket" JBIS 1975 vol 28 p 556-562.

Powell, Conley. "Flight time minimisation for an energy-limited flyby star probe" 1974 JBIS vol 27 p 267-272.

Powell, Conley. "Heating and drag at relativistic speeds" JBIS 1975 vol 28 p 546-552.

Redding, J. L. "Interstellar vehicle propelled by terrestrial laser beam" *Nature* 11 Feb 1967.

Roberts, William. "Relativistic dynamics of a sublight speed interstellar ramjet probe" JBIS 1976 vol 29 p 795-812; also AAS 1975 #75-020.

Sagan, Carl. "Direct contact among galactic civilizations by relativistic interstellar spaceflight" *Plan. & Space Sc.* 1963 vol 11 p 485-498; also in Goldsmith p 205-213.

Sagdeev, Roald, & Kennel, Charles. "Collisionless shock waves" ScAm Apr 1991.

Segre, Emilio, & Wiegand, Clyde. "The antiproton" ScAm Jun 1956.

Sussman, M. "Elementary diffraction theory of zone plates" *Am. J. Phys.* 1960 vol 28 p 394.

von Hoerner, Sebastian. "General limits of space travel" *Science* 6 Jul 1962 vol 137 p 18-23.

Vulpetti, G. "Antimatter propulsion for space exploration" JBIS 1986 vol 39 p 391-409.

Vulpetti, G. "Approach to modeling matter-antimatter propulsion systems" JBIS 1984 vol 37 p 403-409.

Vulpetti, G. "Multiple propulsion concept" JBIS 1979 vol 32 p 209-214.

Vulpetti, G. "Propulsion oriented synthesis of antiproton-nucleon annihilation experimental results" JBIS 1984 vol 37 p 124-134.

Waldrop, M. "Superconductor's critical current at a new high" *Science* 18 Dec 1987.

Whitmire, D. P. "Relativistic spaceflight and the catalytic nuclear ramjet" *Acta Astron.* 1975 vol 2 p 497-509.

Whitmire, D. P., & Jackson, A. A. "Laser powered interstellar ramjet" JBIS 1977 vol 30 p 223-226.

Winston, Roland. "Nonimaging optics" ScAm Mar 1991.

Zito, Richard. "Chain reaction in a hydrogen-antimatter pile" JBIS 1983 vol 36 p 308-310.

Zito, R. R. "Cryogenic confinement of antiprotons for space propulsion systems" JBIS 1982 vol 35 p 411-421.

CHAPTER 4: BOOKS ON RELATIVISTIC PROPULSION AND INTERSTELLAR SPACE:

Hecht, Eugene, & Zajac, Alfred. *Optics* 1974 Addison-Wesley.

Forward, Robert. *Future Magic* 1988 Avon.

Forward, Robert, & Davis, Joel. *Mirror Matter: Pioneering antimatter physics* 1988 Wiley.

French, A. P. *Special Relativity* 1968 Norton.

Friedman, Louis. *Starsailing* 1988 Wiley.

Hollenbach, David, et al. *Interstellar Processes* 1987 Reidel.

Jackson, J. D. *Classical Electrodynamics* 1962 Wiley.

Mauldin, John. *Light, Lasers, and Optics* 1988 TAB.

Nicolson, Ian. *The Road to the Stars* 1978 Wm Morrow.

Sagan, Carl. *Communication with Extraterrestrial Intelligence* 1973 MIT Pr.

Spitzer, Lyman. *Physical Processes in the Interstellar Medium* 1978 Wiley.

Spitzer, Lyman. *Searching between the Stars* 1982 Yale UP.

CHAPTER 5: ARTICLES ON STARSHIP SYSTEMS AND GENERAL ENGINEERING:

---. "How color affects people and work" *Science* 10 Nov 1989 news from NASA.

Aston, Graeme. "Electric propulsion" JBIS 1986 vol 39 p 503-507.

Bond, Alan. "Auxiliary power supplies" JBIS PD 1978 p S126-S129.

Bond, Alan, & Martin, Anthony. "Project Daedalus" (introduction) & "Mission profile" JBIS PD 1978 p S5-S8 & p S37-S42.

Bond, Alan, & Martin, Anthony. "Worldships: assessment of engineering feasibility" JBIS 1984 vol 37 p 254-266.

Dawe, Russell, et al. "Structuring the outer-planet spacecraft" *Astron. & Aeron.* Sep 1970 p 89-93.

Jaffe, L. D., et al. "Interstellar precursor mission" JBIS 1980 vol 33 p 3-26.

Kleinknecht, Kenneth. "Design principles stressing simplicity" *Astron. & Aeron.* Mar 1970 p 46-49.

Martin, Anthony. "Worldships: concept, cause, cost, construction, colonization" JBIS 1984 vol 37 p 243-253.

Matloff, Gregory. "Faster non-nuclear world ships" JBIS 1986 vol 39 p 475-485.

Matloff, Gregory. "Utilization of O'Neill's Model 1 Lagrange point colony as interstellar ark" JBIS 1976 vol 29 p 775-785.

Matloff, G. L., & Fennelly, A. J. "Interstellar application of several electrostatic/ electromagnetic ion collection techniques" JBIS 1977 vol 30 p 213-222.

Matloff, G. L., & Mallove, E. S. "Solar sail starships: clipper ships of the galaxy" JBIS 1981 vol 34 p 371-380.

O'Neill, Gerard. "Colonization of space" *Phys. Today* Sep 1974.

Parkinson, R. C. "Propellent acquisition techniques" JBIS PD 1978 S83-S89.

Powell, Conley. "Flight time minimisation for an energy-limited flyby star probe" 1974 JBIS vol 27 p 267-272.

CHAPTER 5: BOOKS ON STARSHIP SYSTEMS AND GENERAL ENGINEERING

Johnson, Richard, & Holbrow, Charles, eds. *Space Settlements* 1977 NASA #SP-413.

Krick, E. V. *Introduction to Engineering and Engineering Design* 1969 Wiley.

McAleer, Neil. *The Omni Space Almanac* 1987 Omni/World Almanac.

CHAPTER 6: ARTICLES ON INTERSTELLAR MISSIONS, STARS, AND PROBES:

Abt, Helmut. "Companions of sunlike stars" ScAm Apr 1977.

Archer, J. L., & O'Donnell, A. J. "Scientific exploration of near stellar systems" in AASA 1971 vol 29 p 691-724.

Arrhenius, Svante. "Propagation of life in space" *Die Umschau* 1903 vol 7 p 481 (German); in Goldsmith p 32-33 (translated).

Aston, Graeme. "Electric propulsion" JBIS 1986 vol 39 p 503-507.

Aumann, Hartmut. "Protoplanetary material around nearby stars" in P85 p 43-50.

Bahcall, John, & Spitzer, Lyman. "The Space telescope" ScAm Jul 1982.

Black, David. "Worlds around other stars" ScAm Jan 1991.

Black, David, & Suffolk, Graham. "Concerning the planetary system of Barnard's Star" *Icarus* 1973 vol 19 p 353-356, and in Goldsmith p 158-161.

Bond, Alan, & Martin, Anthony. "Project Daedalus" (introduction) & "Mission profile" JBIS PD 1978 p S5-S8 & p S37-S42.

Bond, Alan, & Martin, Anthony. "Worldships: assessment of engineering feasibility" JBIS 1984 vol 37 p 254-266.

Boss, Alan. "Collapse and formation of stars" ScAm Jan 1985.

Bracewell, R. N. "Communications from superior galactic communities" *Nature* 28 May 1960; in Goldsmith p 105-107.

Bracewell, R. N. "Interstellar probes" in P&C p 102-116.

Burke, Bernard. "Astrophysics from the moon" *Science* 7 Dec 1990.

Burns, Jack, et al. "Observatories on the moon" ScAm Mar 1990.

Cameron, A. G. W. "Origin and evolution of the Solar System" ScAm Sep 1975.

Cesarone, Robert, et al. "Prospects for Voyager" JBIS 1984 vol 37 p 99-116.

Crick, F. H. C., & Orgel, L. E. "Directed panspermia" *Icarus* 1973 vol 14 p 341-; in Goldsmith p 34-37.

Croswell, Ken. "Looking for a few good stars" *Spaceworld* Jul 1987.

Doyle, Laurence. "Assisting extrasolar planetary detection through determination of stellar space orientations" in P85 p 97-99.

Finney, B. R., & Jones, E. M. "From Africa to the stars" in AASA 1983 vol 53 #83-205.

Fogg, Martyn. "Feasibility of intergalactic colonisation and its relevance to SETI" JBIS 1988 vol 41 p 491-496.

Forward, R. L. "Program for interstellar exploration" JBIS 1976 vol 29 p 611-632.

Forward, R. L. "Roundtrip interstellar travel using laser-pushed lightsails" *J. Spacecraft* 1984 vol 21 p 187-195.

Forward, Robert. "Starwisp: an ultralight interstellar probe" *J. Spacecraft* 1985 vol 22 p 345-350.

Freitas, Robert. "Case for interstellar probes" JBIS 1983 vol 36 p 490-495.

Freitas, Robert. "Interstellar probes: new approach to SETI" JBIS 1980 vol 33 p 95-100.

Freitas, Robert. "Observable characteristics of extraterrestrial technological civilizations" JBIS 1985 vol 38 p 106-112.

Freitas, Robert. "Self-reproducing interstellar probe" JBIS 1980 vol 33 p 251-264.

Gatley, Ian, et al. "Astronomical imaging with infrared array detectors" *Science* 2 Dec 1988.

Greenberg, J. M., & Weber, Peter. "Panspermia: a modern astrophysical and biological approach" in P85 p. 157-164.

Hart, Michael. "Interstellar migration, biological revolution, and future of the galaxy" in F&J.

Hodges, William. "Division of labor and interstellar migration..." in F&J p 134-151.

Jones, E. M. "Interstellar colonization" JBIS 1978 vol 31 p 103-107.

Jones, Eric, & Finney, Ben. "Fastships and nomads: two roads to the stars" in F&J.

Jones, E. M., & Finney, B. R. "Interstellar nomads" AASA 1983 vol 53 #83-238.

Kemeny, John. "Man viewed as a machine" ScAm Apr 1955.

Marochnik, Leonid, et al. "Estimates of mass and angular momentum in the Oort cloud" *Science* 28 Oct 1988.

Martin, Anthony. "Ranking of nearby stellar systems for exploration" JBIS PD 1978 p S33-S36.

Martin, Anthony. "Worldships: concept, cause, cost, construction, colonization" JBIS 1984 vol 37 p 243-253.

Matloff, Gregory. "Faster non-nuclear world ships" JBIS 1986 vol 39 p 475-485.

Matloff, Gregory. "On potential performance of non-nuclear interstellar arks" JBIS 1985 vol 38 p 113-119.

Matloff, Gregory. "Utilization of O'Neill's Model 1 Lagrange point colony as interstellar ark" JBIS 1976 vol 29 p 775-785.

Matloff, Gregory, & Mallove, Eugene. "First interstellar colonization mission" JBIS 1980 vol 33 p 84-88.

Matloff, G. L., & Mallove, E. S. "Solar sail starships: clipper ships of the galaxy" JBIS 1981 vol 34 p 371-380.

Matloff, G., & Ubell, C. "Worldships: prospects for non-nuclear propulsion and power sources" JBIS 1985 vol 38 p 253-261.

Mattinson, H. R. "Astronomical data on nearby stellar systems" JBIS PD 1978 p S8-S18.

Meot-ner, M., & Matloff, G. L. "Directed panspermia: technical and ethical evaluation of seeding nearby solar systems" JBIS 1979 vol 32 p 419-423.

Moore, Edward. "Artificial living plants" ScAm Oct 1956.

Moore, Edward. "Mathematics in biological sciences" ScAm Sep 1964.

Newman, William, & Sagan, Carl. "Galactic civilizations: population dynamics and interstellar diffusion" *Icarus* 1981 vol 46 p 293-327.

Oliver, Bernard. "Search strategies" in Billingham p 351-376.

O'Neill, Gerard. "Colonization of space" *Phys. Today* Sep 1974.

Paresce, Francesco, & Bowyer, Stuart. "The Sun and the interstellar medium" ScAm Sep 1986.

Penrose, L. S. "Self-reproducing machines" ScAm Jun 1959.

Seward, F. D. "A Trip to the Crab Nebula" JBIS 1978 vol 31 p 83-92.

Singer, Cliff. "Settlements in space and interstellar travel" in H&Z p 46-61.

Stahler, Steven. "Early life of stars" ScAm Jul 1991.

Stephenson, David. "Comets and interstellar travel" JBIS 1983 vol 36 p 210-214.

Tipler, Frank. "Extraterrestrial intelligent beings do not exist" QJRAS 1980 vol 21 p 267- & vol 22; in *Phys. Today* Apr 1981; in Regis p 133-150.

Valdes, Francisco, & Freitas, Robert. "Comparison of reproducing and non-reproducing starprobes..." JBIS 1980 vol 33 p 402-408.

Van Allen, James. "Interplanetary particles and fields" ScAm Sep 1975.

van de Damp, Peter. "Stars nearer than 5 parsecs" *Sky & Telescope* Oct 1955.

von Hoerner, Sebastian. "Likelihood of interstellar colonization and absence of its evidence" in H&Z p 29-35.

Webb, J. A. "Detection of intelligent signals from space" in Cameron p 178-.

Zuckerman, B. "Space telescopes, interstellar probes, and directed panspermia" JBIS 1981 vol 34 p 367-370.

CHAPTER 6: BOOKS ON INTERSTELLAR MISSIONS, STARS, AND PROBES:

Abell, George. *Realm of the Universe* 1976 Holt Rinehart Winston.

Baker, Robert. *Astronomy* 1964 Van Nostrand.

Bernal, J. D. *The World the Flesh and the Devil* 1929 Methuen, 1969 Indiana UP.

Brand, Stewart, ed. *Space Colonies* 1977 Co-Evolution/Penguin.

DOE & NASA. *Satellite Power System* 1978 #DOE/ER-0023, DOE & NASA.

Eicher, David. *The Universe from Your Backyard* 1988 Cambridge UP.

Gilmore, Gerald, et al. *The Milky Way as a Galaxy* 1990 Univ. Sc. Bks.

Moore, Patrick. *New Concise Atlas of the Universe* 1978 Rand McNally.

O'Neill, Gerard. *The High Frontier: Human Colonies in Space* 1976 Wm Morrow.

Ridpath, Ian. *Messages from the Stars* 1978 Harper & Row.

Spitzer, Lyman. *Searching between the Stars* 1982 Yale UP.

von Neumann, John, & Burks, Arthur. *Theory of Self-Reproducing Automata* 1966 Univ. of Illinois Pr.

CHAPTER 7: ARTICLES ON ASTROGATION, OBSERVATION, AND COMMUNICATION:

Anderson, Gerald. "Some problems in communications with relativistic interstellar rockets" JBIS 1975 vol 28 p 168-174.

Bond, Alan, & Martin, Anthony. "Project Daedalus reviewed" JBIS 1986 vol 39 p 385-390.

Cocconi, Giuseppe, & Morrison, Philip. "Searching for interstellar communication" *Nature* 19 Sep 1959.

Dressler, Alan. "Large scale streaming of galaxies" ScAm Sep 1987 (our motions).

Harris, M. J. "On detectability of antimatter propulsion spacecraft" *Astrophys. & Space Sc.* 1986 vol 123 p 297-303.

Jaffe, L. D., et al. "Interstellar precursor mission" JBIS 1980 vol 33 p 3-26.

Malin, David. "A Universe of color" *Kodak Tech Bits* 1990 #3.

McKinley, J. M., & Doherty, P. "In search of the starbow: appearance of starfields from a relativistic spaceship" *Am. J. Phys.* 1979 vol 47 p 309-316.

Moskowitz, Saul. "Visual aspects of trans-stellar space flight" *Sky & Tele.* May 1967.

Moskowitz, Saul, & Devereux, William. "Trans-stellar space navigation" *AIAA J.* 1968 vol 6 p 1021-1029.

Oliver, R. M. "View from the starship bridge..." *IEEE Spectrum* Jan 1964.

Richards, G. R. "The Navigation problem" JBIS PD 1978 p S143-S148.

Ross, Monte. "Likelihood of finding extraterrestrial laser signals" JBIS 1979 vol 32 p 203-208.

Schwartz, R. N., & Townes, C. H. "Interstellar and interplanetary communication by optical masers" *Nature* 1961 vol 190 p 205-.

Sheldon, E., & Giles, R. H. "Celestial views from non-relativistic and relativistic interstellar spacecraft" JBIS 1983 vol 36 p 99-114.

Stimets, R. W., & Sheldon, E. "Celestial view from a relativistic starship" JBIS 1981 vol 34 p 83-99.

Tarter, Jill. "Planned observational strategy for NASA's first systematic SETI" in F&J p 314-330.

Viewing, D. R., et al. "Detection of starships" JBIS 1977 vol 30 p 99-104.

Vulpetti, G. "Problem in relativistic navigation: three-dimensional rocket equation" JBIS 1978 vol 31 p 344-351.

Wertz, J. R. "Interstellar navigation" *Spaceflight* 1972 vol 14 p 206-216.

Wilkinson, David. "Anisotropy of the cosmic blackbody radiation" *Science* 20 Jun 1986.

CHAPTER 7: BOOKS ON ASTROGATION, OBSERVATION, AND COMMUNICATION:

Abell, George. *Realm of the Universe* 1976 Holt Rinehart Winston.

Pierce, J. R. *Symbols, Signals, and Noise* 1961 Harper.

Sagan, Carl. *Communication with Extraterrestrial Intelligence* 1973 MIT Pr.

Smithsonian. *Smithsonian Astrophysical Observatory Star Catalog* 1966.

CHAPTER 8: ARTICLES ON TECHNOLOGICAL REQUIREMENTS AND HAZARDS:

Bak, Per, & Chen, Kan. "Self-organized criticality" ScAm Jan 1991 (failures).

Birch, Paul. "Orbital ring systems and jacob's ladders" 3 parts JBIS 1982 vol 35 p 475- & 1983 vol 36 p 115- & p 231-.

Birch, Paul. "Radiation shields for ships and settlements" JBIS 1982 vol 35 p 515-519.

Bond, Alan. "Target system encounter protection" JBIS PD 1978 p S123-S125 (dust).

Bond, Alan, & Martin, Anthony. "Project Daedalus reviewed" JBIS 1986 vol 39 p 385-390.

Corcoran, Elizabeth. "Calculating reality" ScAm Jan 1991 (computers).

Freitas, Robert. "Case for interstellar probes" JBIS 1983 vol 36 p 490-495.

Freitas, Robert. "Interstellar probes: new approach to SETI" JBIS 1980 vol 33 p 95-100.

Freitas, Robert. "Self-reproducing interstellar probe" JBIS 1980 vol 33 p 251-264.

Grant, T. J. "Need for onboard repair" JBIS PD 1978 p S172-S179.

Hannah, Eric. "Radiation protection for space colonies" JBIS 1977 vol 30 p 310-313.

Hayes, Thomas. "Superplastic rare-metal legacy of Stealth" NYT 25 Nov 1990 sec 3.

Heacock, Raymond. "The Voyager Spacecraft" IME Proceedings 1980 vol 194 p 211-224 (advanced technology).

Heppenheimer, T. A. "Launching the Aerospace Plane" *High Tech.* Jul 1986.

Hodges, William. "Division of labor and interstellar migration..." in F&J p 134-151.

Isaacs, John, et al. "Satellite elongation into a true sky-hook" *Science* 11 Feb 1966 (cable from Earth to synchronous orbit and beyond).

Jaffe, L. D., et al. "Interstellar precursor mission" JBIS 1980 vol 33 p 3-26.

Jones, C. P. "Engineering challenges of inflight spacecraft: Voyager case study" JBIS 1985 vol 38 p 465-471 (reliability).

Kodak. "Optical Disk System" 1989 Eastman Kodak Co. sales brochures.

Kryder, Mark. "Data storage technologies for advanced computing" ScAm Oct 1987.

Linsley, John. "Highest energy cosmic rays" ScAm Jul 1978.

Logsdon, John. "Space shuttle program: policy failure?" *Science* 30 May 1986.

Marshall, Eliot. "Academy panel faults NASA's safety analysis" *Science* 11 Mar 1988 news.

Marshall, Eliot. "Feynman issues his own shuttle report attacking NASA's risk estimate" *Science* 27 Jun 1986 news.

Martin, Anthony. "Bombardment by interstellar material and its effects on the vehicle" JBIS PD 1978 p S116-S121.

Matloff, Gregory. "Cosmic ray shielding for manned interstellar arks and mobile habitats" JBIS 1977 vol 30 p 96-98.

Matloff, Gregory. "Utilization of O'Neill's Model 1 Lagrange point colony as interstellar ark" JBIS 1976 vol 29 p 775-785.

Meyer, Peter, et al. "Cosmic rays" *Phys. Today* Oct 1974.

Morrison, David. "Testing the limits of mach 25" *Science* 20 May 1988 news.

Parfitt, J. A., & White, A. G. A. "Structural material selection" JBIS PD 1978 p S97-S103.

Pearson, Jerome. "Orbital tower: spacecraft launcher using Earth's rotational energy" *Acta Astron.* 1975 vol 2 p 785-799.

Pollack, Andrew. "Hologram computers of tomorrow" NYT 9 Jun 1991.

Smith, Alexander. "Failures, setbacks, and compensations in interstellar expansion" JBIS 1985 vol 38 p 265-269.

Steinberg, Morris. "Materials for aerospace" ScAm Oct 1986 (incl. spaceplane).

Stone, E. C., & Miner, E. D. "Voyager 2 encounter with the Neptunian system" *Science* 15 Dec 1989.

Tipler, Frank. "Extraterrestrial intelligent beings do not exist" QJRAS 1980 vol 21 p 267- & vol 22; in *Phys. Today* Apr 1981; in Regis p 133-150.

Waldrop, M. "Challenger disaster: assessing implications" *Science* 14 Feb 1986 news.

CHAPTER 8: BOOKS ON TECHNOLOGICAL REQUIREMENTS AND HAZARDS:

Churchland, Paul & Patricia. "Could a machine think?" ScAm Jan 1990.

Dennett, Daniel. *Brainstorms* 1978 Bradford Bks.

Dyson, Freeman. *Disturbing the Universe* 1979 Harper & Row.

Forward, Robert. *Future Magic* 1988 Avon.

Friedlander, Michael. *Cosmic Rays* 1989 Harvard UP.

Hofstadter, Douglas. *Metamagical Themas* 1985 Basic Bks.

Johnson, Richard, & Holbrow, Charles, eds. *Space Settlements* 1977 NASA #SP-413.

McAleer, Neil. *The Omni Space Almanac* 1987 Omni/World Almanac.

Morse, Philip. *Thermal Physics* 1964 Benjamin.

O'Neill, Gerard. *2081* 1981 Simon & Schuster.

Rees, D. J. *Health Physics* 1967 MIT Pr.

von Neumann, John, & Burks, Arthur. *Theory of Self-Reproducing Automata* 1966 Univ. of Illinois Pr.

CHAPTER 9: ARTICLES ON BIOLOGICAL SYSTEMS AND HAZARDS:

Bittles, Alan, et al. "Reproductive behavior and health in consanguineous marriages" *Science* 10 May 1991.

Bond, Alan, & Martin, Anthony. "Worldships: assessment of engineering feasibility" JBIS 1984 vol 37 p 254-266.

Briefings. "New cancer data from Oak Ridge" *Science* 5 Apr 1991.

Briefings. "NASA vegematic" *Science* 6 Jul 1990.

Crosby, Alfred. "Life with all its problems in space" in F&J p 210-219.

Drake, Frank. "On hands and knees in search of Elysium" *Tech. Rev.* Jun 1976.

Gibbons, Ann. "Gerontology research comes of age" *Science* 2 Nov 1990 news.

Hart, Michael. "Interstellar migration, biological revolution, and future of the galaxy" in F&J p 278-291.

Janick, Jules, et al. "Cycles of plant and animal nutrition" ScAm Sep 1976.

Leggett, N., & Fielder, J. "Space greenhouse design" JBIS 1984 vol 37 p 495-498.

Macklin, Martin. "Waste-combustion and water-recovery system" *J. Spacecraft* 1964 vol 1 p 349-350.

Matloff, Gregory. "Cosmic ray shielding for manned interstellar arks and mobile habitats" JBIS 1977 vol 30 p 96-98.

Matloff, Gregory. "Utilization of O'Neill's Model 1 Lagrange point colony as interstellar ark" JBIS 1976 vol 29 p 775-785.

Moffat, Anne. "Nitrogenase structure revealed" *Science* 14 Dec 1990 news.

Moore, Berrien, & MacElroy, R. D. eds. "Controlled ecological life support system" NASA 1982 #2233.

Murphy, J. R. "Medical considerations for manned interstellar flight" JBIS 1981 vol 39 p 466-475.

Olshansky, S. Jay, et al. "In search of methuselah: estimating upper limits to human longevity" *Science* 2 Nov 1990.

Richards, I. R. "Closed ecosystem for space colonies" JBIS 1981 vol 34 p 392-399.

Richards, I. R., & Parker, P. J. "Estimates of crop areas for large space colonies" JBIS 1976 vol 29 p 769-774.

Scrimshaw, Nevin, & Young, Vernon. "Requirements of human nutrition" ScAm Sep 1976.

Storey, Kenneth & Janet. "Frozen alive" ScAm Dec 1990.

Wurtman, Richard & Judith. "Carbohydrates and depression" ScAm Jan 1989.

CHAPTER 9: BOOKS ON BIOLOGICAL SYSTEMS AND HAZARDS:

Brand, Stewart, ed. *Space Colonies* 1977 Co-Evolution/Penguin.

Heppenheimer, T. A. *Colonies in Space* 1977 Stackpole.

Johnson, Richard, & Holbrow, Charles, eds. *Space Settlements* 1977 NASA #SP-413.

Lorr, David, et al. *Working in Orbit and Beyond: Challenges for space medicine* AASST 1989 vol 72.

O'Neill, Gerard. *The High Frontier: Human Colonies in Space* 1976 Wm Morrow.

Rees, D. J. *Health Physics* 1967 MIT Pr.

CHAPTER 10: ARTICLES ON PSYCHOLOGICAL, SOCIAL, POLITICAL, AND ECONOMIC MATTERS:

Bainbridge, William. "Religions for a galactic civilization" AASH 1982 vol 5 p 187-201.

Bloomfield, Masse. "Sociology of an interstellar vehicle" JBIS 1986 vol 39 p 116-120.

Bond, Alan, & Martin, Anthony. "Project Daedalus" (introduction) & "Mission profile" JBIS PD 1978 p S5-S8 & p S37-S42.

Criswell, David. "Solar system industrialization..." in F&J p 50-87.

Chernow, Ron. "Colonies in space...nice places to live" *Smithsonian* Feb 1976.

Emme, Eugene. "An Eclectic bibliography on the history of space futures" AASH 1982 vol 5 p 213-245.

Freitas, Robert. "Self-reproducing interstellar probe" JBIS 1980 vol 33 p 251-264.

Glaser, Peter. "Power from the sun" *Science* 22 Nov 1968 (via satellite).

Hart, Michael. "Interstellar migration, biological revolution, and future of the galaxy" in F&J p 278-291.

Hartmann, William. "Resource base in our solar system" in F&J p 26-41.

Hodges, William. "Division of labor and interstellar migration..." in F&J p 134-151.

Holmes, D. L. "Worldships: a sociological view" JBIS 1984 vol 37 p 296-304.

Jones, Eric. "Where are they? Implications of ancient and future migrations" in P85 p 465-476.

Lacey, David. "Some social implications of a generation starship" JBIS 1984 vol 37 p 499-501.

Martin, Anthony. "Worldships: concept, cause, cost, construction, colonization" JBIS 1984 vol 37 p 243-253.

Molton, Peter. "On the likelihood of human interstellar civilization" JBIS 1978 vol 31 p 203-208.

Murphy, J. R. "Medical considerations for manned interstellar flight" JBIS 1981 vol 39 p 466-475.

O'Leary, Brian. "Mining Apollo and Amor asteroids" *Science* 22 Jul 1977.

O'Neill, Gerard. "Colonization of space" *Phys. Today* Sep 1974.

O'Neill, Gerard. "Space colonies and energy supply to earth" *Science* 5 Dec 1975.

Parkinson, R. C. "Propellant acquisition techniques" JBIS PD 1978 p S83-S89.

Parkinson, B. "Starship as exercise in economics" JBIS 1974 vol 27 p 692-696.

Regis, Edward. "Moral status of multi-generational interstellar exploration" in F&J p 248-259.

Tanner, Nancy. "Interstellar migration" in F&J p 220-233.

Wachter, Kenneth. "Predicting demographic contours of an interstellar future" in F&J p 120-133.

CHAPTER 10: BOOKS ON PSYCHOLOGICAL, SOCIAL, POLITICAL, AND ECONOMIC MATTERS:

Asimov, Isaac. *A Choice of Catastrophes* 1979 Doubleday.

Brand, Stewart, ed. *Space Colonies* 1977 Co-Evolution/Penguin.

DOE & NASA. *Satellite Power System* 1978 #DOE/ER-0023, DOE & NASA.

Golden, F. *Colonies in Space* 1977 Harcourt Brace Jovanovich.

Heppenheimer, T. A. *Colonies in Space* 1977 Stackpole.

Johnson, Richard, & Holbrow, Charles, eds. *Space Settlements* 1977 NASA #SP-413.

Lewis, John & Ruth. *Space Resources* 1987 Columbia UP.

O'Neill, Gerard. *The High Frontier: Human Colonies in Space* 1976 Wm Morrow.

Pournelle, Jerry, ed. *The Endless Frontier* 1979 Ace.

Reiber, Duke, ed. *The NASA Mars Conference* AASST 1988 vol 71.

Verschuur, G. *Cosmic Catastrophes* 1978 Addison-Wesley.

CHAPTER 11: ARTICLES ON OTHER LIFE, SETI, AND CIVILIZATIONS:

Archer, J. L., & O'Donnell, A. J. "Scientific exploration of near stellar systems" in AASA 1971 vol 29 p 691-724.

Argyle, Edward. "Chance and the origin of life" in H&Z p 100-112.

Aumann, Hartmut. "Protoplanetary material around nearby stars" in P85 p 43-50.

Ball, John. "Extraterrestrial intelligence: where is everybody?" in P85 p 483-486.

Ball, John. "Universal aspects of biological evolution" in P85 p 251-253.

Ball, John. "Zoo hypothesis" *Icarus* 1973 vol 19 p 347-; in Goldsmith p 241-242.

Beck, Lewis. "Extraterrestrial intelligent life" in Regis p 3-18.

Benford, Gregory. "Alien technology" in B&P p 165-177 (placed near us).

BIS. "Fermi paradox: forum" JBIS 1979 vol 32 p 424-434.

Black, David. "Prospects for detecting other planetary systems" in Billingham p 163-176.

Black, David. "Worlds around other stars" ScAm Jan 1991.

Bracewell, R. N. "Communications from superior galactic communities" *Nature* 28 May 1960; in Goldsmith p 105-107.

Bracewell, R. N. "An extended Drake equation..." *Acta Astron.* 1979 vol 6 p 67-69.

Bracewell, R. N. "Interstellar probes" in P&C p 102-116.

Bracewell, R., & MacPhie, R. "Searching for nonsolar planets" *Icarus* 1979 vol 38 p 136-; in Goldsmith p 162-169.

Brin, David. "Mystery of the great silence" in B&P p 118-139.

Cameron, A. G. W. "Origin and evolution of the Solar System" ScAm Sep 1975.

Carr, Bernard, & Rothman, Tony. "Coincidences in nature and the hunt for the anthropic principle" in Rothman p 107-130.

Clark, Benton. "Sulfur: fountainhead of life in the universe?" in Billingham p 47-60.

Clark, D. H., et al. "Frequency of nearby supernovae and climatic and biological catastrophes" *Nature* 1977 vol 265 p 318-319.

Clement, Hal. "Alternative life designs" in B&P p 41-54.

Cocconi, Giuseppe, & Morrison, Philip. "Searching for interstellar communication" *Nature* 19 Sep 1959.

Cohen, Martin. "Stellar influences on emergence of intelligent life" in Billingham p 115-118.

Cox, Laurence. "An Explanation for absence of extraterrestrials on earth" QJRAS 1976 vol 17 p 201-; in Goldsmith p 232-235.

Criswell, David. "Solar system industrialization..." in F&J p 50-87.

Deardorff, J. W. "Examination of embargo hypothesis as explanation for the great silence" JBIS 1987 vol 40 p 373-379.

Dickerson, Richard. "Chemical evolution and origin of life" ScAm Sep 1978.

Drake, Frank. "Comparative analysis of space colonization enterprises" in P85 p 443-448.

Drake, Frank. "N is neither very small nor very large" in P80 p 27-34.

Drake, Frank. "On hands and knees in search of Elysium" *Tech. Rev.* Jun 1976.

Dunlop, Alistair. "Existence of extra-terrestrial life" JBIS 1984 vol 37 p 513-514.

Dyson, Freeman. "Search for artificial stellar sources of infrared radiation" *Science* 1 Jun 1960 (& letters 22 Jul 1960).

Dyson, Freeman. "Time without end: physics and biology in an open universe" *Rev. Mod. Phys.* 1979 vol 51 p 447-460.

Ferris, Timothy. "Seeking an end to cosmic loneliness" NYT Mag. 23 Oct 1977.

Finney, Ben. "SETI and interstellar migration" JBIS 1985 vol 38 p 274-275.

Finney, B. R., & Jones, E. M. "From Africa to the stars" in AASA 1983 vol 53 #83-205.

Fogg, Martyn. "Feasibility of intergalactic colonisation and its relevance to SETI" JBIS 1988 vol 41 p 491-496.

Frautschi, Steven. "Entropy in an expanding universe" *Science* 13 Aug 1982.

Freeman, J., & Lampton, M. "Interstellar archeology and prevalence of intelligence" *Icarus* 1975 vol 25 p 368-369.

Freitas, Robert. "Case for interstellar probes" JBIS 1983 vol 36 p 490-495.

Freitas, Robert. "Extraterrestrial intelligence in the solar system" JBIS 1983 vol 36 p 496-500.

Freitas, Robert. "Interstellar probes: new approach to SETI" JBIS 1980 vol 33 p 95-100.

Freitas, Robert. "Observable characteristics of extraterrestrial technological civilizations" JBIS 1985 vol 38 p 106-112.

Freitas, Robert. "Search for extraterrestrial artifacts" JBIS 1983 vol 36 p 501-506.

Gale, George. "Anthropic principle" ScAm Dec 1981.

Harrington, J. P. "Frequency of planetary systems in the galaxy" in H&Z p 142-153.

Harrington, R. S. "Planetary orbits in multiple star systems" in Billingham p 119-124.

Harris, M. J. "On detectability of antimatter propulsion spacecraft" *Astrophys. & Space Sc.* 1986 vol 123 p 297-303.

Hart, Michael. "Atmospheric evolution, Drake equation, and DNA" in H&Z p 154-165.

Hart, Michael. "An Explanation for the absence of extraterrestrials on Earth" QJRAS 1975 vol 16 p 128-; in H&Z p 1-8.

Hart, M. H. "Habitable zones around main sequence stars" *Icarus* 1979 vol 37 p 351-357; in Goldsmith p 236-240.

Horgan, John. "In the beginning" ScAm Feb 1991 (origins of life).

Horowitz, Norman. "Search for life on Mars" ScAm Nov 1977.

Huang, Su-Shu. "Life-supporting regions in vicinity of binary systems" *Astron. Soc. Pacific* 1960 vol 72 p 489-; in Cameron p 93-102.

Jones, E. M. "Discrete calculations of interstellar migration and settlement" *Icarus* 1981 vol 46 p 328-336.

Jones, Eric. "Estimate of expansion time scales" in H&Z p 66-76.

Jones, E. M. "Interstellar colonization" JBIS 1978 vol 31 p 103-107.

Jones, Eric. "Where are they? Implications of ancient and future migrations" in P85 p 465-476.

Jones, Eric, & Finney, Ben. "Fastships and nomads: two roads to the stars" in F&J.

Kardashev, Nikolai. "On inevitability and possible structures of super-civilizations" in P85 p 497-504.

Kardashev, N. S. "Transmission of information by extraterrestrial civilizations" *Sov. Astron.* 1964 vol 8 p 217-; in Goldsmith p 136-139.

Kuiper, T. B. H., & Morris, M. "Searching for extraterrestrial civilizations" *Science* 6 May 1977.

Lederberg, Joshua. "Exobiology: approaches to life beyond earth" *Science* 1960 vol 132 p 393-.

Mallove, Eugene. "Renaissance in the search for galactic civilization" *Tech. Rev.* Jan 1984.

Markowitz, William. "Physics and metaphysics of unidentified flying objects" *Science* 15 Sep 1967 vol 157 p 1274-1279.

Margulis, Lynn, & Lovelock, James. "Biological modulation of Earth's atmosphere" *Icarus* 1974 vol 21 p 471-489.

Margulis, Lynn, & Lovelock, James. "Atmospheres and evolution" in Billingham p 79-100.

Martin, A., & Bond, A. "Is mankind unique?" JBIS 1983 vol 36 p 223-225.

McKay, Christopher, & Haynes, Robert. "Should we implant life on Mars" ScAm Dec 1990 essay.

Michaud, Michael. "Final question: paradigms for intelligent life in the universe" JBIS 1982 vol 35 p 131-134.

Michaud, Michael. "Unique moment in human history" in B&P p 243-261 (contact).

Miller, S. L. "Production of some organic compounds under possible primitive earth conditions" *J. Am. Chem. Soc.* 1955 vol 77 p 2351.

Miller, S. L., & Urey, H. C. "Organic compound synthesis on primitive earth" *Science* 31 Jul 1959.

Minsky, Marvin. "Why intelligent aliens will be intelligible" in Regis p 117-125.

Morrison, Philip. "Interstellar communication" *Bull. Phil. Soc. Wash.* 1962 vol 16 p 58-; in Cameron p 249-271; in Goldsmith p 122-131.

Morrison, Philip. "25 years of search for extraterrestrial communications" in P85 p 13-20.

Murray, Bruce, et al. "Extraterrestrial intelligence: observational approach" *Science* 3 Feb 1978.

Newman, William, & Sagan, Carl. "Galactic civilizations: population dynamics and interstellar diffusion" *Icarus* 1981 vol 46 p 293-327.

Newman, William, & Sagan, Carl. "Nonlinear diffusion and population dynamics" in F&J p 302-312.

Niven, Larry. "Bigger than worlds" in Pournelle, Jerry, ed. *The Endless Frontier* 1979 Ace.

Oliver, Bernard. "Proximity of galactic civilizations" *Icarus* 1975 vol 25 p 360-; in Goldsmith p 178-183.

Oliver, Bernard. "Search strategies" in Billingham p 351-376.

Papagiannis, Michael. "Colonies in the asteroid belt or a missing term in the Drake equation" in H&Z p 77-86.

Papagiannis, Michael. "Historical introduction to search for extraterrestrial life" in P85 p 5-12.

Papagiannis, Michael. "Infrared search in our solar system..." in P85 p 505-511.

Papagiannis, Michael. "Recent progress and future plans on SETI" *Nature* 14 Nov 1985.

Papagiannis, Michael. "Retention by planets of liquid water..." JBIS 1989 vol 42 p 401-405.

Pollard, W. G. "Prevalence of earth-like planets" *Amer. Sc.* 1979 vol 67 p 653-659.

Ponnamperuma, Cyril. "Primordial organic chemistry" in H&Z p 87-99.

Purcell, Edward. "Radioastronomy and communication through space" USAEC 1961 #BNL-658; in Goldsmith p 188-196; in Cameron p 121-143.

Raup, David. "ETI without intelligence" in Regis p 31-42.

Regis, Edward. "SETI debunked" in Regis p 231-244.

Rescher, Nicholas. "Extraterrestrial science" in Regis p 83-116.

Ross, Monte. "Likelihood of finding extraterrestrial laser signals" JBIS 1979 vol 32 p 203-208.

Rothman, Tony. "What you see is what you beget theory" *Discover* May 1987 (anthropic principle).

Ruse, Michael. "Is rape wrong on Andromeda? An introduction to extraterrestrial evolution, science, and morality" in Regis p 43-78.

Sagan, Carl. "Direct contact among galactic civilizations by relativistic interstellar spaceflight" *Plan. & Space Sc.* 1963 vol 11 p 485-498; also in Goldsmith p 205-213.

Sagan, Carl. "Message from earth" *Science* 25 Feb 1972 (plaque).

Sagan, Carl. "On the detectivity of advanced galactic civilizations" *Icarus* 1973 vol 19 p 350-; in Goldsmith p 140-141.

Sagan, Carl. "Search for extraterrestrial intelligence" ScAm May 1975.

Sagan, Carl. "The solar system" ScAm Sep 1975.

Sagan, Carl, & Newman, William. "Solipsist approach to extraterrestrial intelligence" QJRAS 1982; in Regis p 151-161.

Sagan, C., & Walker, R. G. "Infrared detectability of Dyson civilizations" *Astrophys. J.* 1966 vol 144 p 1216-1218.

Schwartz, R. N., & Townes, C. H. "Interstellar and interplanetary communication by optical masers" *Nature* 1961 vol 190 p 205- (lasers).

Seeger, Charles. "Fermi question, Fermi paradox..." in P85 p 487-492.

Shapiro, Robert, & Feinberg, Gerald. "Possible forms of life in environments different from Earth" in H&Z p 113-121.

Simpson, George. "Nonprevalence of humanoids" *Science* 21 Feb 1964.

Slysh, V. I. "Search in infrared to microwave for astro-engineering activity" in P85 p 315-320.

Smith, Alexander. "Settlers and metals: industrial supplies in barren planetary systems" JBIS 1982 vol 35 p 209-217.

Soule, Michael. "Conservation: tactics for a constant crisis" *Science* 16 Aug 1991. Earth eco-crisis.

staff. "Arecibo message of Nov. 1974" *Icarus* 1975 vol 26 p 462-; in Goldsmith p 293-296.

Stahler, Steven. "Early life of stars" ScAm Jul 1991.

Stephenson, David. "Classification of extraterrestrial cultures" JBIS 1981 vol 34 p 486-490.

Stephenson, David. "Comets and interstellar travel" JBIS 1983 vol 36 p 210-214.

Stephenson, David. "Extraterrestrial cultures within the solar system?" QJRAS 1979 vol 20 p 422-; in Goldsmith p 246-249.

Sullivan, Woodruff. "Eavesdropping mode and radio leakage from Earth" in Billingham p 377-390.

Sullivan, Woodruff, & Knowles, Stephen. "Lunar reflections of terrestrial radio leakage" in P85 p 327-334.

Sullivan, W. T., et al. "Eavesdropping: radio signature of Earth" *Science* 27 Jan 1978.

Tang, Tong. "Fermi paradox and CETI" JBIS 1982 vol 35 p 236-240.

Tarter, Jill. "Planned observational strategy for NASA's first systematic SETI" in F&J p 314-330.

Tarter, Jill. "Searching for extraterrestrials" in Regis p 167-190.

Tarter, J., & Zuckerman, B. "Microwave searches in US and Canada" in P80 p 81-92.

Tipler, Frank. "Extraterrestrial intelligent beings do not exist" QJRAS 1980 vol 21 p 267- & vol 22; in *Phys. Today* Apr 1981; in Regis p 133-150.

Trimble, Virginia. "Nucleosynthesis and galactic evolution: implications for the origin of life" in H&Z p 135-141.

Tucker, Wallace. "Astrophysical crises in evolution of life in the galaxy" in Billingham p 287-296.

Turner, Edwin. "Galactic colonization and competition in young galactic disk" in P85 p 477-482.

Valdes, F., & Freitas, R. *Icarus* 1980 vol 42 p 442- & 1983 vol 55 p 453- (search L4 and L5).

Verschuur, Gerrit. "Search for narrow band 21-cm wavelength signals from ten nearby stars" *Icarus* 1973 vol 19 p 329-; in Goldsmith p 142-151.

Viewing, D. R., et al. "Detection of starships" JBIS 1977 vol 30 p 99-104.

von Hoerner, Sebastian. "Search for signals from other civilizations" *Science* 8 Dec 1961.

Wetherill, George. "Occurrence of earth-like bodies in planetary systems" *Science* 2 Aug 1991.

Wolfe, John. "On the question of interstellar travel" in P85 p 449-454.

Worden, Simon. "Detecting planets in binary systems with speckle interferometry" in Billingham p 177-192.

CHAPTER 11: BOOKS ON OTHER LIFE, SETI, AND CIVILIZATIONS:

Asimov, Isaac. *A Choice of Catastrophes* 1979 Doubleday.

Asimov, Isaac. *Extraterrestrial civilizations* 1979 Crown.

Averner, M. M., & MacElroy, R. D., eds. *On the Habitability of Mars* NASA 1976 #SP-414 (redoing Mars).

Barrow, John, & Tipler, Frank. *The Anthropic Cosmological Principle* 1986 Oxford UP.

Berendzen, Richard, ed. *Life Beyond Earth and the Mind of Man* 1973 NASA.

Billingham, John, ed. *Life in the Universe* 1981 MIT Pr.

Bova, Ben, & Preiss, Byron, eds. *First Contact* 1990 New American Lib.

Bracewell, Ronald. *The Galactic Club* 1974 Freeman.

Cameron, A. G. W., ed. *Interstellar Communication* 1963 Benjamin.

Christian, James, ed. *Extra-Terrestrial Intelligence: First encounter* 1976 Prometheus (philosophy).

Dyson, Freeman. *Infinite in All Directions* 1988 Harper & Row.

Engdahl, Sylvia. *The Planet-Girded Suns* 1974 Atheneum (history).

Feinberg, Gerald, & Shapiro, Robert. *Life Beyond Earth* 1980 Wm Morrow.

Finney, Ben, & Jones, Eric, eds. *Interstellar Migration and the Human Experience* 1985 Univ. of California Pr.

Folsome, Clair. *The Origin of Life* 1979 Freeman.

Freudenthal, Hans. *Lincos: Design of a language for cosmic intercourse* 1960 North-Holland.

Gale, W. A., ed. *Life in the Universe* 1979 Westview.

Goldsmith, Donald, ed. *The Quest for Extraterrestrial Life* 1980 University Science Bks.

Hart, Michael, & Zuckerman, Ben, eds. *Extraterrestrials* 1982 Pergamon.

Lovelock, James. *The Ages of Gaia* 1988 Norton.

Mallove, E. F. *The Quickening Universe: Cosmic evolution and human destiny* 1987 St. Martin's.

McDonough, Thomas. *Search for Extraterrestrial Intelligence* 1987 Wiley.

Morrison, Philip, et al, eds. *The Search for Extraterrestrial Intelligence* 1977 NASA, 1979 DOver.

O'Neill, Gerard. *2081* 1981 Simon & Schuster.

Papagiannis, Michael, ed. *The Search for Extraterrestrial Life* 1985 Reidel.

Papagiannis, M. D., ed. *Strategies for the Search for Life in the Universe* 1980 Reidel.

Ponnamperuma, Cyril, & Cameron, A. G. W., eds. *Interstellar Communication* 1974 Houghton Mifflin.

Poynter, Margaret, & Klein, Michael. *Cosmic Quest: searching for intelligent life among the stars* 1985 Atheneum.

Regis, Edward, ed. *Extraterrestrials* 1985 Cambridge UP.

Ridpath, Ian. *Messages from the Stars* 1978 Harper & Row.

Rood, R. T., & Trefil, J. S. *Are We Alone?* 1981 Scribner's.

Sagan, Carl. *Communication with Extraterrestrial Intelligence* 1973 MIT Pr.

Sagan, Carl, & Agel, Jerome. *The Cosmic Connection* 1973 Doubleday.

Sagan, Carl, et al. *Murmurs of Earth* 1978 Random House.

Shklovskii, I. S., & Sagan, Carl. *Intelligent Life in the Universe* 1966 Dell.

Sullivan, Walter. *We Are Not Alone* 1964 McGraw-Hill.

Swift, David. *SETI Pioneers* 1990 Univ. Arizona Pr.

Verschuur, G. *Cosmic Catastrophes* 1978 Addison-Wesley.

CHAPTER 12: ARTICLES ON LONGER-TERM, MORE SPECULATIVE IDEAS:

Abbott, Larry. "Mystery of the cosmological constant" ScAm May 1988 (vacuum energy).

Amato, Ivan. "Lab of rising microsuns" *Science* 25 Oct 1991 new on fusion.

Bilaniuk, Olexa-Myron, & Sudarshan, George. "Beyond the light barrier" *Phys. Today* May 1969 (tachyons).

Birrell, N. D., & Davies, P. C. W. "On falling through a blackhole into another universe" *Nature* 2 Mar 1978.

Boyer, Timothy. "The Classical vacuum" ScAm Aug 1985 (casimir force).

Corcoran, E., & Beardsley, Tim. "New space race" ScAm Jul 1990.

Broad, William. "Critics say NASA decline threatens agency goals" NYT 9 Sep 1990 p 1.

Broad, William. "How $8 billion space station became $120 billion showpiece" NYT 10 Jun 1990 p 1.

DeWitt, Bryce. "Quantum gravity" ScAm Dec 1983.

Dyson, Freeman. "Gravitational machines" in Cameron p 115-120.

Feinberg, Gerald. "Particles that go faster than light" ScAm Feb 1970.

Forward, Robert. "Extracting electrical energy from vacuum by cohesion of charged foliated conductors" *Phys. Rev.* 1984 vol 30B p 1700-1702.

Forward, Robert. "Guidelines to antigravity" *Am. J. Phys.* 1963 vol 31 p 166-170.

Forward, Robert. "Space warps: review of one form of propulsionless transport" JBIS 1989 vol 42 p 533-542.

Froning, H. D. "Investigation of quantum ramjet for interstellar flight" AIAA 1981 #81-1534.

Froning, H. D. "Propulsion requirements for quantum interstellar ramjet" JBIS 1980 vol 33 p 265-270.

Froning, H. D. "Use of vacuum energies for interstellar space flight" JBIS 1986 vol 39 p 410-415.

Gribbin, John. article on white holes, gateways *Smithsonian* Nov 1977.

Hawking, Stephen. "Quantum mechanics of black holes" ScAm Jan 1977.

Jackson, J. D., et al. "Superconducting supercollider" ScAm Mar 1986. Higgs.

Logsdon, John, & Williamson, Ray. "US access to space" ScAm Mar 1989.

Marshall, Eliot. "Space station science: up in the air" *Science* 1 Dec 1989.

Matzner, Richard, et al. "Demythologizing the black hole" in Rothman p 21-48.

Matzner, Richard, et al. "Grand illusions: further conversations on the edge of spacetime" in Rothman p 49-86 (black holes).

Norman, Michael. "His head in the stars" NYT Mag. 20 May 1990. JPL's Ed Stone.

Penrose, Roger. "Black holes" ScAm May 1972.

Pool, Robert. "Brookhaven chemists find new fusion method" *Science* 29 Sep 1989.

Ruderman, Malvin. "Solid stars" ScAm Feb 1971 (degenerate matter, neutron stars).

ScAm. 13 articles on advanced materials ScAm Oct 1986.

Thorne, Kip. "Search for black holes" ScAm Dec 1974.

Veltman, Martinus. "Higgs boson" ScAm Nov 1986.

Whitman, L. J., et al. "Manipulation of adsorbed atoms and creation of new structures...with scanning tunneling microscope" *Science* 8 Mar 1991.

CHAPTER 12: BOOKS ON LONGER-TERM, MORE SPECULATIVE IDEAS:

Brown, Lester. *Building a Sustainable Society* 1981 W. W. Norton.

Clarke, Arthur. *Profiles of the Future* 1977 Fawcett.

Dyson, Freeman. *Infinite in All Directions* 1988 Harper & Row.

Emme, Eugene, ed. *Science Fiction and Space Futures* AASH 1982 vol 5.

Forward, Robert. *Future Magic* 1988 Avon.

Forward, Robert, & Davis, Joel. *Mirror Matter: Pioneering antimatter physics* 1988 Wiley.

Hawking, Stephen. *A Brief History of Time* 1988 Bantam (cosmology, gravity).

Kaufmann, William. *The Cosmic Frontiers of General Relativity* 1977 Little, Brown.

Lederman, Leon, & Schramm, David. *From Quarks to the Cosmos* 1989 Freeman.

Pagels, Heinz. *Perfect Symmetry* 1985 Simon & Schuster (particles & cosmology).

Index

This index covers key ideas and terms only as they apply to the problem of interstellar travel. It includes the Appendices but not the Bibliography.

About the Author

John H. Mauldin has a bachelor's degree in engineering physics (Cornell University), master's in physics (Purdue University), and Ph.D. in science education (University of Texas). He has four books published in science and technology, covering mathematical graphics in *Perspective Design* (1985; second edition now being prepared), physics in *Particles in Nature* (1986), solar energy in *Sunspaces,* (1987), and optics in *Light, Lasers, and Optics* (1988). He has taught physics and engineering at several colleges and universities, done education research and development at MIT and University of Texas, and worked at NASA in electronic power engineering on an early phase of the Voyager missions. Astronomy, space travel, and science fiction were his earliest interests. For lack of funds to build a space colony, he and his wife have designed and built the lowest-cost superinsulated circular solar house on Earth, for which they purchase or burn no fuel for heat and which received a state conservation award.

**Author, partner Susan, and their superinsulated solar house on Earth
that needs no other heat (Photo credit: Juan Espinosa)**